CALCULUS
SUPER REVIEW®

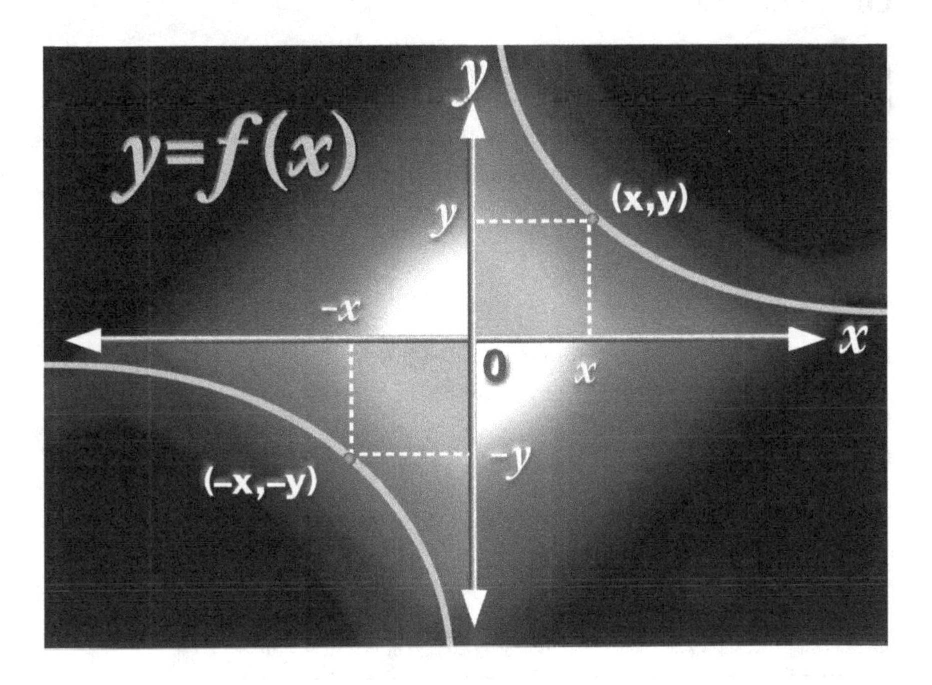

Staff of
Research & Education Association

Research & Education Association
Visit our website at: www.rea.com

Research & Education Association
61 Ethel Road West
Piscataway, New Jersey 08854
E-mail: info@rea.com

SUPER REVIEW®
OF CALCULUS

Published 2016

Printed in the United States of America

Library of Congress Control Number 2012945108

ISBN-13: 978-0-7386-1106-8
ISBN-10: 0-7386-1106-9

G15

Need help with calculus? Want a quick review or refresher for class? This is the book for you!

REA's *Calculus Super Review®* gives you everything you need to know!

This *Super Review®* can be used as a supplement to your high school or college textbook, or as a handy guide for anyone who needs a fast review of the subject.

- **Comprehensive, yet concise coverage** – review covers the material that is typically taught in a beginning-level calculus course. Each topic is presented in a clear and easy-to-understand format that makes learning easier.
- **Questions and answers for each topic** – let you practice what you've learned and build your calculus skills
- **End-of-chapter quizzes** – gauge your understanding of the important information you need to know, so you'll be ready for any calculus problem you encounter on your next quiz or test

Whether you need a quick refresher on the subject, or are prepping for your next test, we think you'll agree that REA's *Super Review®* provides all you need to know!

Available Super Review® Titles

ARTS/HUMANITIES

Basic Music
Canadian History
Classical Mythology
European History
History of Architecture
History of Greek Art
United States History

BUSINESS/ACCOUNTING

Accounting
Macroeconomics
Microeconomics

COMPUTER SCIENCE

C++
Java

LANGUAGES

English
French
French Verbs
Italian
Japanese for Beginners
Japanese Verbs
Latin
Spanish

MATHEMATICS

Algebra & Trigonometry
Basic Math & Pre-Algebra
Calculus
Geometry
Linear Algebra
Pre-Calculus
Statistics

SCIENCES

Anatomy & Physiology
Biology
Chemistry
Entomology
Geology
Microbiology
Organic Chemistry I & II
Physics

SOCIAL SCIENCES

Psychology I & II
Sociology

WRITING

College & University Writing

About REA

Founded in 1959, Research & Education Association (REA) is dedicated to publishing the finest and most effective educational materials—including study guides and test preps—for students in middle school, high school, college, graduate school, and beyond.

Today, REA's wide-ranging catalog is a leading resource for students, teachers, and other professionals. Visit www.rea.com to see a complete listing of all our titles.

Acknowledgements

We would like to thank Pam Weston, Publisher, for setting the quality standards for production integrity and managing the publication to completion; Larry B. Kling, Vice President, Editorial, for supervision of revisions and overall direction; Alice Leonard, Senior Editor, for coordinating development of this edition; Mel Friedman, REA's Lead Mathematics Editor for his editorial review and revisions; Claudia Petrilli, Graphic Designer, for typesetting this edition; and Christine Saul, Senior Graphic Designer, for designing our cover.

Contents

CHAPTER 1

Fundamentals

1.1 Number Systems

The real number system can be broken down into several parts and each of these parts has certain operations which can be performed on it. First, let us define the components of the real number system.

The natural numbers, denoted N, are 1,2,3,4,.... The integers, denoted Z, are ...-3,-2,-1,0,1,2,3,.... The rational numbers, denoted Q, are all numbers of the form p/q where p and q are integers and $q \neq 0$. A real number x is a non-terminating decimal (with a sign + or -).

Six basic algebraic properties of rational numbers:

a) The closure property: If x and y are rational numbers, then $x+y$ and $x \cdot y$ are also rational numbers.

b) Additive and multiplicative identity elements: If x is a rational number, then $x+0=x$ and $x \cdot 1=x$.

c) Associative property: If x, y and z are rational numbers, then $x+(y+z)=(x+y)+z$, $x\cdot(y\cdot z)=(x\cdot y)\cdot z$.

d) Additive and multiplicative inverses : For each rational number, x, there exists a rational number, denoted - x such that $x+(-x)=0$; if $x\neq 0$, there exists a rational number x^{-1} such that $x\cdot x^{-1}=1$.

e) **Commutative property:** If x and y are rational numbers, then $x+y=y+x$, $x \cdot y=y \cdot x$.

f) **Distributive property:** If x, y and z are rational numbers, then

$$x \cdot (y+z) = (x \cdot y) + (x \cdot z)$$

If q and p are rational numbers and $p-q$ is negative, then q is greater than p, $(q>p)$ or p is less than q, $(p<q)$.

Problem Solving Examples:

Find all real numbers satisfying the inequality $|3x + 2| > 5$.

A

Writing the given equation into the form which can be dealt with algebraically, we note that the given inequality expressed with an absolute number is equal to the following inequalities:

$$3x + 2 > 5 \text{ or } 3x + 2 < -5.$$

Considering, now, the first inequality, we have:

$$3x + 2 > 5,$$

or

$$x > 1.$$

Therefore, the interval $(1, +\infty)$ is a solution.

From the second inequality, we have:

$$3x + 2 < -5,$$

or

$$x < -\frac{7}{3}.$$

Hence, the interval $(-\infty, -\frac{7}{3})$ is a solution.

The solution of the given inequality consists of the two intervals $(-\infty, -\frac{7}{3})$ and $(1, +\infty)$ or, expressed in another way, all x not in the closed interval $[-\frac{7}{3}, 1]$.

1.1.1 Properties of Rational Numbers

a) Trichotomy property: If p and q are rational numbers, then one and only one of the relations $q=p$, $q>p$ or $q<p$ is true.

b) Transitive property: If p, q and r are rational numbers, and if $p<q$ and $q<r$, then $p<r$.

c) If p, q and r are rational numbers and $p<q$, then $p+r<q+r$.

d) If p, q and r are rational numbers and if $p<q$ and $0<r$, then $pr<qr$.

Problem Solving Examples:

Show that the set Q of rational numbers x such that $0 < x < 1$ is countably infinite.

A set A is countably infinite if it is in a one to one correspondence with the natural numbers (i.e., $\{1, 2, 3, \ldots\}$). Construct a table of Q in the following manner:

$$\frac{1}{2} \quad \frac{1}{3} \quad \frac{1}{4} \quad \frac{1}{5} \quad \frac{1}{6} \quad \ldots$$

$$\frac{2}{3} \quad \frac{2}{5} \quad \frac{2}{7} \quad \frac{2}{9} \quad \frac{2}{11} \quad \ldots$$

$$\frac{3}{4} \quad \frac{3}{5} \quad \frac{3}{7} \quad \frac{3}{8} \quad \frac{3}{10} \quad \ldots$$

$$\ldots$$

The numerators in successive rows of this table are 1, 2, 3 … . The denominators in each row are increasing but so that each fraction is proper (i.e., in Q) and in lowest terms (i.e., appears only once in the table). Now match the above table of Q with this table of the natural numbers:

```
1 → 2   6 → 7  15 → 16 ...
  ↙   ↗   ↙   ↗
3    5   8     14 ...
↓ ↗  ↙   ↗
4    9  13 ...
  ↙   ↗
10  12 ...
↓ ↗
11 ...

...
```

Consequently the set Q is in a one–to–one correspondence with the natural numbers.

Remark: Let $A_n = \{p/n : p \text{ an integer}\}$ for $n \geq 1$, i.e., A_n is the set of all rational numbers with denominator n. Then

$$Q = \bigcup_{n=1}^{\infty} A_n$$

and by construction each A_n is countable. A countable union of countable sets is countable. Hence Q is countable. This shows that Q is countably infinite.

1.2 Inequalities

To solve a linear inequality

$$ax+b>0 \text{ or } x>-b/a, \text{ where } a>0$$

draw a number line, dashed for $x<-b/a$ and solid for $x>-b/a$.

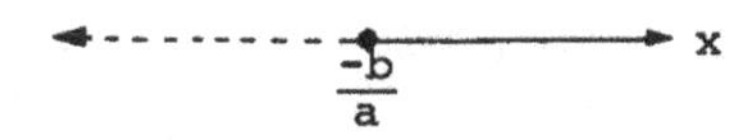

To solve $(ax+b)(cx+d) > 0$ graphically, where $a>0$ and $c>0$:

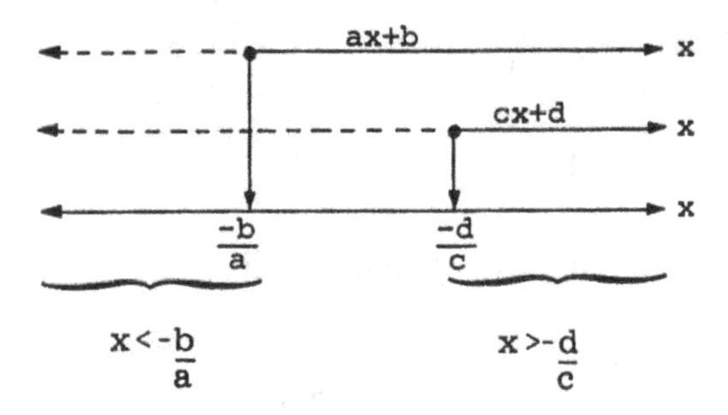

The solution lies in the interval where both lines are dashed and both lines are solid.

Hence, $x>-d/c$, $x<\frac{-b}{a}$ is the solution to the above inequality.

Problem Solving Examples:

Solve the inequality $|3 - 2x| < 1$.

$|3 - 2x| < 1$ can be represented as $-1 < 3 - 2x < 1$
By subtracting 3 from all the terms, we have

$$-4 < -2x < -2.$$

By multiplying all terms by $-\frac{1}{2}$ and remembering to reverse the signs of the inequalities, we have

$$2 > x > 1$$

or all values of x in the open interval (1, 2).

Graph the following two inequalities and show where the two graphs coincide:

$$2 \le x < 3 \text{ and } |y - 2| < \frac{1}{2}.$$

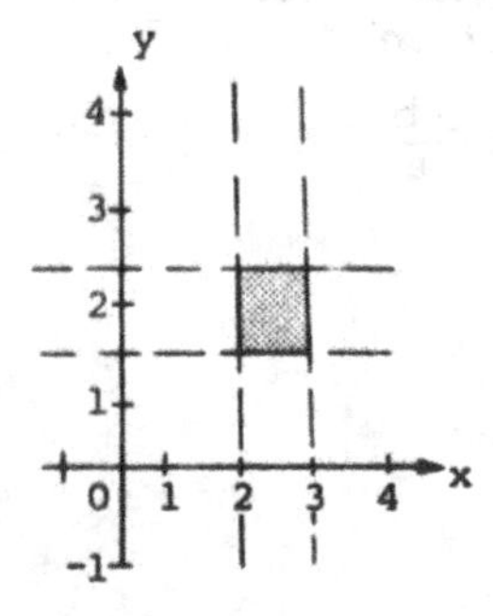

The first inequality consists of an infinite strip between the lines $x = 2$ and $x = 3$. Note that the points on the line $x = 2$ are included, but the points on the $x = 3$ are not.

For the second inequality, two cases must be considered, depending on whether (y - 2) is positive or negative. This may be expressed as: $-\frac{1}{2} < y - 2 < \frac{1}{2}$.

Adding 2 to each term to simplify, gives:

$$\frac{3}{2} < y < \frac{5}{2} .$$

This is an infinite strip which intersects the first one in a rectangle, with vertices at the points $\left(2,\frac{3}{2}\right)$, $\left(3,\frac{3}{2}\right)$, $\left(3,\frac{5}{2}\right)$ and $\left(2,\frac{5}{2}\right)$. The result is actually a square. Therefore, all points inside this square satisfy both inequalities. However, in addition, the points on the boundary of the square along the line x = 2 (left boundary), except for the corners, also satisfy both inequalities.

1.3 Absolute Value

Definition: The absolute value of a real number x is defined as

$$|x| = \begin{cases} x \text{ if } x \geq 0 \\ -x \text{ if } x<0 \end{cases}$$

For real numbers a and b:

a) $|a| = |-a|$

b) $|ab| = |a| \cdot |b|$

c) $-|a| \leq a \leq |a|$

d) $ab \leq |a||b|$

e) $|a+b|^2 = (a+b)^2$

f) $|a+b| \leq |a| + |b|$ (Triangle Inequality)

g) $|a-b| \geq |a| - |b|$

For positive values of b

a) $|a| < b$ if and only if $-b < a < b$

b) $|a| > b$ if and only if $a > b$ or $a < -b$

c) $|a| = b$ if and only if $a = b$ or $a = -b$

Problem Solving Examples:

Solve for x when $|5 - 3x| = -2$.

This problem has no solution, since the absolute value can never be negative and we need not proceed further.

Solve for x when $|3x + 2| = 5$.

First we write expressions which replace the absolute symbols in forms of equations that can be manipulated algebraically. Thus this equation will be satisfied if either

$$3x + 2 = 5 \text{ or } 3x + 2 = -5.$$

Considering each equation separately, we find

$$x = 1 \text{ and } x = -\frac{7}{3}.$$

Accordingly the given equation has two solutions.

1.4 Set Notation

A set is a collection of objects called elements. Let A and B be sets.

$x \in A$: indicates that x is an element of A
$x \notin B$: indicates that x is not an element of B

A is a subset of B, $(A \subset B)$, means that A is contained in another set B and each element of A is also an element of B.

A is equal to B (A=B), if and only if $A \subset B$ and $B \subset A$.

$A \cup B$: The union of A and B; the set consists of all elements of A and B.

$A \cap B$: the intersection of A and B; the set consists of elements, common to both A and B.

$A \cap B = \emptyset$: It is the set that has no elements common to both A and B; thus it is an empty set. In this case A and B are said to be disjoint.

These notations may be used to describe intervals of numbers such as:

The open interval $(a,b) = \{x : a < x < b\}$
The closed interval $[a,b] = \{x : a \leq x \leq b\}$
The half-open intervals $[a,b) = \{x : a \leq x < b\}$
and $(a,b] = \{x : a < x \leq b\}$

1.5 Summation Notation

If we are given a set or collection of numbers $\{a_1, a_2, a_3, \ldots, a_n\}$, the sum of these numbers can be represented by the symbol $\sum_{i=1}^{n} a_i$, that is,

$$\sum_{i=1}^{n} a_i = a_1 + a_2 + a_3 + \ldots + a_n.$$

In general, $\sum_{i=1}^{n} c = nc$ for every real number c, and

$$\sum_{i=1}^{n} c\,a_i = c\left(\sum_{i=1}^{n} a_i\right).$$

For any positive integer n and the sets of numbers $\{a_1, a_2, a_3, \ldots, a_n\}$ and $\{b_1, b_2, b_3, \ldots, b_n\}$,

$$\sum_{i=1}^{n} (a_i + b_i) = \sum_{i=1}^{n} a_i + \sum_{i=1}^{n} b_i$$

$$\sum_{i=1}^{n} (a_i - b_i) = \sum_{i=1}^{n} a_i - \sum_{i=1}^{n} b_i$$

Problem Solving Examples:

Find the numerical value of the following:

a) $\sum_{j=1}^{7} (2j + 1)$ b) $\sum_{j=1}^{21} (3j - 2)$.

If A(r) is some mathematical expression and n is a positive integer, then the symbol $\sum_{r=0}^{n} A(r)$ means

"Successively replace the letter r in the expression A(r) with the numbers 0,1,2,...,n and add up the terms. The symbol Σ is the Greek letter sigma and is a shorthand way to denote "the sum". It avoids having to write the sum A(0) + A(1) + A(2) + ... + A(n).

a) For a) successively replace j by 1,…,7 and add up the terms

$$\begin{aligned}\sum_{j=1}^{7} (2j + 1) &= (2(1)+1) + (2(2)+1) + (2(3)+1) + \\ &\quad (2(4)+1) + (2(5)+1) + (2(6)+1) + (2(7)+1) \\ &= (2+1) + (4+1) + (6+1) + (8+1) + \\ &\quad (10+1) + (12+1) + (14+1) \\ &= 3 + 5 + 7 + 9 + 11 + 13 + 15 \\ &= 63.\end{aligned}$$

b) For b) successively replace j by 1,2,3,…,21 and add up the terms.

$$\sum_{j=1}^{21} (3j - 2) = (3(1)-2) + (3(2)-2) + (3(3)-2) + (3(4)-2) + (3(5)-2) + (3(6)-2) + (3(7)-2) + (3(8)-2) + (3(9)-2) + (3(10)-2) + (3(11)-2) + (3(12)-2) + (3(13)-2) + (3(14)-2) + (3(15)-2) + (3(16)-2) + (3(17)-2) + (3(18)-2) + (3(19)-2) + (3(20)-2) + (3(21)-2)$$

$$= (3-2) + (6-2) + (9-2) + (12-2) + (15-2) + (18-2) + (21-2) + (24-2) + (27-2) + (30-2) + (33-2) + (36-2) + (39-2) + (42-2) + (45-20) + (48-2) + (51-2) + (54-2) + (57-2) + (60-2) + (63-2)$$

$$= 1 + 4 + 7 + 10 + 13 + 16 + 19 + 22 + 25 + 28 + 31 + 34 + 37 + 40 + 43 + 46 + 49 + 52 + 55 + 58 + 61$$

$$= 651 .$$

Functions

2.1 Functions

Definition: function is a correspondence between two sets, the domain and the range, such that for each value in the domain there corresponds exactly one value in the range.

A function has three distinct features:

a) the set x which is the domain,

b) the set y which is the co-domain or range,

c) a functional rule, f, that assigns only one element $y \in Y$ to each $x \in X$. We write $y = f(x)$ to denote the functional value y at x.

Consider Figure 2.1. The "machine" f transforms the domain X, element by element, into the co-domain Y.

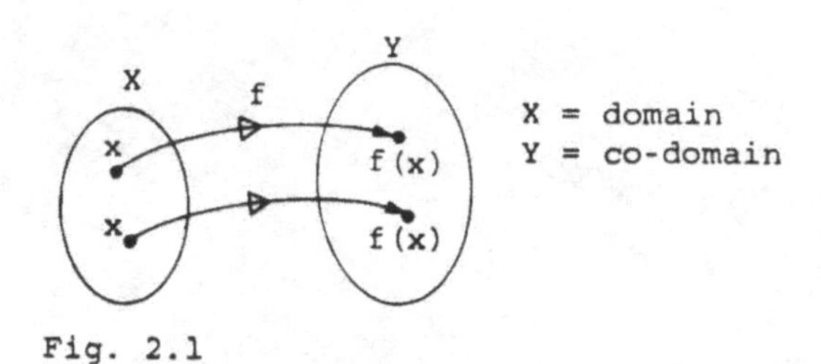

Fig. 2.1

Problem Solving Examples:

Determine the domain of the function:

$$y = \frac{2x}{(x-2)(x+1)}.$$

$$y = \frac{2x}{(x-2)(x+1)}$$

is defined only if the denominator is not equal to zero.

The denominator is equal to zero when $(x - 2)(x + 1) = 0$, or, equivalently, when $(x - 2) = 0$ or $(x + 1) = 0$. $(x - 2) = 0$ when $x = 2$, and $(x + 1) = 0$ when $x = -1$.

Therefore, the domain of the function is all values of x where $x \neq -1$ and $x \neq 2$.

Determine the domain of the function:

$$y = \sqrt{\frac{x}{2-x}}$$

Since we are restricted to the real number system, there are two cases to be considered:

Case 1. When the denominator equals 0, the fraction is undefined. This occurs when $x = 2$.

Case 2. Since we are restricted to real values of y, it is necessary that $\frac{x}{2-x} \geq 0$, because, if $\frac{x}{2-x} < 0$, the square root gives an imaginary number. Therefore,

$\frac{x}{2-x} < 0$ only if $x < 0$, and $x > 2$.

Combining the conditions in cases 1 and 2, the domain of x is the interval $0 \leq x < 2$.

2.2 Combination of Functions

Let f and g represent functions, then

a) the sum $(f+g)(x) = f(x) + g(x)$,
b) the difference $(f-g)(x) = f(x) - g(x)$,
c) the product $(fg)(x) = f(x)g(x)$,

d) the quotient $(\frac{f}{g})(x) = \frac{f(x)}{g(x)}$, $g(x) \neq 0$,

e) the composite function $(g \circ f)(x) = g(f(x))$ where $f(x)$ must be in the domain of g.

A polynomial function of degree n is denoted as

$$f(x) = a_n x^n + a_{n-1} x^{n-1} + a_{n-2} x^{n-2}$$

$$+ \ldots + a_1 x + a_0$$

where a_n is the leading coefficient and not equal to zero and $a_k x^k$ is the kth term of the polynomial.

2.3 Properties of Functions

A) A function F is one to one if for every range value there corresponds exactly one domain value of x.

B) A function is even if $f(-x) = f(x)$ or

$$f(x) + f(-x) = 2f(x).$$

C) A function is said to be odd if $f(-x) = -f(x)$ or $f(x) + f(-x) = 0$.

D) Periodicity

A function f with domain X is periodic if there exists a positive real number p such that $f(x+p) = f(x)$ for all $x \in X$.

The smallest number p with this property is called the period of f.

Over any interval of length p, the behavior of a periodic function can be completely described.

E) Inverse of a function

Assuming that f is a one-to-one function with domain X and range Y, then a function g having domain Y and range X is called the inverse function of f if:

$f(g(y)) = y$ for every $y \in Y$ and

$g(f(x)) = x$ for every $x \in X$.

The inverse of the function f is denoted f^{-1}.

To find the inverse function f^{-1}, you must solve the equation $y = f(x)$ for x in terms of y.

Be careful: This solution must be a function.

F) The identity function $f(x) = x$ maps every x to itself.

G) The constant function $f(x) = c$ for all $x \in R$.

The "zeros" of an arbitrary function $f(x)$ are particular values of x for which $f(x) = 0$.

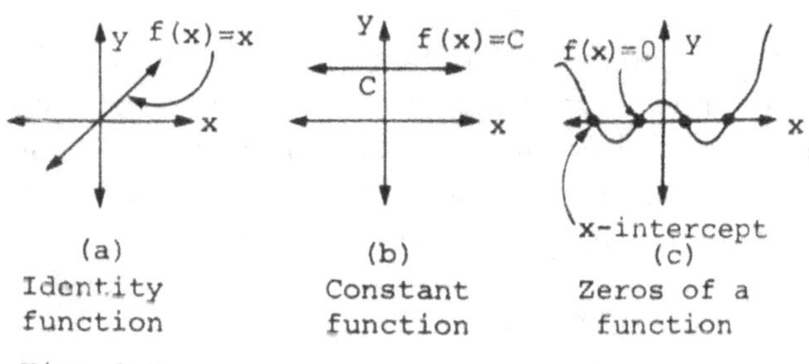

(a) Identity function

(b) Constant function

(c) Zeros of a function

Fig. 2.2

Problem Solving Examples:

Find the domain D and the range R of the function $x, \frac{x}{|x|}$.

Note that the y-value of any coordinate pair (x,y) is $\frac{x}{|x|}$.

We can replace x in the formula $\frac{x}{|x|}$ · with any number except 0, since the denominator, $|x|$, cannot equal 0, (i.e. $|x| \neq 0$) which is the equivalent to $x \neq 0$. This is because division by 0 is undefined. Therefore, the domain D is the set of all real numbers except 0. If x is negative, i.e., $x < 0$, then $|x| = -x$ by definition. Hence, if x is negative, then $\frac{x}{|x|} = \frac{x}{-x} = -1$. If x is positive, i.e., $x > 0$, then $|x| = x$ by definition. Hence, if x is positive, then $\frac{x}{|x|} = \frac{x}{x} = 1$. (The case where $x = 0$ has already been found to be undefined.) Thus, there are only two numbers -1 and 1 in the range R of the function; that is, $R = \{-1,1\}$.

If $f(x) = 3x + 4$ and $D = \{x \mid -1 \leq x \leq 3\}$, find the range of f(x).

A We first prove that the value of $3x + 4$ increases when x increases. If $X > x$, then we may multiply both sides of the inequality by a positive number to obtain an equivalent inequality. Thus, $3X > 3x$. We may also add a number to both sides of the inequality to obtain an equivalent inequality. Thus,

$$3X + 4 > 3x + 4.$$

Hence, if x belongs to D, the function value $f(x) = 3x + 4$ is least when $x = -1$ and greatest when $x = 3$. Consequently, since

$f(-1) = -3 + 4 = 1$ and $f(3) = 9 + 4 = 13$, the range is all y from 1 to 13; that is,

$$R = \{y \mid 1 \leq y \leq 13\}.$$

2.4 Graphing a Function

2.4.1 The Cartesian Coordinate System

Consider two lines x and y drawn on a plane region called R.

Let the intersection of x and y be the origin and let us impose a coordinate system on each of the lines.

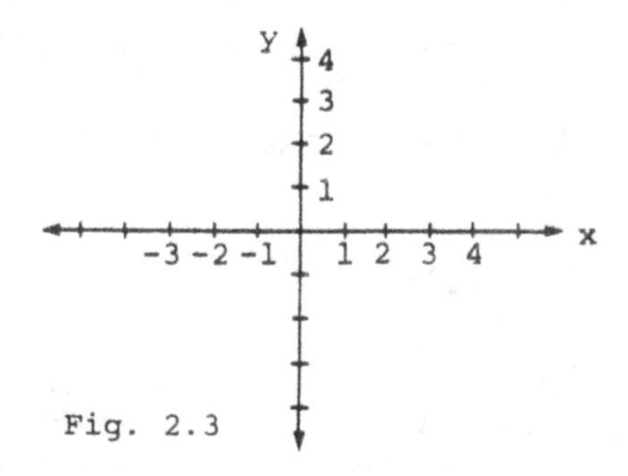

Fig. 2.3

If (x,y) is a point or ordered pair on the coordinate plane R then x is the first coordinate and y is the second coordinate.

To locate an ordered pair on the coordinate plane simply measure the distance of x units along the x-axis, then measure vertically (parallel to the y-axis) y units.

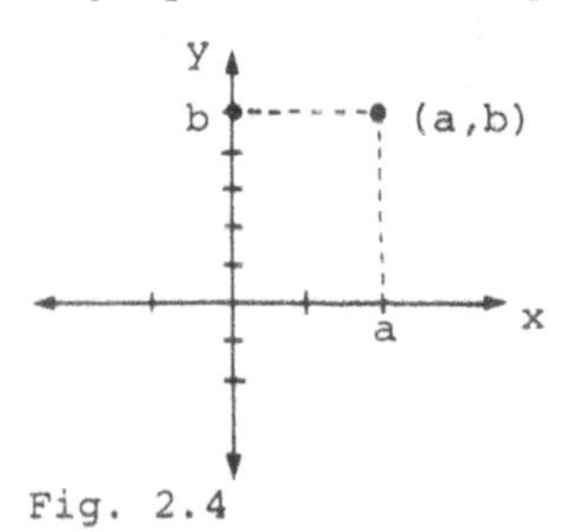

Fig. 2.4

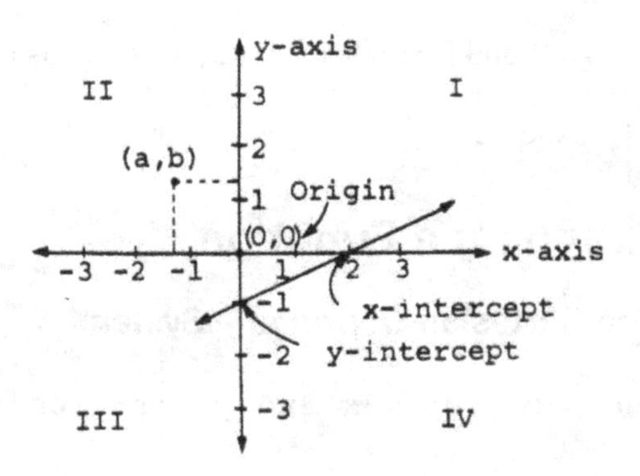

I, II, III, IV are called quadrants in the COORDINATE PLANE.

(a,b) is an ordered pair with x-coordinate a and y-coordinate b.

Fig. 2.5 CARTESIAN COORDINATE SYSTEM

2.4.2 Drawing the Graph

There are several ways to plot the graph of a function. The process of computing and plotting points on the graph is always an aid in this endeavor. The more points we locate on the graph, the more accurate our drawing will be.

It is also helpful if we consider the symmetry of the function. That is,

a) A graph is symmetric with respect to the x-axis if whenever a point (x,y) is on the graph, then $(x,-y)$ is also on the graph.

b) Symmetry with respect to the y-axis occurs when both points $(-x,y)$ and (x,y) appear on the graph for every x and y in the graph.

c) When the simultaneous substitution of -x for x and -y for y does not change the solution of the equation, the graph is said to be symmetric about the origin.

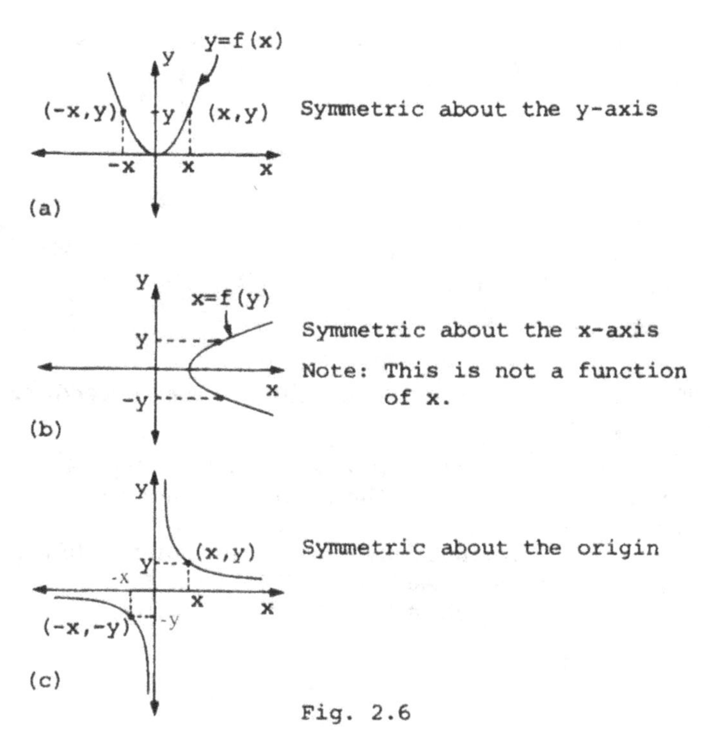

Fig. 2.6

Another aid in drawing a graph is locating any vertical asymptotes.

A vertical asymptote is a vertical line $x = a$, such that the functional value $|f(x)|$ grows indefinitely large as x approaches the fixed value a.

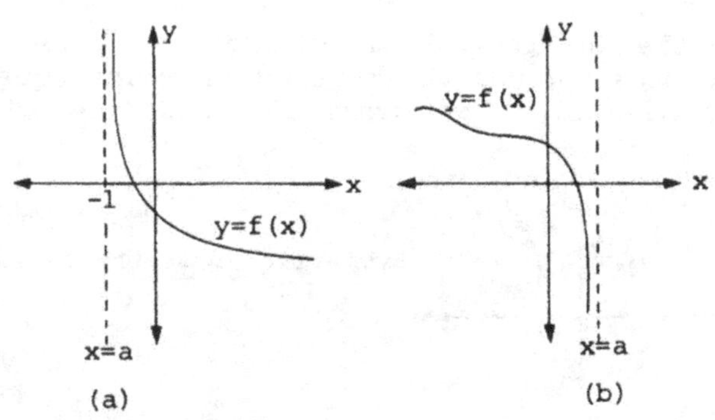

x=a is a vertical asymptote
for these functions

Fig. 2.7

The following steps encapsulate the procedure for drawing a graph:

a) Determine the domain and range of the function.
b) Find the intercepts of the graph and plot them.
c) Determine the symmetries of the graph.
d) Locate the vertical asymptotes and plot a few points on the graph near each asymptote.
e) Plot additional points as needed.

Problem Solving Examples:

Construct the graph of the function defined by $y = 3x - 9$.

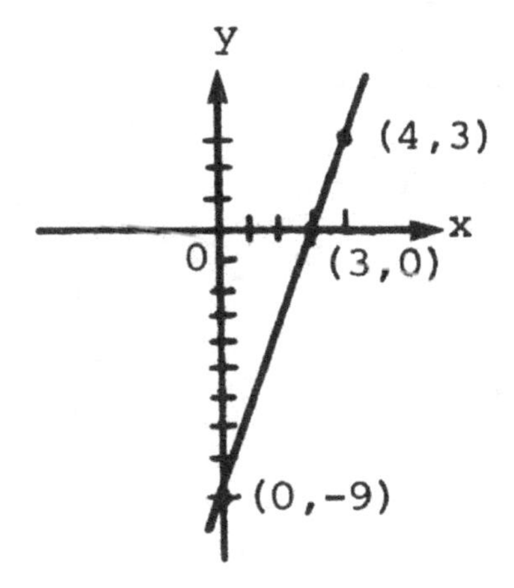

An equation of the form $y = mx + b$ is a linear equation; that is, the equation of a line.

A line can be determined by two points. Let us choose the intercepts. The x-intercept lies on the x-axis and the y-intercept on the y-axis.

We find the intercepts by assigning 0 to x and solving for y and by assigning 0 to y and solving for x. It is helpful to have a third point. We find the third point by assigning 4 to x and solving for y. Thus we get the following table of corresponding numbers:

x	y = 3x –9	y
0	y = 3(0) – 9 = 0 – 9 =	–9
4	y = 3(4) –9 = 12 - 9 =	3

Solving for x to get the x-intercept:

$$y = 3x - 9$$
$$y + 9 = 3x$$
$$x = \frac{y + 9}{3}$$

When y = 0, $x = \frac{9}{3} = 3$. The three points are (0,–9), (4,3), and (3,0). Draw a line through them (see sketch).

Are the following points on the graph of the equation 3x – 2y = 0?
a) point (2,3)? b) point (3,2)? c) point (4,6)?

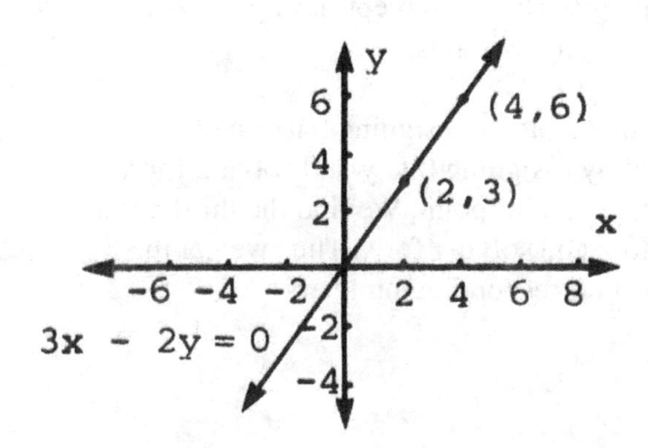

The point (a,b) lies on the graph of the equation $3x - 2y = 0$ if replacement of x and y by a and b, respectively, in the given equation results in an equation which is true.

a) Replacing (x,y) by (2,3):

$$3x - 2y = 0$$
$$3(2) - 2(3) = 0$$
$$6 - 6 = 0$$
$$0 = 0, \text{ which is true.}$$

Therefore (2,3) is a point on the graph.

b) Replacing (x,y) by (3,2):

$$3x - 2y = 0$$
$$3(3) - 2(2) = 0$$
$$9 - 4 = 0$$
$$5 = 0, \text{ which is not true.}$$

Therefore (3,2) is not a point on the graph.

c) Replacing (x,y) by (4,6):

$$3x - 2y = 0$$
$$3(4) - 2(6) = 0$$
$$12 - 12 = 0$$
$$0 = 0, \text{ which is true.}$$

Therefore (4,6) is a point on the graph.

This problem may also be solved geometrically as follows: draw the graph of the line $3x - 2 = 0$ on the coordinate axes. This can be done by solving for y:

$$3x - 2y = 0$$
$$-2y = -3x$$
$$y = \frac{-3}{-2}x = \frac{3}{2}x,$$

and plotting the points shown in the following table:

x	$y = \frac{3}{2}x$
0	0
1	$\frac{3}{2} = 1\frac{1}{2}$
2	3
–2	–3

(See accompanying figure.)

Observe that we obtain the same result as in our algebraic solution. The points (2,3) and (4,6) lie on the line $3x - 2y = 0$, whereas (3,2) does not.

2.5 Lines and Slopes

Each straight line in a coordinate plane has an equation of the form $Ax + By + C = 0$, where A and B are not zero.

If we consider only a portion or a segment of the line we can find both, the length of the segment and its midpoint.

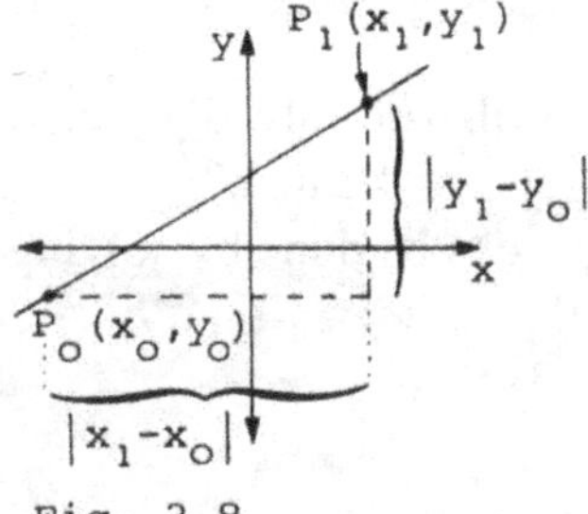

Fig. 2.8

The distance between two points P_0 and P_1 in a coordinate plane is $d(P_0,P_1) = \sqrt{(x_1-x_0)^2+(y_1-y_0)^2}$.

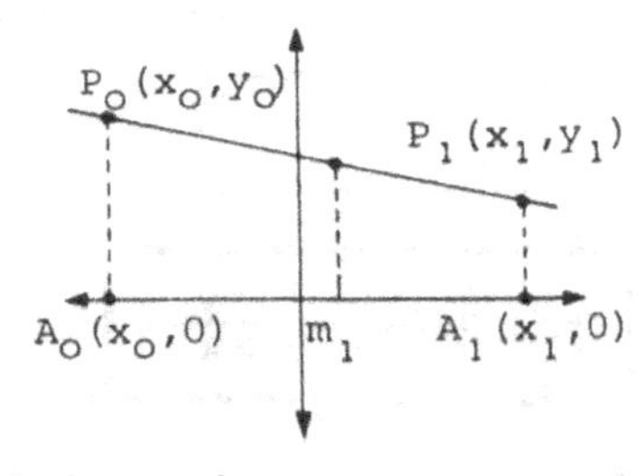

Fig. 2.9

The midpoint of a line segment from P_0 to P_1 is the point

$$\left(\frac{x_1+x_0}{2}, \frac{y_1+y_0}{2}\right)$$

However we are more often concerned with finding the slope of the line.

If given two points (x_1,y_1) and (x_0,y_0) the ratio

$$\frac{y_1-y_0}{x_1-x_0}$$

is the slope of the line.

Any two segments of the same line must have the same slope. Therefore, looking at Fig. 2.10, we see

$$\frac{y_3-y_2}{x_3-x_2} = \frac{y_1-y_0}{x_1-x_0}.$$

It is easy to show that if two line segments have the same slopes and a common endpoint, then they must be the same line.

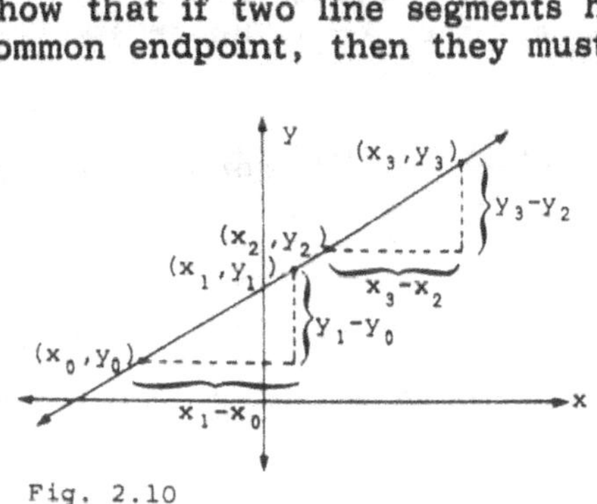

Fig. 2.10

The equation for a line can be conveniently written as

$$y = mx + b$$

where

$$m = \text{slope} = \frac{y_1 - y_0}{x_1 - x_0}$$

and b = y-intercept; where the line intersects the y-axis.

The value of m will help us determine the position of the line on a graph.

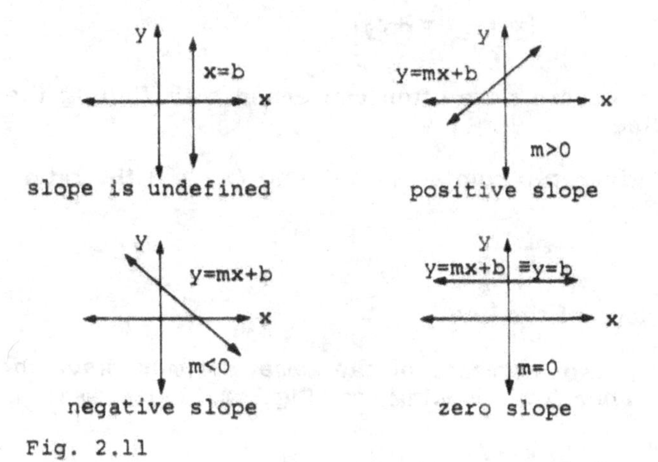

Fig. 2.11

The slope of a line can be used to determine whether or not several points are collinear. Given n points (a_1,b_1) $(a_2,b_2),\ldots,(a_n,b_n)$ they are collinear if and only if

$$\frac{b_i - b_{i-1}}{a_i - a_{i-1}} = \frac{b_2 - b_1}{a_2 - a_1} \quad \text{for } i = 3,4,\ldots n.$$

Two lines are parallel if and only if their slopes are equal.

Two lines having slopes m_1 and m_2 are perpendicular if and only if $m_1 m_2 = -1$.

Problem Solving Examples:

Find the slope f(x) = 3x + 4.

Two points on the line determined by f(x) = 3x + 4 are A(0,4) and B(1,7).

$$\frac{\text{difference of ordinates}}{\text{difference of abscissas}} = \frac{7-4}{1-0} = 3$$

Note that the ordinates are the y-coordinates and the abscissas are the x-coordinates. The slope determined by points A and B is 3. Hence, the slope of f(x) = 3x + 4 is 3. In general, the slope of a linear function of the form
f(x) = mx + b is m.

Show that the slope of the segment joining (1,2) and (2,6) is equal to the slope of the segment joining (5,15) and (10,35).

The slope of the line segment, m, joining the points (x_1,y_1) and (x_2,y_2) is given by the formula

$$m = \frac{(y_2 - y_1)}{(x_2 - x_1)}$$

Therefore, the slope of the segment joining (1,2) and (2,6) is

$$\frac{6-2}{2-1} = \frac{4}{1} = 4.$$

The slope of the segment joining (5,15) and (10,35) is

$$\frac{35-15}{10-5} = \frac{20}{5} = 4.$$

Thus, the slopes of the two segments are equal.

2.6 Parametric Equations

If we have an equation $y = f(x)$, and the explicit functional form contains an arbitrary constant called a parameter, then it is called a parametric equation. A function with a parameter represents not one but a family of curves.

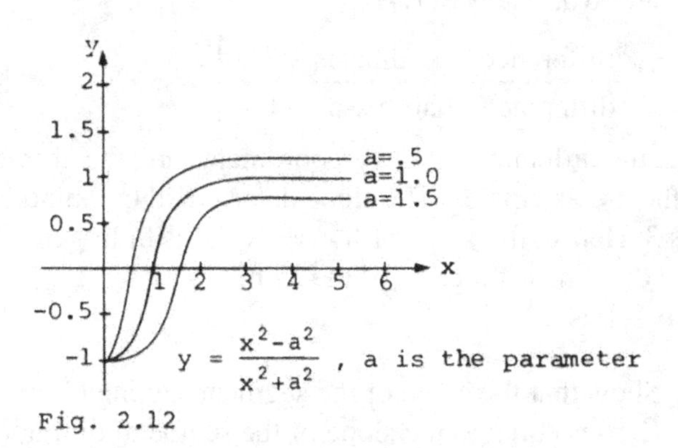

Fig. 2.12

Often the equation for a curve is given as two functions of a parameter t, such as

$$X = x(t) \text{ and } Y = y(t).$$

Corresponding values of x and y are calculated by solving for t and substituting.

Transcendental Functions

3.1 Trigonometric Functions

The trigonometric functions are defined in terms of a point P which moves in a circular track of unit radius.

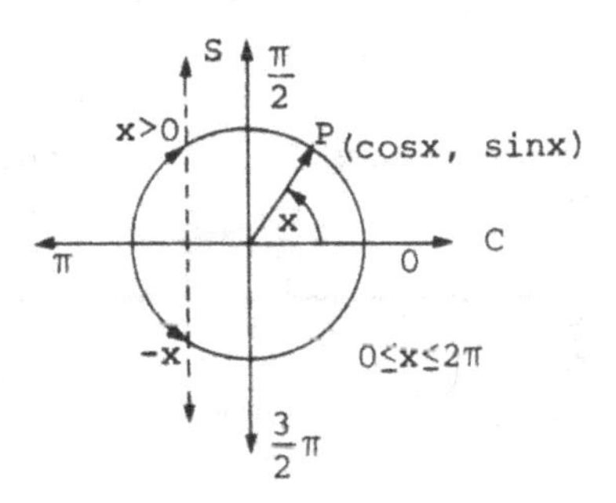

Fig. 3.1

If we let P(x) = (cos x, sin x), then for any x such that $0 \leq x \leq 2\pi$, the points P(x) and P(-x) lie on the same vertical axis. They are symmetrically located with respect to the c-axis.

This implies that cos(-x) = cos x and sin(-x) = -sin x

Another fundamental identity to remember is

$$\boxed{\cos^2 x + \sin^2 x = 1}$$

For any $x \in R$, $-1 \le \cos x \le 1$ and $-1 \le \sin x \le 1$, therefore, both the sine and cosine functions are continuous for all real numbers.

If PQR is an angle t and P has coordinates (x,y) on the unit circle, then by joining PR we get angle PRQ = 90° (Fig. 3.2), and then we can define all the trigonometric functions in the following way:

sine of t, $\sin t = y$

cosine of t, $\cos t = x$

tangent of t, $\tan t = \frac{y}{x}$, $x \neq 0$

cotangent of t, $\tan t = \frac{x}{y}$, $y \neq 0$

secant of t, $\sec t = \frac{1}{x}$, $x \neq 0$

cosecant of t, $\csc t = \frac{1}{y}$, $y \neq 0$.

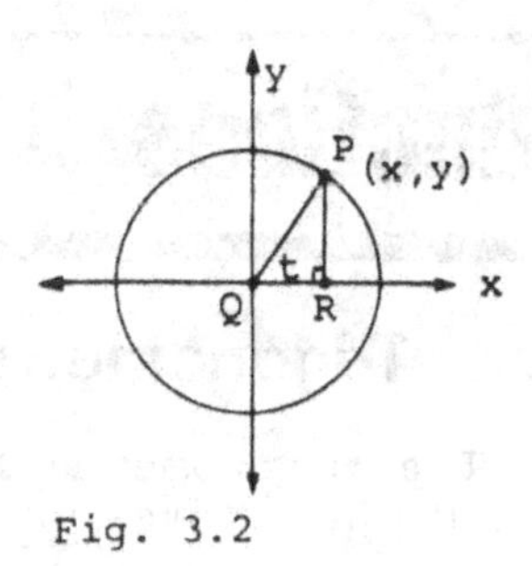

Fig. 3.2

Provided the denominators are not zero, the following relationships exist:

$$\sin t = \frac{1}{\csc t} \qquad \tan t = \frac{\sin t}{\cos t}$$

$$\cos t = \frac{1}{\sec t} \qquad \cot t = \frac{\cos t}{\sin t}$$

$$\tan t = \frac{1}{\cot t}$$

Figures 3.3 and 3.4 show the graphs of each of the trigonometric functions. Notice that the x-axis is measured in radians.

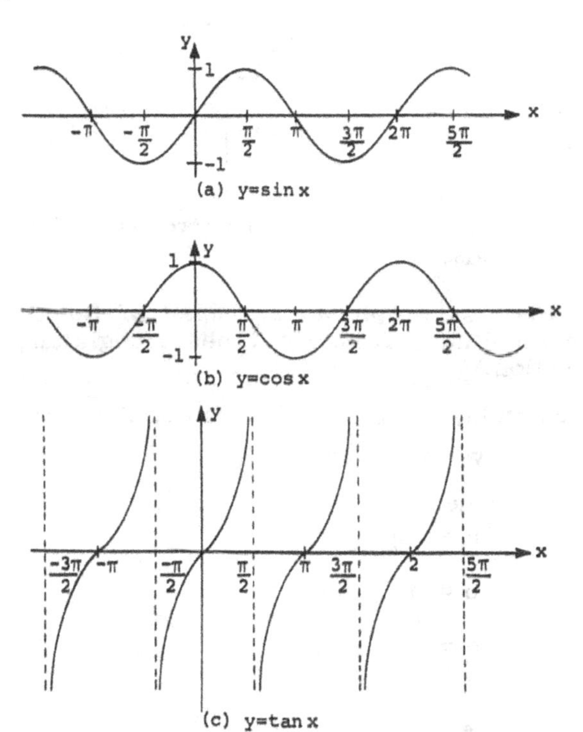

(a) y=sin x

(b) y=cos x

(c) y=tan x

Fig. 3.3

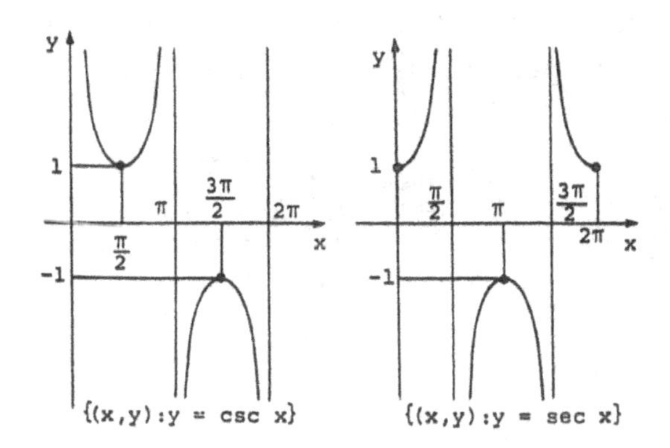

{(x,y):y = csc x}

{(x,y):y = sec x}

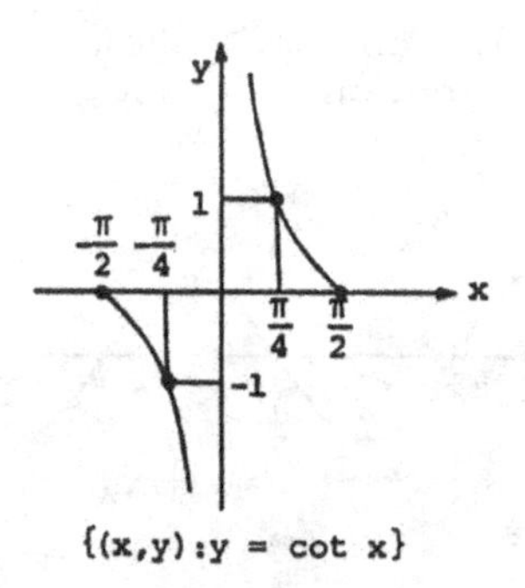

$\{(x,y): y = \cot x\}$

Fig. 3.4

In order to graph a trigonometric function we must know the amplitude, frequency, phase angle and the period of the function.

For example, to graph a function of the form

$$y = a \sin(bx+c)$$

we must determine:

$$a = \text{amplitude}$$

$$b = \text{frequency}$$

$$\frac{c}{b} = \text{phase angle}$$

and $\frac{2\pi}{b} = \text{period}.$

Let us graph the function $y = 2\sin(2x + \frac{\pi}{4})$. Amplitude = 2, period = $\frac{2\pi}{2} = \pi$, phase $\not{x} = \frac{\pi}{8}$.

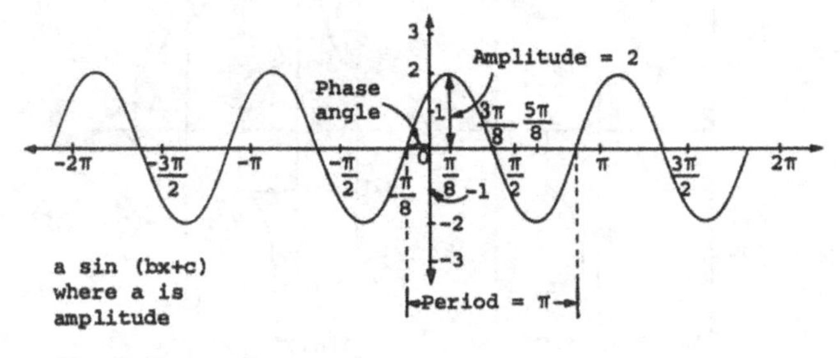

a sin (bx+c) where a is amplitude

Fig. 3.5

Problem Solving Examples:

Given the right triangle with a = 3, b = 4 and c = 5, find the values of the trigonometric functions of α.

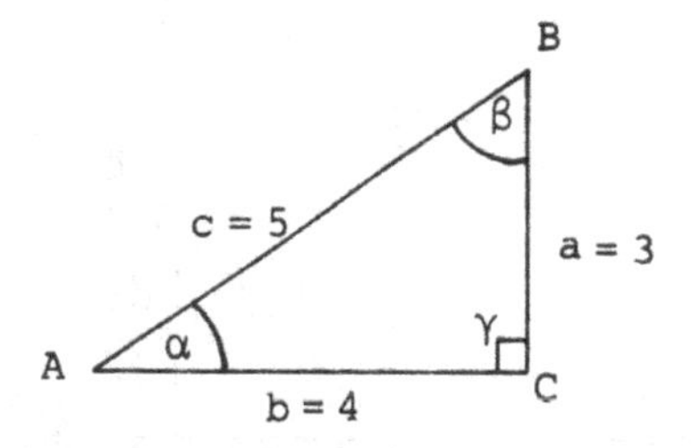

In the accompanying figure, a is the side opposite of angle α, b is the side opposite angle β, and c is the opposite angle γ. The values of the trigonometric functions of α are:

$$\cos\alpha = \frac{\text{adjacent side}}{\text{hypotenuse}}, \quad \sin\alpha = \frac{\text{opposite side}}{\text{hypotenuse}},$$

$$\tan\alpha = \frac{\text{opposite side}}{\text{adjacent side}}, \quad \cot\alpha = \frac{1}{\tan\alpha}$$

$$\sec\alpha = \frac{1}{\cos\alpha} \quad \text{and} \quad \csc\alpha = \frac{1}{\sin\alpha}.$$

Therefore:

$$\cos\alpha = \frac{4}{5}, \qquad \sin\alpha = \frac{3}{5},$$

$$\tan\alpha = \frac{3}{4}, \qquad \cot\alpha = \frac{1}{\frac{3}{4}}, = \frac{4}{3},$$

$$\sec\alpha = \frac{1}{\frac{4}{5}} = \frac{5}{4}, \quad \csc\alpha = \frac{1}{\frac{3}{5}} = \frac{5}{3}.$$

Q Calculate the values of the six trigonometric functions at the point $\frac{1}{3}\pi$.

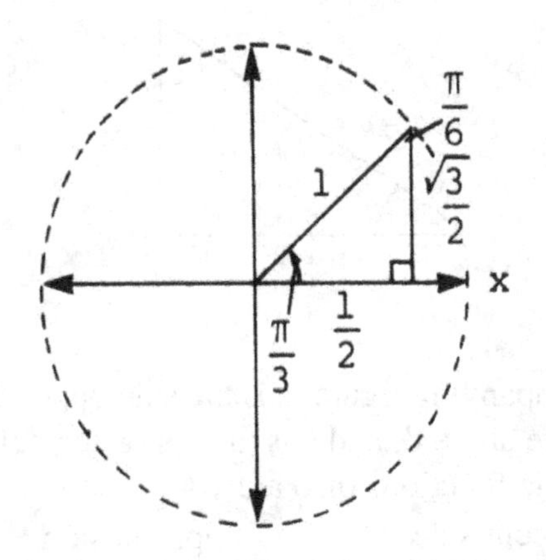

A To find the trigonometric point $p\left(\frac{1}{3}\right)\pi$, proceed around the unit circle in the counterclockwise direction, since $\frac{\pi}{3}$ is a positive angle. Recall that sin 60° i.e., $\sin\left(\frac{\pi}{3}\right) = \frac{\sqrt{3}}{2}$. Now using the Pythagorean theorem and the fact that the hypotenuse is unity because it is a unit circle we can compute the third side, which we find to be $\frac{1}{2}$ (see figure). Therefore, the coordinates of the trigonometric point $P\frac{1}{3}\pi$ are $\frac{1}{2}$, $\frac{3}{\sqrt{2}}$. Hence, we apply the following equations:

$$\cos\theta = \frac{\text{adjacent side}}{\text{hypotenuse}} \qquad \sec\theta = \frac{1}{\cos\theta} = \frac{\text{hypotenuse}}{\text{adjacent side}}$$

$$\sin\theta = \frac{\text{opposite side}}{\text{hypotenuse}} \qquad \csc\theta = \frac{1}{\sin\theta} = \frac{\text{hypotenuse}}{\text{opposite side}}$$

$$\tan\theta = \frac{\text{opposite side}}{\text{adjacent side}} \qquad \cot\theta = \frac{\cos\theta}{\sin\theta} = \frac{\text{adjacent side}}{\text{opposite side}}$$

Thus,

$$\cos\frac{1}{3}\pi = \frac{1}{2}, \qquad \sec\frac{1}{3}\pi = 2,$$

$$\sin\frac{1}{3}\pi = \frac{1}{2}\sqrt{3}, \qquad \csc\frac{1}{3}\pi = \frac{2}{\sqrt{3}} = \frac{2}{\sqrt{3}}\cdot\frac{\sqrt{3}}{\sqrt{3}} = \frac{2}{3}\sqrt{3}$$

$$\tan\frac{1}{3}\pi = \sqrt{3}, \qquad \cot\frac{1}{3}\pi = \frac{1}{\sqrt{3}} = \frac{1}{\sqrt{3}}\cdot\frac{\sqrt{3}}{\sqrt{3}} = \frac{\sqrt{3}}{3}.$$

3.2 Inverse Trigonometric Functions

By restricting the domain of the trigonometric functions we can define their inverse functions.

The inverse sine function, denoted $\sin^{-1}$, is defined to be $\sin^{-1}x = y$ if and only if $\sin y = x$ where $-1 \leq x \leq 1$ and $-\pi/2 \leq y \leq \frac{\pi}{2}$.

In a similar manner we define:

$-\frac{\pi}{2} \leq \sin^{-1}x \leq \frac{\pi}{2}$	$-1 \leq x \leq 1$	monotone increasing
$0 \leq \cos^{-1}x \leq \pi$	$-1 \leq x \leq 1$	monotone decreasing
$\frac{-\pi}{2} < \tan^{-1}x < \frac{\pi}{2}$	for all $x \in R$	
$0 < \cot^{-1}x < \pi$	for all $x \in R$	
$0 \leq \sec^{-1}x \leq \pi$	$\lvert x \rvert \geq 1, \sec^{-1}x \neq \frac{\pi}{2}$	
$-\frac{\pi}{2} \leq \csc^{-1}x \leq \frac{\pi}{2}$	$\lvert x \rvert \geq 1, \csc^{-1}x \neq 0$	

Problem Solving Examples:

Evaluate: a) $\sin^{-1}\frac{\sqrt{3}}{2}$, b) $\tan^{-1}(-\sqrt{3})$.

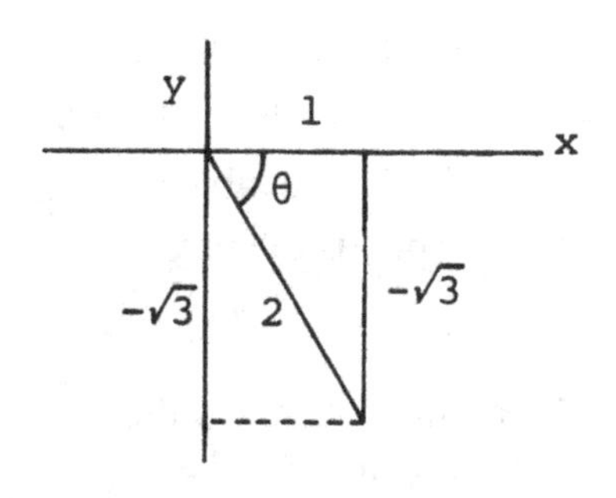

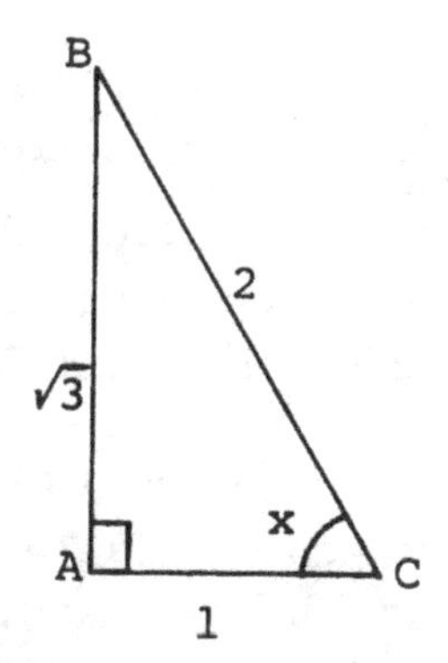

a) Recall that inverse sines are angles. Thus we are looking for the angle whose sin is $\frac{\sqrt{3}}{2}$. $\sin^{-1} \frac{\sqrt{3}}{2} = x$ means $\sin x = \frac{\sqrt{3}}{2}$ where $\sin = \frac{\text{opposite}}{\text{hypotenuse}}$.

We note that triangle ABC is a 30-60 right triangle, and angle $x = 60^\circ$. Since $\sin 60^\circ = \frac{\sqrt{3}}{2}$, $\sin^{-1} \frac{\sqrt{3}}{2} = 60^\circ$.

b) Recall that inverse tangents are angles. Thus we are looking for the angle whose tangent is $-\sqrt{3}$. $\text{Tan}^{-1}(-\sqrt{3}) = \theta$ means $\tan\theta = -\sqrt{3}$ where $\tan = \frac{\text{opposite}}{\text{adjacent}}$.

Since tangent is negative in the 4th quadrant, we draw our triangle there, and note it is a 30-60 right triangle, and angle $\theta = (-60^\circ)$.

Since $\tan(-60^\circ) = \frac{\sqrt{3}}{1}$, $\tan^{-1}(-\sqrt{3}) = -60^\circ$.

Calculate the following numbers, in radians.
a) Arctan $\sqrt{3}$ b) $\text{Tan}^{-1}.2027$ c) $\text{Tan}^{-1}1.871$.

a) The expression tan y = x is equivalent to arctan x = $\tan^{-1}x$ = y. Let the expression arctan $\sqrt{3}$ = y. Hence, the expression arctan $\sqrt{3}$ = y is equivalent to tan y = $\sqrt{3}$ = 1.7321. In a table of trigonmetric functions, the number y that corresponds to tan y = 1.732 is approximately 1.05.

b) Note that the expression $\tan^{-1}.2027$ = arctan .2027. Let the expression $\tan^{-1}.2027$ = y. Hence, the expression $\tan^{-1}.2027$ = arctan .2027 = y is equivalent to tan y = .2027. In a table of trigonometric functions, the number y that corresponds to tan y = .2027 is .20.

c) Note that the expression $\tan^{-1}1.871$ = arctan 1.871. Let the expression $\tan^{-1} 1.871$ = y. Hence, the expression $\tan^{-1}1.871$ = arctan 1.871 = y is equivalent to tan y = 1.871. In a table of trigonometric functions, the number y that corresponds to tan y = 1.871 is 1.08.

3.3 Exponential and Logarithmic Functions

If f is a nonconstant function that is continuous and satisfies the functional equation $f(x+y) = f(x) \cdot f(y)$, then $f(x) = a^x$ for some constant a. That is, f is an exponential function.

Consider the exponential function a^x, $a > 0$ and the logarithmic function $\log_a x$, $a > 0$. Then a^x is defined for all $x \in R$, and $\log_a x$ is defined only for positive $x \in R$.

These functions are inverses of each other,

$$a^{\log_a x} = x; \log_a(a^y) = y.$$

Let a^x, $a > 0$ be an exponential function. Then for any real numbers x and y

$$\text{a) } a^x \cdot a^y = a^{x+y}$$

$$\text{b) } (a^x)^y = a^{xy}$$

Let $\log_a x$, $a > 0$ be a logarithmic function. Then for any positive real numbers x and y

$$\text{a) } \log_a(xy) = \log_a(x) + \log_a(y)$$

$$\text{b) } \log_a(x^y) = y \log_a(x)$$

Let $h > -1$ be any real number. Then for any natural number $n \in N$,

$$(1+h)^n \geq 1 + nh.$$

3.3.1 The Natural Logarithmic Function

A) To every real number y there corresponds a unique positive real number x such that the natural logarithm, ln, of x is equal to y. That is $\ln x = y$.

B) The natural exponential function, denoted by exp, is defined by

$$\exp x = y \text{ if and only if } \ln y = x$$

for all x, where $y > 0$.

C) The natural log and natural exponential are inverse functions. $\ln(\exp x) = x$ and $\exp(\ln y) = y$.

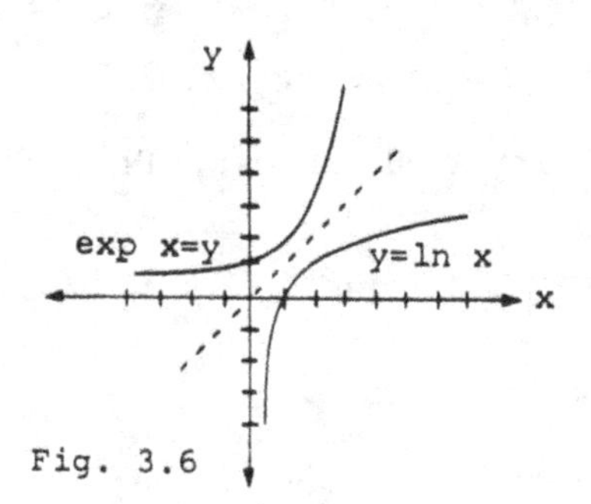

Fig. 3.6

D) The letter e denotes the unique positive real number such that $\ln e = 1$.

E) If x is a real number then e^x is the unique real number y such that

$e^x = y$ if and only if $\ln y = x$.

F) If p and q are real numbers and r is rational then

a) $e^p e^q = e^{p+q}$

b) $\frac{e^p}{e^q} = e^{p-q}$

c) $(e^p)^r = e^{pr}$

Quiz: Fundamentals, Functions and Transcendental Functions

1. The domain of the function defined by $f(x) = \ln(x^2 - x - 6)$ is the set of all real numbers x such that

   (A) $x > 0$

   (B) $-2 \leq x \leq 3$

   (C) $-2 \leq x$ or $x \geq 3$

   (D) $-2 < x < 3$

   (E) $-2 > x$ or $x > 3$

2. If $\dfrac{f(x_1)}{f(x_2)} = f\left(\dfrac{x_1}{x_2}\right)$ for all real numbers x_1 and x_2, where $x_2 \neq 0$ and $f(x_2) \neq 0$, which of the following could define f?

   (A) $f(x) = \dfrac{1}{x}$

   (B) $f(x) = x^2 + 3$

   (C) $f(x) = x + 1$

   (D) $f(x) = \ln x$

   (E) $f(x) = e^x$

3. $\dfrac{\ln(x^3 e^x)}{x} =$

   (A) $\dfrac{3(\ln x + e^x)}{x}$

   (B) $\ln(x^3 e^x - x)$

   (C) $\ln x^2 + 1$

   (D) $\dfrac{3\ln x + x}{x}$

   (E) $\dfrac{3\ln x}{x}$

4. If $\frac{r^2}{r-1} \geq r$, then

(A) $r \geq 0$

(B) $r \leq 0$

(C) $r \leq 0$ or $r > 1$

(D) $r \leq 0$ or $r \geq 1$

(E) $0 \leq r < 1$

5. If $(x) = \frac{\sqrt{x+2}}{x+2}$ and $g(x) = \frac{1}{x} - 2$, then $f[g(x)] =$

(A) $\frac{\sqrt{\frac{1}{x} - 2}}{\frac{1}{x} - 2}$

(B) $\frac{\sqrt{1 - 2x}}{x}$

(C) $\frac{\sqrt{\frac{1}{x-2} + 2}}{\frac{1}{x-2} + 2}$

(D) $\sqrt{x}$

(E) $\frac{\sqrt{x}}{x}$

6. If $\tan x = 2$, then $(\sin x)^2 =$

(A) $\frac{2}{5}$

(B) $\frac{4\sqrt{5}}{5}$

(C) $\frac{4}{5}$

(D) $\frac{4}{3}$

(E) $\frac{2}{3}$

7. The domain of $f(x) = \sqrt{4 - x^2}$ is

(A) $-2 \le x \le 2$

(B) $-2 \le x$ or $x \ge 2$

(C) $-2 < x$ or $x > 2$

(D) $-2 < x < 2$

(E) $x \ge 2$

8. If $f(x) = \sqrt{1 - x^2}$, which of the following is NOT true?

(A) Domain of $f = [-1, 1]$

(B) $[f(x)]^2 + x^2 = 1$

(C) Range of f is $[0, 1]$

(D) $f(x) = f(-x)$

(E) The line $y = 1$ intersects the graph of f at two points.

9. Which of the following represents a function?

(A) $x^2 + y^2 = 1$

(B) $y^2 = x$

(C) $x^2 - y = 0$

(D) $y = \pm \sqrt{1 - x^2}$

(E) None of the above

10. If A = $\{x{:}x \leq 2\}$ and B = $\{x{:}x > -1\}$, which of the following intervals describes $A \cap B$?

(A) $(-1, \infty)$

(B) $(-\infty, 2)$

(C) $[-1, 2]$

(D) $(-1, 2]$

(E) $[-1, 2)$

ANSWER KEY

1. (E)
2. (A)
3. (D)
4. (C)
5. (D)
6. (C)
7. (A)
8. (E)
9. (C)
10. (D)

CHAPTER 4

Limits

4.1 Definition

Let f be a function that is defined on an open interval containing a, but possibly not defined at a itself. Let L be a real number. The statement

$$\lim_{x \to a} f(x) = L$$

defines the limit of the function f(x) at the point a. Very simply, L is the value that the function has as the point a is approached.

Problem Solving Examples:

Find $\lim_{x \to 2} f(x) = 2x + 1$

As $x \to 2$, $f(x) \to 5$. Therefore, $\lim_{x \to 2} (2x + 1) = 5$

Find $\lim_{x \to 3} f(x) = \dfrac{x^2-9}{x+1}$

$$\lim_{x \to 3} \frac{x^2-9}{x+1} = \frac{0}{4} = 0.$$

4.2 Theorems on Limits

The following are important properties of limits:

Consider $\lim_{x \to a} f(x) = L$ and $\lim_{x \to a} g(x) = K$, then

A) Uniqueness – If $\lim_{x \to a} f(x)$ exists then it is unique.

B) $\lim_{x \to a} [f(x)+g(x)] = \lim_{x \to a} f(x) + \lim_{x \to a} g(x) = L+K$

C) $\lim_{x \to a} [f(x)-g(x)] = \lim_{x \to a} f(x) - \lim_{x \to a} g(x) = L-K$

D) $\lim_{x \to a} [f(x) \cdot g(x)] = \lim_{x \to a} f(x) \cdot \lim_{x \to a} g(x) = L \cdot K$

E) $\lim_{x \to a} \frac{f(x)}{g(x)} = \frac{\lim_{x \to a} f(x)}{\lim_{x \to a} g(x)} = \frac{L}{K}$ provided $K \neq 0$

F) $\lim_{x \to a} \frac{1}{g(x)} = \frac{1}{K}$, $K \neq 0$

G) $\lim_{x \to a} [f(x)]^n = [\lim_{x \to a} f(x)]^n$ for $n > 0$

H) $\lim_{x \to a} [cf(x)] = c[\lim_{x \to a} f(x)]$, $c \in R$

I) $\lim_{x \to a} cx^n = c \lim_{x \to a} x^n = ca^n$, $c \in R$

J) If f is a polynomial function then $\lim_{x \to a} f(x) = f(a)$ for all $a \in R$.

K) $\lim_{x \to a} \sqrt[n]{x} = \sqrt[n]{a}$ when $a \geq 0$ and n is a positive integer or when $a \leq 0$ and n is an odd positive integer.

L) $\lim_{x \to a} \sqrt[n]{f(x)} = \sqrt[n]{\lim_{x \to a} f(x)}$ when n is a positive integer

M) If $f(x) \leq h(x) \leq g(x)$ for all x in an open interval containing a, except possibly at a, and if $\lim_{x \to a} f(x) = L = \lim_{x \to a} g(x)$ then $\lim_{x \to a} h(x) = L$.

Problem Solving Examples:

Find $\lim_{x \to 0} (x\sqrt{x-3})$.

In checking the function by simple substitution, we see that:

$$x\sqrt{x-3} = 0$$

if $x = 0$.

However, this function does not have real values for values of x less than 3. Therefore, since x cannot approach 0, f(x) does not approach 0 and the limit does not exist. This example illustrates that we cannot properly find

$$\lim_{x \to a} f(x)$$

by finding f(a), even though they are equal in many cases. We must consider values of x near a, but not equal to a.

4.3 One-Sided Limits

Suppose f is a function such that it is not defined for all values of x. Rather, it is defined in such a way that it "jumps" from one y value to the next instead of smoothly going from one y value to the next. Examples are shown in Fig. 4.1 and 4.2.

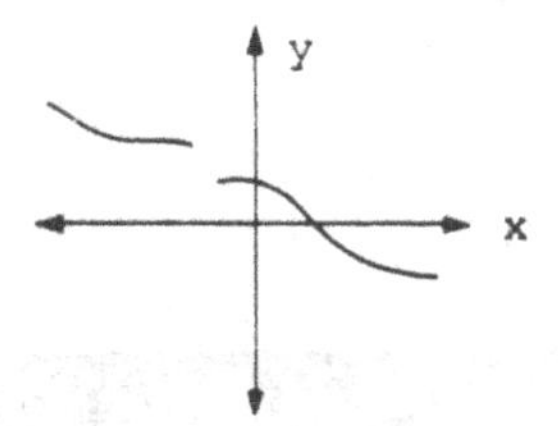

Fig. 4.1 y=f(x) is not defined for all x values.

Fig. 4.2 y=f(x) "jumps" from a positive value to a negative one.

The statement $\lim_{x \to a^+} f(x) = R$ tells us that as x approaches "a" from the right or from positive infinity, the function f has the limit R.

Similarly, the statement $\lim_{x \to a^-} f(x) = L$ says that as x approaches "a" from the left-hand side or from negative infinity, the function f has the limit L.

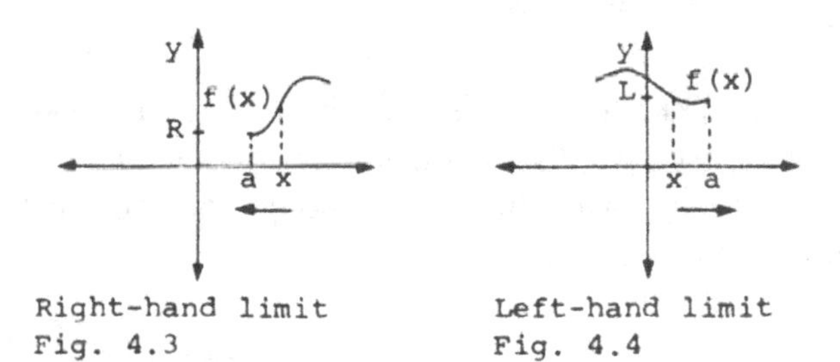

Right-hand limit
Fig. 4.3

Left-hand limit
Fig. 4.4

If f is defined in an open interval containing a, except possibly at a, then

$\lim_{x \to a} f(x) = L$ if and only if

$$\lim_{x \to a^+} f(x) = L = \lim_{x \to a^-} f(x).$$

Notice that in Fig. 4.2 the right-hand limit is not the same as the left-hand limit, as it is in Fig. 4.5.

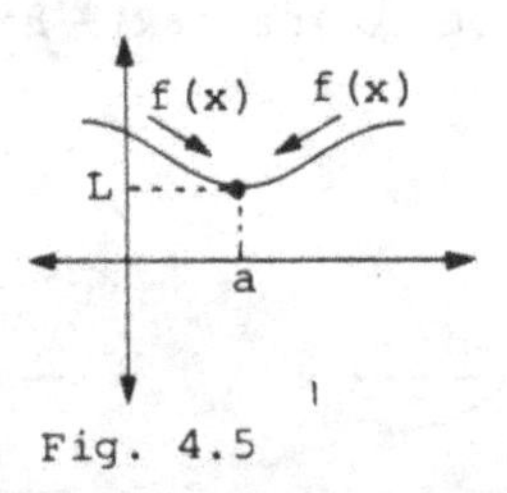

Fig. 4.5

Problem Solving Examples:

Let f(x) and g(x) be defined by:

$$f(x) = \begin{cases} x^2 + 2x, & x \leqq 1, \\ 2x, & x > 1, \end{cases}$$

$$g(x) = \begin{cases} 2x^3, & x \leqq 1, \\ 3, & x > 1. \end{cases}$$

Find $\lim_{x \to 1} [f(x) \cdot g(x)]$ if it exists.

A Neither f(x) not g(x)have limits as $x \to 1$, but one-sided limits exist for both functions. It is possible that the product of two functions may have a limit, even though the two functions do not have limits individually.

$$\lim_{x \to 1^-} f(x) = 3, \qquad \lim_{x \to 1^+} f(x) = 2,$$

$$\lim_{x \to 1^-} g(x) = 2, \qquad \lim_{x \to 1^+} g(x) = 3.$$

Therefore,

$$\lim_{x \to 1^-} [f(x) \cdot g(x)] = 6$$

and

$$\lim_{x \to 1^+} [f(x) \cdot g(x)] = 6.$$

Consequently,

$$\lim_{x \to 1} [f(x) \cdot g(x)] = 6.$$

Given that f is the function defined by:

$$f(x) = \begin{cases} x - 3 & \text{if } x \neq 4 \\ 5 & \text{if } x = 4, \end{cases}$$

Find $\lim_{x \to 4} f(x)$.

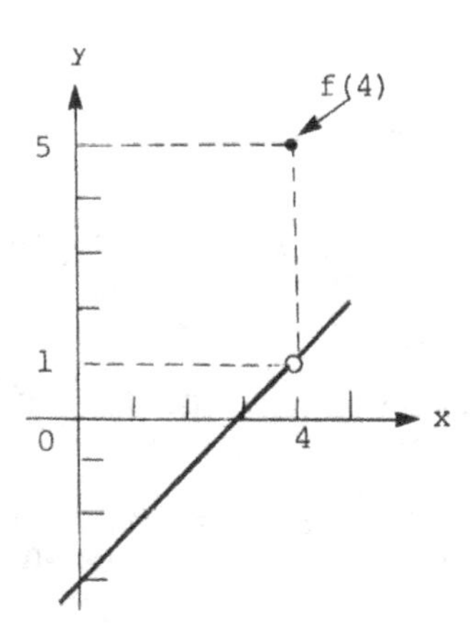

When plotting f(x) to obtain a visual representation, it is seen that $f(x) = x - 3$ is a straight line which has a break or discontinuity at the point $x = 4$. At $x = 4$, the value of f(x) is given as 5, and not as 1—the value that f(x) would assume if the line were continuous. However, when evaluating $\lim_{x \to 4} f(x)$, we are considering values of x close to 4 but not equal to 4. Thus we have

$$\lim_{x \to 4} f(x) = \lim_{x \to 4} (x-3)$$
$$= 1.$$

In this example,

$\lim_{x \to 4} f(x) = 1$ but $f(4) = 5$; therefore,

$$\lim_{x \to 4} f(x) \neq f(4).$$

4.4 Special Limits

A) $\lim_{x \to 0} \frac{\sin x}{x} = 1, \qquad \lim_{x \to 0} \frac{1-\cos x}{x} = 0$

B) $\lim_{n \to \infty} (1 + \frac{1}{n})^n = e, \qquad \lim_{n \to 0} (1+n)^{1/n} = e$

C) For $a > 1$ $\lim_{x \to +\infty} a^x = +\infty, \lim_{x \to -\infty} a^x = 0$

$$\lim_{x \to +\infty} \log_a x = +\infty, \lim_{x \to 0} \log_a x = -\infty$$

D) For $0 < a < 1$, $\lim_{x \to +\infty} a^x = 0, \lim_{x \to -\infty} a^x = +\infty$

$$\lim_{x \to +\infty} \log_a x = -\infty, \lim_{x \to 0} \log_a x = +\infty$$

Some nonexistent limits which are frequently encountered are:

A) $\lim_{x \to 0} \frac{1}{x^2}$, as x approaches zero, x^2 gets very small and also becomes zero therefore $\frac{1}{0}$ is undefined and the limit does not exist.

B) $\lim_{x \to 0} \frac{|x|}{x}$ does not exist.

Proof:

If $x > 0$, then $\frac{|x|}{x} = \frac{x}{x} = 1$ and hence lies to the right of the y-axis, the graph of f coincides with the line $y = 1$. If $x < 0$ then $\frac{-x}{x} = -1$ and the graph of f coincides with the line $y = -1$ to the left of the y-axis.

If it were true that $\lim_{x \to 0} \frac{|x|}{x} = L$ for some L, then the preceding remarks imply that $-1 \leq L \leq 1$.

If we consider any pair of horizontal lines $y = L \pm \varepsilon$, where $0 < \varepsilon < 1$, then there exists points on the graph which are not between these lines for some non-zero x in every interval $(-\delta, \delta)$ containing 0. It follows that the limit does not exist.

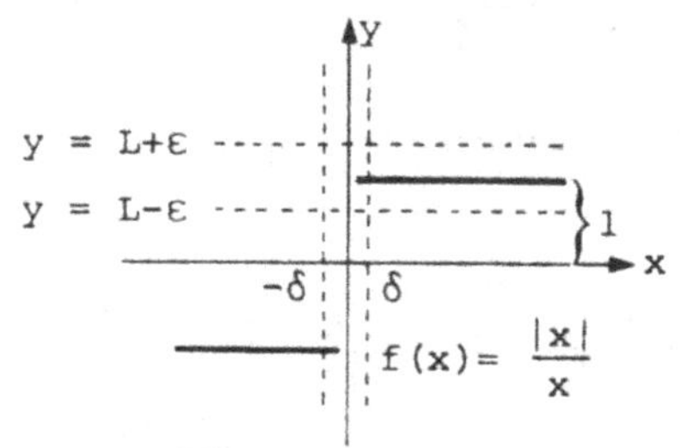

Fig. 4.6

Problem Solving Examples:

Find $\lim_{x \to 3} f(x) = \frac{1}{(x-3)^2}$, $x \neq 3$.

Sketching the graph of this function about $x = 3$, we see that it increases without bound as x tends to 3.

Using the method of simple substitution, we find that $\lim_{x \to 3} f(x) = \infty$. There is no limit.

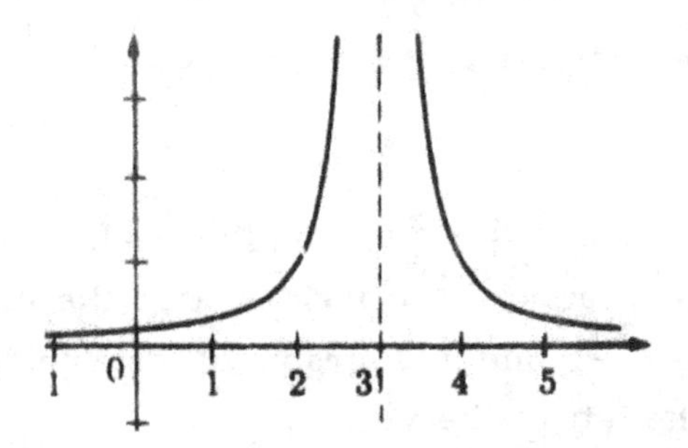

Find $\lim_{x \to 3} \frac{5x}{6-2x}$.

$$\lim_{x \to 3} 5x = 15$$

and

$$\lim_{x \to 3} [6-2x] = 0$$

Therefore,

$$\lim_{x \to 3} \frac{5x}{6-2x} = \frac{15}{0} = \infty .$$

The function has no limit.

4.5 Continuity

A function f is continuous at a point a if

$$\lim_{x \to a} f(x) = f(a).$$

This implies that three conditions are satisfied:

a) f(a) exists, that is, f is defined at a
b) $\lim_{x \to a} f(x)$ exists, and
c) the two numbers are equal.

To test continuity at a point x = a we test whether

$$\lim_{x \to a^+} F(x) = \lim_{x \to a^-} F(x) = F(a)$$

Problem Solving Examples:

Investigate the continuity of the function:

$$h(x) = \begin{cases} 3 + x & \text{if } x \leq 1 \\ 3 - x & \text{if } 1 < x . \end{cases}$$

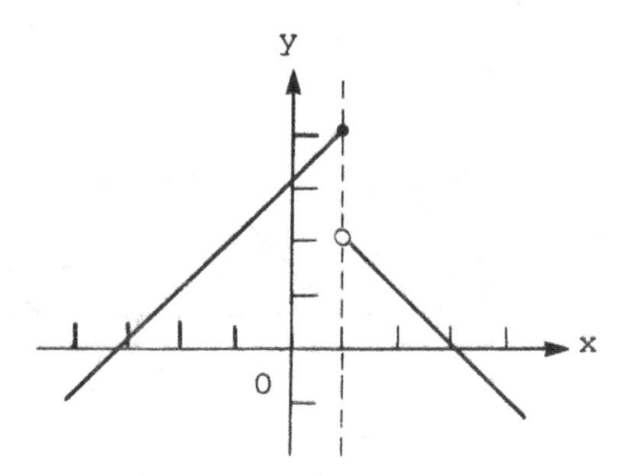

Because there is a break in the graph at the point x = 1, we investigate the three conditions for continuity at the point x = 1. The three conditions are: (1) $f(x_0)$ is defined, (2) $\lim_{x \to x_0} f(x)$ exists,

(3) $\lim_{x \to x_0} f(x) = f(x_0)$. At x = 1, h(1) = 4; therefore, condition (1) is satisfied.

$$\lim_{x \to 1^-} h(x) = \lim_{x \to 1^-} (3 + x) = 4$$

$$\lim_{x \to 1^+} h(x) = \lim_{x \to 1^+} (3 - x) = 2.$$

Because $\lim_{x \to 1^-} h(x) \neq \lim_{x \to 1^+} h(x)$, we conclude that $\lim_{x \to 1} h(x)$ does not exist. Therefore, condition (2) fails to hold at 1.

Hence, h is discontinuous at 1.

Investigate continuity of:

$$F(x) = \begin{cases} |x - 3| & \text{if } x \neq 3 \\ 2 & \text{if } x = 3. \end{cases}$$

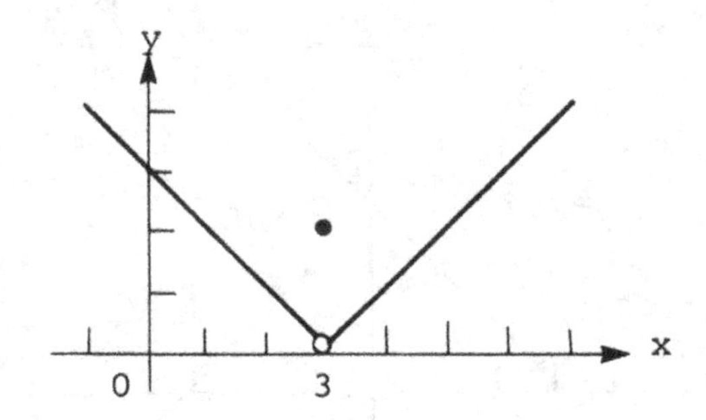

We investigate the three conditions for continuity at the point $x = 3$. The three conditions are: (1) $f(x_0)$ is defined,

(2) $\lim_{x \to x_0} f(x)$ exists, (3) $\lim_{x \to x_0} f(x) = f(x_0)$. At $x = 3$ we have

$F(3) = 2$; therefore, condition (1) is satisfied.

$\lim_{x \to 3^-} F(x) = 0$ and $\lim_{x \to 3^+} F(x) = 0$. Therefore, $\lim_{x \to 3} F(x)$

exists and is 0; therefore, condition (2) is satisfied.

$\lim_{x \to 3} F(x) = 0$ but $F(3) = 2$. Therefore, condition (3) is not

satisfied. F is thus discontinous at 3.

4.5.1 Theorems of Continuity

A) A function defined in a closed interval [a,b] is continuous in [a,b] if and only if it is continuous in the open interval (a,b), as well as continuous from the right at "a" and from the left at "b".

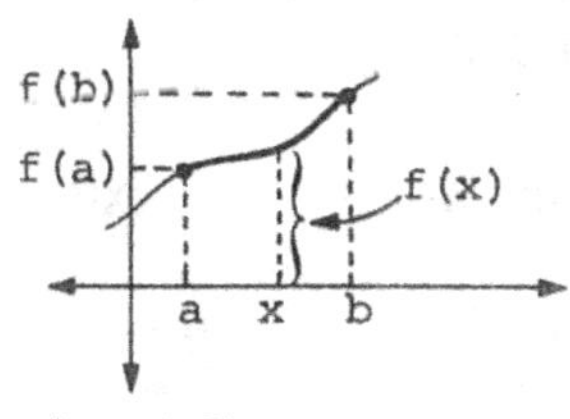

Fig. 4.7

B) If f and g are continuous functions at a, then so are the functions f+g, f-g, fg and f/g where $g(a) \neq 0$.

C) If $\lim_{x \to a} g(x) = b$ and f is continuous at b,

$$\text{then } \lim_{x \to a} f(g(x)) = f(b) = f[\lim_{x \to a} g(x)].$$

D) If g is continuous at a and f is continuous at $b = g(a)$, then

$$\lim_{x \to a} f(g(x)) = f[\lim_{x \to a} g(x)] = f(g(a)).$$

E) Intermediate Value Theorem. If f is continuous on a closed interval [a,b] and if $f(a) \neq f(b)$, then f takes on every value between f(a) and f(b) in the interval [a,b].

F) $f(x) = k$, $k \in R$ is continuous everywhere.

G) $f(x) = x$, the identity function is continuous everywhere.

H) If f is continuous at a, then $\lim_{n \to \infty} f(a + \frac{1}{n}) = f(a)$.

I) If f is continuous on an interval containing a and b, $a < b$, and if $f(a) \cdot f(b) < 0$ then there exists at least one point c, $a < c < b$ such that $f(c) = 0$.

Problem Solving Examples:

Let h be defined by:

$$h(x) = \begin{cases} 4 - x^2 & \text{if } x \le 1 \\ 2 + x^2 & \text{if } 1 < x . \end{cases}$$

Find each of the following limits if they exist:

$$\lim_{x \to 1^-} h(x) , \lim_{x \to 1^+} h(x), \lim_{x \to 1} h(x).$$

It is desirable to sketch the given function to aid in visualizing the problem.

Now,

$$\lim_{x \to 1^-} h(x) = \lim_{x \to 1^-} (4 - x^2) = 3$$

$$\lim_{x \to 1^+} h(x) = \lim_{x \to 1^+} (2 + x^2) = 3$$

Therefore, $\lim_{x \to 1} h(x)$ exists and is equal to 3. Note that $h(1) = 3$. This holds because the function is continuous.

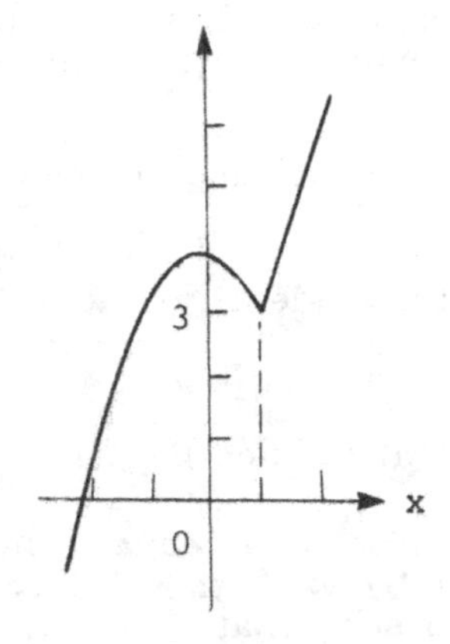

If $h(x) = \sqrt{4 - x^2}$, prove that h(x) is continuous in the closed interval [–2, 2].

A

To prove continuity we employ the following definition: A function defined in the closed interval [a,b] is said to be continuous in [a,b] if and only if it is continuous in the open interval (a,b), as well as continuous from the right at a and continuous from the left at b. The function h is continuous in the open interval (–2,2). We must show that the function is continuous from the right at –2 and from the left at 2. Therefore, we must show that f(–2) is defined and $\lim_{x \to -2^+} f(x)$ exists and that these are equal. Also, we must show that $f(2) = \lim_{x \to 2^-} f(x)$. We have:

$$\lim_{x \to -2^+} \sqrt{4 - x^2} = 0 = h(-2),$$

and

$$\lim_{x \to 2^-} \sqrt{4 - x^2} = 0 = h(2).$$

Thus, h is continuous in the closed interval [–2,2].

Quiz: Limits

1. The equation of each horizontal asymptote for $f(x) = \frac{1 - |x|}{x}$ is

   (A) $y = 1$

   (B) $y = -1$

   (C) $x = 0, x = 1, x = -1$

   (D) $y = 0$

   (E) $y = 1, y = -1$

2. $\lim_{x \to 1} \frac{\frac{1}{x+1} - \frac{1}{2}}{x - 1} =$

   (A) $-\frac{1}{4}$

   (B) -1

   (C) $\frac{1}{4}$

   (D) 0

   (E) does not exist

3. $\lim_{x \to 9} \frac{4x^2 - 35x - 9}{x - 9} =$

   (A) 35

   (B) 37

   (C) 0

   (D) 1

   (E) $+\infty$

4. If $f(x) = \begin{cases} \dfrac{2x-6}{x-3} & x \neq 3 \\ 5 & x = 3 \end{cases}$, then $\lim_{x \to 3} f(x) =$

(A) 5

(B) 1

(C) 2

(D) 6

(E) 0

5. The vertical asymptote and horizontal asymptote for $f(x) = \dfrac{\sqrt{x}}{x+4}$ is

(A) $x = -4, y = 0$

(B) no vertical asymptote, $y = 0$

(C) no vertical or horizontal asymptote

(D) $x = -4$, no horizontal asymptote

(E) $x = -4, y = 1$

6. $\lim_{n \to \infty} \left(1 + \dfrac{1}{n}\right)^{n+2} =$

(A) e^2

(B) $e + 2$

(C) $2e$

(D) e

(E) $e + e^2$

7. If $f(x) = \frac{1}{x-2}$ and $\lim_{x \to (-k+1)} f(x)$ does not exist, then $k =$

(A) 2

(B) 3

(C) 1

(D) –2

(E) –1

8. If $2 \leq f(x) \leq (1 - x)^2 + 2$ for $x \neq 1$, then $= \lim_{x \to 1} f(x) =$

(A) 3

(B) 2

(C) 4

(D) $\frac{5}{2}$

(E) 1

9. $\lim_{x \to -1} \frac{x + x^2}{x^2 - 1} =$

(A) $-\frac{1}{2}$

(B) 1

(C) –1

(D) $\frac{1}{2}$

(E) Does not exist

10. $\lim_{x \to +\infty} \left[\frac{1}{x} - \frac{x}{x-1}\right] = ?$

(A) -1

(B) 0

(C) 1

(D) 2

(E) None of these

ANSWER KEY

1. (E)	6. (D)
2. (A)	7. (E)
3. (B)	8. (B)
4. (C)	9. (D)
5. (B)	10. (A)

CHAPTER 5

The Derivative

5.1 The Definition and Δ-Method

The derivative of a function expresses its rate of change with respect to an independent variable. The derivative is also the slope of the tangent line to the curve.

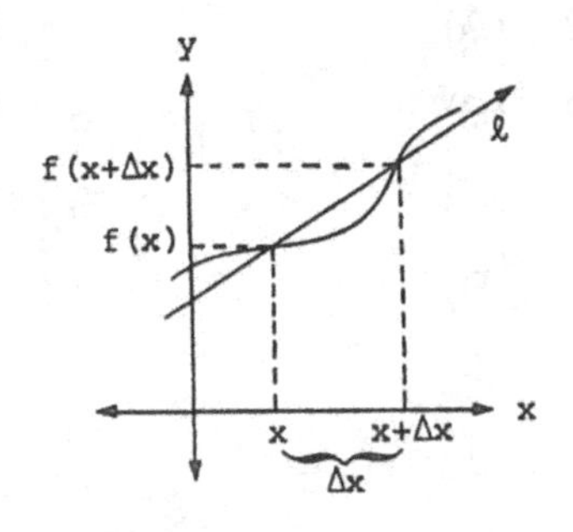

Fig. 5.1

Consider the graph of the function f in Fig. 5.1. Choosing a point x and a point $x + \Delta x$ (where Δx denotes a small distance on the x-axis) we can obtain both, f(x) and $f(x+\Delta x)$. Drawing a tangent line, ℓ, of the curve through the points f(x) and $f(x+\Delta x)$, we can measure the rate of change of this line. As we let the distance, Δx, approach zero, then

$$\lim_{\Delta x \to 0} \frac{f(x+\Delta x)-f(x)}{\Delta x}$$

becomes the instantaneous rate of change of the function or the derivative.

We denote the derivative of the function f to be f'. So we have

$$f'(x) = \lim_{\Delta x \to 0} \frac{f(x+\Delta x)-f(x)}{\Delta x}$$

If y = f(x), some common notations for the derivative are

$$y' = f'(x)$$

$$\frac{dy}{dx} = f'(x)$$

$$D_x y = f'(x) \text{ or } Df = f'$$

Problem Solving Examples:

Find the slope of each of the following curves at the given point, using the Δ-method.

a) $y = 3x^2 - 2x + 4$ at (1,5)

b) $y = x^3 - 3x + 5$ at (–2,3).

The slope of a given curve at a specified point is the derivative, in this case $\frac{\Delta y}{\Delta x}$, evaluated at that point.

a) From the Δ-method we know that:

$$\frac{\Delta y}{\Delta x} = \frac{f(x+\Delta x) - f(x)}{\Delta x}.$$

For the curve $y = 3x^2 - 2x + 4$, we find:

$$\frac{\Delta y}{\Delta x} = \frac{3(x + \Delta x)^2 - 2(x + \Delta x) + 4 - (3x^2 - 2x + 4)}{\Delta x}$$

$$= \frac{3x^2 + 6x\Delta x + 3(\Delta x)^2 - 2x - 2\Delta x + 4 - 3x^2 + 2x - 4}{\Delta x}$$

$$= \frac{6x\Delta x + 3(\Delta x)^2 - 2\Delta x}{\Delta x}$$

$$= 6x + 3\Delta x - 2.$$

$$\lim_{\Delta x \to 0} \frac{\Delta y}{\Delta x} = \lim_{\Delta x \to 0} 6x + 3\Delta x - 2 = 6x - 2.$$

At (1,5) $\frac{\Delta y}{\Delta x} = 4$ is the required slope.

b) Again using the Δ-method, $\frac{\Delta y}{\Delta x}$ for the curve:

$y = x^3 - 3x + 5$, can be found as follows:

$$\frac{\Delta y}{\Delta x} = \frac{f(x+\Delta x) - f(x)}{\Delta x}.$$

$$\frac{\Delta y}{\Delta x} = \frac{(x + \Delta x)^3 - 3(x\ \Delta x) + 5 - (x^3 - 3x + 5)}{\Delta x}.$$

$$\frac{\Delta y}{\Delta x} = \frac{(x + \Delta x)^3 - 3(x + \Delta x) + 5 - (x^3 - 3x + 5)}{\Delta x}$$

$$= \frac{x^3 + 3x^2\Delta x + 3x(\Delta x)^2 + (\Delta x)^3 - 3x - 3\Delta x + 5 - x^3 + 3x - 5}{\Delta x}$$

$$= \frac{3x^2\Delta x + 3x(\Delta x)^2 + (\Delta x)^3 - 3\Delta x}{\Delta x}$$

$$= 3x^2 + 3x\Delta x + (\Delta x)^2 - 3.$$

$$\lim_{\Delta x \to 0} \frac{\Delta y}{\Delta x} = \lim_{\Delta x \to 0} 3x^2 + 3x\Delta x + (\Delta x)^2 - 3 = 3x^2 - 3.$$

At $(-2,3), \frac{\Delta y}{\Delta x} = 9$ is the required slope.

Find the average rate of change, by the Δ process, for:

$$y = \frac{1}{x} \ .$$

$$y = f(x) = \frac{1}{x}$$

The average rate of change is defined to be

$$\frac{\Delta y}{\Delta x} \text{ with } \Delta y = f(x + \Delta x) - f(x).$$

Since

$$f(x) = \frac{1}{x}, f(x + \Delta x) = \frac{1}{x + \Delta x},$$

and

$$\Delta y = \frac{1}{x + \Delta x} - \frac{1}{x} = \frac{x - (x + \Delta x)}{x(x + \Delta x)}$$

$$= \frac{-\Delta x}{x(x + \Delta x)} \ .$$

Now,

$$\frac{\Delta y}{\Delta x} = \frac{-\Delta x}{x(x + \Delta x)\Delta x} = -\frac{1}{x(x + \Delta x)} \ .$$

Therefore, the average rate of change is $\frac{-1}{x(x + \Delta x)}$.

5.1.1 The Derivative at a Point

If f is defined on an open interval containing "a", then

$$f'(a) = \lim_{x \to a} \frac{f(x)-f(a)}{x-a},$$

provided the limit exists.

Problem Solving Examples:

Find the instantaneous rate of the function:

$$y = \frac{2x}{x+1}$$

for any value of x and for x = 2.

The instantaneous rate of change of a function is defined as,

$$\lim_{\Delta x \to 0} \frac{\Delta y}{\Delta x} = \lim_{\Delta x \to 0} \frac{f(x+\Delta x)-f(x)}{\Delta x}.$$

Therefore,

$$\Delta y = f(x + \Delta x) - f(x).$$

In this case,

$$f(x) = \frac{2x}{x+1}, \text{ therefore,}$$

$$f(x + \Delta x) = \frac{2(x + \Delta x)}{x + \Delta x + 1}.$$

Substituting, we have:

$$\Delta y = \frac{2x + 2 \cdot \Delta x}{x + \Delta x + 1} - \frac{2x}{x+1}$$

$$= \frac{(2x + 2 \cdot \Delta x)(x + 1) - 2x(x + \Delta x + 1)}{(x + \Delta x + 1)(x + 1)}$$

$$= \frac{2x^2 + 2x \cdot \Delta x + 2x + 2 \cdot \Delta x - 2x^2 - 2x \cdot \Delta x - 2x}{(x + \Delta x + 1)(x + 1)}$$

$$= \frac{2 \cdot \Delta x}{(x + \Delta x + 1)(x + 1)} .$$

$$\frac{\Delta y}{\Delta x} = \frac{2 \cdot \Delta x}{(x + \Delta x + 1)(x + 1)(\Delta x)}$$

$$= \frac{2}{(x + \Delta x + 1)\ (x + 1)}$$

Now,

$$\lim_{\Delta x \to 0} \frac{\Delta y}{\Delta x} = \lim_{\Delta x \to 0} \frac{2}{(x + \Delta x + 1)\ (x + 1)} .$$

Substituting 0 for Δx we have,

$$\lim_{\Delta x \to 0} \frac{\Delta y}{\Delta x} = \frac{2}{(x + 1)^2} ,$$

the instantaneous rate of change for any value of x.

For $x = 2$, we have,

$$\frac{2}{(x + 1)^2} = \frac{2}{(2 + 1)^2} = \frac{2}{9} .$$

Find the instantaneous rate of change of y with respect to x at the point $x = 5$, if

$$2y = x^2 + 3x - 1.$$

Instantaneous Rate of change is defined as

$$\lim_{\Delta x \to 0} \frac{\Delta y}{\Delta x}, \text{ with}$$

$$\Delta y = f(x + \Delta x) - f(x).$$

We have:

$$2\Delta y = (x + \Delta x)^2 + 3(x + \Delta x) - 1 - (x^2 + 3x - 1)$$

$$= x^2 + 2x \cdot \Delta x + (\Delta x)^2 + 3x + 3\Delta x - 1 - x^2 - 3x + 1$$

$$= 2x \cdot \Delta x + (\Delta x)^2 + 3\Delta x.$$

Dividing by Δx,

$$\frac{2\,\Delta y}{\Delta x} = \frac{2x \cdot \Delta x}{\Delta x} + \frac{(\Delta x)^2}{\Delta x} + \frac{3\,\Delta x}{\Delta x}$$

$$= 2x + \Delta x + 3$$

and

$$\frac{\Delta y}{\Delta x} = x + \frac{\Delta x}{2} + \frac{3}{2}.$$

Now,

$$\lim_{\Delta x \to 0} \frac{\Delta y}{\Delta x} = \lim_{\Delta x \to 0} x + \frac{\Delta x}{2} + \frac{3}{2} = x + \frac{3}{2}.$$

For $x = 5$,

$$\lim_{\Delta x \to 0} \frac{\Delta y}{\Delta x} = 5 + \frac{3}{2} = 6\frac{1}{2}.$$

5.2 Rules for Finding the Derivatives

General rule:

A) If f is a constant function, $f(x) = c$, then $f'(x) = 0$.

B) If $f(x) = x$, then $f'(x) = 1$.

C) If f is differentiable, then $(cf(x))' = cf'(x)$

D) Power Rule If $f(x) = x^n$, $n \in Z$, then

$f'(x) = nx^{n-1}$; if $n < 0$ then x^n is not defined at $x = 0$.

E) If f and g are differentiable on the interval (a,b) then:

a) $(f+g)'(x) = f'(x) + g'(x)$

b) Product Rule. $(fg)'(x) = f(x)g'(x) + g(x)f'(x)$

Example: Find $f'(x)$ if $f(x) = (x^3+1)(2x^2+8x-5)$.

$$f'(x) = (x^3+1)(4x+8)+(2x^2+8x-5)(3x^2)$$

$$= 4x^4 + 8x^3 + 4x + 8 + 6x^4 + 24x^3 - 15x^2$$

$$= 10x^4 + 32x^3 - 15x^2 + 4x + 8$$

c) Quotient Rule: $\left(\frac{f}{g}\right)'(x) = \frac{g(x)f'(x)-f(x)g'(x)}{[g(x)]^2}$

Example: Find $f'(x)$ if $f(x) = \frac{3x^2-x+2}{4x^2+5}$

$$f'(x) = \frac{-(3x^2-x+2)(8x)+(4x^2+5)(6x-1)}{(4x^2+5)^2}$$

$$= \frac{-(24x^3-8x^2+16x)+(24x^3-4x^2+30x-5)}{(4x^2+5)^2}$$

$$= \frac{4x^2+14x-5}{(4x^2+5)^2}$$

F) If $f(x) = x^{m/n}$, then $f'(x) = \frac{m}{n} x^{\frac{m}{n}-1}$

where $m, n \in Z$ and $n \neq 0$

G) Polynomials. If $f(x) = (a_0+a_1x+a_2x^2+...+a_nx^n)$

then $f'(x) = a_1+2a_2x+3a_3x^2+...+na_nx^{n-1}$

This employs the power rule and rules concerning constants.

H) Chain Rule. Let f(u) be a composite function, where $u=g(x)$.

Then $f'(x) = f'(u)g'(x)$ or if $y=f(u)$ and $u=g(x)$ then $D_xy = (D_uy)(D_xu) = f'(u)g'(x)$

Problem Solving Examples:

Find the derivative of: $y = x^{3b}$.

Applying the theorem for $d(u^n)$,

$$\frac{dy}{dx} = 3b \cdot x^{3b-1}.$$

Find the derivative of: $y = (x^2 + 2)^3$.

Method 1. We may expand the cube and write:

$$\frac{dy}{dx} = \frac{d}{dx}[(x^2+2)^3] = \frac{d}{dx}(x^6 + 6x^4 + 12x^2 + 8)$$

$$= 6x^5 + 24x^3 + 24x.$$

Method 2. Let $u = x^2 + 2$, then $y = (x^2 + 2)^3 = u^3$;

Using the chain rule we have:

$$\frac{dy}{dx} = \frac{dy}{du} \cdot \frac{du}{dx} = \frac{d(u^3)}{du} \cdot \frac{d(x^2 + 2)}{dx} = 3u^2(2x)$$

$$= 3(x^2 + 2)^2 \cdot (2x) = 3(x^4 + 4x^2 + 4) \cdot (2x)$$

$$= 6x^5 + 24x^3 + 24x.$$

5.3 Implicit Differentiation

An implicit function of x and y is a function in which one of the variables is not directly expressed in terms of the other. If these variables are not easily or practically separable, we can still differentiate the expression.

Apply the normal rules of differentiation such as the product rule, the power rule, etc. Remember also the chain rule which states $\frac{du}{dx} \times \frac{dx}{dt} = \frac{du}{dt}$.

Once the rules have been properly applied we will be left with as in the example of x and y, some factors of $\frac{dy}{dx}$.

We can then algebraically solve for the derivative $\frac{dy}{dx}$ and obtain the desired result.

Problem Solving Examples:

If $x^2 + y^2 = 16$, find $\frac{dy}{dx}$ as an implicit function of x and y.

Since y is a function of x, we differentiate the equation implicitly in terms of x and y. We have:

$$2x + 2y \cdot \frac{dy}{dx} = 0 \text{ or } 2y\frac{dy}{dx} = -2x.$$

$$\frac{dy}{dx} = -\frac{x}{y}.$$

Find $\frac{dy}{dx}$ for the expression: $2x^4 - 3x^2y^2 + y^4 = 0$.

A The equation $2x^4 - 3x^2y^2 + y^4 = 0$, could be solved for y and then differentiated to obtain $\frac{dy}{dx}$, but an easier method is to differentiate implicitly and then solve for $\frac{dy}{dx}$.

Hence, from $2x^4 - 3x^2y^2 + y^4 = 0$ we obtain:

$$8x^3 - 6x^2y\frac{dy}{dx} - 6xy^2 + 4y^3\frac{dy}{dx} = 0.$$

Solving for $\frac{dy}{dx}$,

$$4y^3\frac{dy}{dx} - 6x^2y\frac{dy}{dx} = 6xy^2 - 8x^3.$$

$$(4y^3 - 6x^2y)\frac{dy}{dx} = 6xy^2 - 8x^3.$$

$$\frac{dy}{dx} = \frac{6xy^2 - 8x^3}{4y^3 - 6x^2y}$$

$$= \frac{3xy^2 - 4x^3}{2y^3 - 3x^2y}.$$

5.4 Trigonometric Differentiation

The three most basic trigonometric derivatives are:

$$\frac{d}{dx}(\sin x) = \cos x,$$

$$\frac{d}{dx}(\cos x) = -\sin x,$$

$$\frac{d}{dx}(\tan x) = \sec^2 x.$$

Given any trigonometric function, it can be differentiated by applying these basics in combination with the general rules for differentiating algebraic expressions.

The following will be most useful if committed to memory:

$D_x \sin u = \cos u \, D_x u$
$D_x \cos u = -\sin u \, D_x u$
$D_x \tan u = \sec^2 u \, D_x u$
$D_x \sec u = \tan u \sec u \, D_x u$
$D_x \cot u = -\csc^2 u \, D_x u$
$D_x \csc u = -\csc u \cot u \, D_x u$

Problem Solving Examples:

Find the derivative of: $y = \sin ax^2$.

Applying the theorem for the derivative of the sine of a function,

$$\frac{dy}{dx} = \cos ax^2 \cdot \frac{d}{dx}(ax^2)$$

$$= 2ax \cos ax^2.$$

Find the derivative of: y tan 3θ.

Let $u = 3\theta$.

Then, $y = \tan u$, and

$$\frac{dy}{d\theta} = \frac{dy}{du} \cdot \frac{du}{d\theta}$$

$$\frac{du}{d\theta} = 3,$$

and $$\frac{dy}{du} = \sec^2 u.$$

Therefore,

$$\frac{dy}{d\theta} = \frac{dy}{du} \cdot \frac{du}{d\theta} = \sec^2 u \cdot 3 = 3 \sec^2 (3\theta).$$

5.5 Inverse Trigonometric Differentiation

Here are the derivatives for the inverse trigonometric functions which can be found in a manner similar to the above function. If u is a differentiable function of x then :

$$D_x \sin^{-1} u = \frac{1}{\sqrt{1-u^2}} D_x u \quad , \quad |u| < 1$$

$$D_x \cos^{-1} u = \frac{-1}{\sqrt{1-u^2}} D_x u \quad , \quad |u| < 1$$

$$D_x \tan^{-1} u = \frac{1}{1+u^2} D_x u \quad ,$$

$$D_x \sec^{-1} u = \frac{1}{|u|\sqrt{u^2-1}} D_x u \ , \ u = f(x), \quad |f(x)| > 1$$

$$D_x \cot^{-1} u = \frac{-1}{1+u^2} D_x u \quad ,$$

$$D_x \csc^{-1}u = \frac{-1}{|u|\sqrt{u^2-1}} D_x u, u = f(x), \quad |f(x)| > 1$$

Problem Solving Examples:

Find the derivative of y=arc sin 4x.

We use the formula for differentiation of the $\sin^{-1}$ or arc sin function, which states:

$$\frac{d}{dx} \sin^{1}u \quad \frac{1}{\sqrt{1 \quad u^2}}.$$

Hence

$$\frac{dy}{dx} \quad \frac{1}{\sqrt{1 \quad 16x^2}} 4 \quad \frac{4}{\sqrt{1 \quad 16x^2}}$$

Given: $y \quad \text{arc } \tan\frac{3}{x}$, find $\frac{dy}{dx}$.

In this example, we use the formula:

$$\frac{d \text{ arc } \tan u}{dx} \quad \frac{1}{1 \quad u^2} J \frac{du}{dx}.$$

For

$$y \quad \text{arc } \tan\frac{3}{x}, \; u \quad \frac{3}{x}, \text{ and } du \quad \frac{3}{x^2}.$$

Therefore,

$$\frac{dy}{dx} \quad \frac{1 \frac{3}{x^2}}{1 \quad \frac{3}{x}^{2}} \quad \frac{\frac{3}{x^2}}{\frac{x^2 \quad 9}{x^2}} \quad \frac{3}{x^2 \quad 9}.$$

5.6 Exponential and Logarithmic Differentiation

The exponential function e^x has the simplest of all derivatives. Its derivative is itself.

$$\frac{d}{dx} e^x = e^x$$

and

$$\frac{d}{dx} e^u = e^u \frac{du}{dx}$$

Since the natural logarithmic function is the inverse of $y = e^x$ and $\ln e = 1$, it follows that

$$\frac{d}{dx} \ln y = \frac{1}{y} \frac{dy}{dx}$$

and

$$\frac{d}{dx} \ln u = \frac{1}{u} \frac{du}{dx}$$

If x is any real number and a is any positive real number, then

$$a^x = e^{x \ln a}$$

From this definition we can obtain the following:

a) $\frac{d}{dx} a^x = a^x \ln a$ and $\frac{d}{dx} a^u = a^u \ln a \frac{du}{dx}$

b) $\frac{d}{dx} (\log_a x) = \frac{1}{x \ln a}$ and $\frac{d}{dx} \log_a |u| = \frac{1}{u \ln a} \frac{du}{dx}$
where $u \neq 0$

Sometimes it is useful to take the logs of a function and then differentiate since the computation becomes easier (as in the case of a product).

Problem Solving Examples:

If $y = e^{\frac{1}{x^2}}$, find $D_x y$.

To find $D_x y = \frac{dy}{dx}$, we use the differentiation formula:

$\frac{d}{dx} e^u = e^u \frac{du}{dx}$, with $u = \frac{1}{x^2}$. We obtain:

$$D_x y = e^{\frac{1}{x^2}} \left(-\frac{2x}{x^4}\right)$$

$$= e^{\frac{1}{x^2}} \left(-\frac{2}{x^3}\right) = -\frac{2e^{\frac{1}{x^2}}}{x^3}.$$

Find the derivative of:

$$y = \left(e^{\frac{1}{x}}\right)^2.$$

We can first rewrite the function as:

$$y = e^{\frac{2}{x}}.$$

Now we use the formula:

$$\frac{d}{dx} e^u = e^u \frac{du}{dx},$$

letting $u = \frac{2}{x}$. Then,

$$\frac{du}{dx} = \frac{(x)(0) - (2)(1)}{x^2} = -\frac{2}{x^2}.$$

Applying the formula, we obtain:

$$\frac{dy}{dx} = e^{\frac{2}{x}} \cdot -\frac{2}{x^2}$$

$$= -\frac{2e^{\frac{2}{x}}}{x^2}.$$

5.6.1 Steps in Logarithmic Differentiation

1. $y = f(x)$ given
2. $\ln y = \ln f(x)$ take logs and simplify
3. $D_x(\ln y) = D_x(\ln f(x))$ differentiate implicitly
4. $\frac{1}{y} D_x y = D_x(\ln f(x))$
5. $D_x y = f(x) D_x(\ln f(x))$ multiply by $y = f(x)$

To complete the solution it is necessary to differentiate $\ln f(x)$. If $f(x) < 0$ for some x then step 2 is invalid and we should replace step 1 by $|y| = |f(x)|$, and then proceed.

Example: $y = (x+5)(x^4+1)$

$$\ln y = \ln[(x+5)(x^4+1)] = \ln(x+5) + \ln(x^4+1)$$

$$\frac{d}{dx} \ln y = \frac{d}{dx} \ln(x+5) + \frac{d}{dx} \ln(x^4+1)$$

$$\frac{1}{y}\frac{dy}{dx} = \frac{1}{x+5} + \frac{4x^3}{x^4+1}$$

$$\frac{dy}{dx} = (x+5)(x^4+1) \left[\frac{1}{x+5} + \frac{4x^3}{x^4+1}\right]$$

$$= (x^4+1) + 4x^3(x+5)$$

This is the same result as obtained by using the product rule.

Problem Solving Examples:

Find the derivative of $y = \ln(1 - 2x)^3$

It is best to rewrite the equation as:

$$y = 3 \ln(1 - 2x).$$

Then we apply the formula:

$$\frac{d}{dx} \ln u = \frac{1}{u} \frac{du}{dx},$$

letting $u = (1 - 2x)$. Then $\frac{du}{dx} = -2$. We obtain:

$$\frac{dy}{dx} = 3\left(\frac{1}{1-2x}\right)(-2) = \frac{-6}{1-2x}.$$

Differentiate: $y = \log \frac{2x}{1 + x^2}$.

To find $\frac{dy}{dx}$, we use the formula, $\frac{d}{dx} \log_a u = \frac{1}{u} \frac{du}{dx} \log_a e$,

letting $u = \frac{2x}{1 + x^2}$, and the a is understood to equal 10. (We recall that when the base 10 is used, it need not be written.) Applying the formula, we obtain:

$$\frac{dy}{dx} = \frac{1 + x^2}{2x} \cdot \frac{(1+x^2)(2) - (2x)(2x)}{(1+x^2)^2} \cdot \log e.$$

Simplifying, we obtain,

$$\frac{dy}{dx} = (\log e)\frac{1 - x^2}{x(1 + x^2)}.$$

Find the derivative of the expression:

$$y = \ln 3x^4.$$

To find $\frac{dy}{dx}$, we use the differentiation formula:

$$\frac{d}{dx} \ln u = \frac{1}{u}\frac{du}{dx}, \text{ letting } u = 3x^4, \frac{du}{dx} = 12x^3.$$

Applying the formula, we obtain:

$$\frac{dy}{dx} = \frac{1}{3x^4}(12x^3) = \frac{4}{x}.$$

5.7 Higher Order Derivatives

The derivative of any function is also a legitimate function which we can differentiate. The second derivative can be obtained by:

$$\frac{d}{dx}\left[\frac{d}{dx} u\right] = \frac{d^2}{dx^2} u = u'' = D^2u,$$

where $u = g(x)$ is differentiable.

The general formula for higher orders and the nth derivative of u is,

$$\underbrace{\frac{d}{dx}\frac{d}{dx}\cdots\frac{d}{dx}}_{\text{n times}} u = \frac{d^{(n)}}{dx^n} u = u^{(n)} = D_x^{(n)}u.$$

The rules for first order derivatives apply at each stage of higher order differentiation (e.g., sums, products, chain rule).

A function which satisfies the condition that its nth derivative is zero, is the general polynomial

$$p_{n-1}(x) = a_{n-1}x^{n-1} + a_{n-2}x^{n-2} + \ldots + a_0.$$

Problem Solving Examples:

Find the sixth derivative of $y = x^6$.

First derivative $= 6x^{6-1} = 6x^5$

Second derivative $= 5 \cdot 6x^{5-1} = 30x^4$

Third derivative $= 4 \cdot 30x^{4-1} = 120x^3$

Fourth derivative $= 3 \cdot 120x^{3-1} = 360x^2$

Fifth derivative $= 2 \cdot 360x^{2-1} = 720x^1 = 720x$

Sixth derivative $= 1 \cdot 720x^{1-1} = 720x^0 = 720$

The seventh derivative is seen to be zero, and therefore the function $y = x^6$ has seven derivatives.

Find the general expression for the nth derivative of:

$$f(x) = \frac{1}{3x + 2}.$$

$$f(x) = \frac{1}{3x + 2} = (3x + 2)^{-1}$$

First derivative:
$f'(x) = -1(3x + 2)^{-2}\ (3) = -3(3x + 2)^{-2}$.

Second derivative:
$f''(x) = 6(3x + 2)^{-3}\ (3) = 18(3x + 2)^{-3}$.

Third derivative:
$f'''(x) = -54(3x + 2)^{-4}\ (3) = 162(3x + 2)^{-4}$.

Fourth derivative:
$f''''(x) = 648(3x + 2)^{-5}\ (3) = 1944(3x + 2)^{-5}$.

To express the nth derivative, a pattern that the sequence of derivatives follows must be found.

1) First, it is noted that the derivatives are alternately negative and positive. The odd order of differentiation results in a negative value, and the even order of differentiation, in a positive value. This property can be expressed by $(-1)^n$ for the nth derivative.

2) The coefficients of $(3x + 2)$ (ignoring the sign) are 3, 18, 162, 1944,...

First derivative:
$n = 1$: $3 = 3^1 \cdot = 3^n \cdot n!$

Second derivative:
$n = 2$: $18 = 3^2\ (2!) = 3^n \cdot n!$

Third derivative:
$n = 3: 162 = 3^3 (3!) = 3^n \cdot n!$

Fourth derivative:
$n = 4: 1944 = 3^4 (4!) = 3^n \cdot n!$

3) The power of $(3x + 2)$ is considered.

First derivative:
$n = 1: (3x + 2)^{-2} = \frac{1}{(3x + 2)^2} = \frac{1}{(3x + 2)^{n+1}}$

Second derivative:
$n = 2: (3x + 2)^{-3} = \frac{1}{(3x + 2)^3} = \frac{1}{(3x + 2)^{n+1}}$

Third derivative:
$n = 3: (3x + 2)^{-4} = \frac{1}{(3x + 2)^4} = \frac{1}{(3x + 2)^{n+1}}$

Fourth derivative:
$n = 4: (3x + 2)^{-5} = \frac{1}{(3x + 2)^5} = \frac{1}{(3x + 2)^{n+1}}$

Combining these results, the nth derivative is:

$$f^n(x) = (-1)^n \frac{3^n \cdot n!}{(3x + 2)^{n+1}}.$$

If $y = a\sqrt{x} + \frac{b}{\sqrt{x}}$, show that $4x^2D_x^2 y + 4xD_x y - y = 0$.

We first rewrite the given function as

$$y = ax^{\frac{1}{2}} + bx^{-\frac{1}{2}}.$$

$$D_x y = \frac{dy}{dx} = \frac{1}{2} ax^{-\frac{1}{2}} - \frac{1}{2} bx^{-\frac{3}{2}}.$$

$$D_x^2 y = \frac{d^2y}{dx^2} = -\frac{1}{4} ax^{-\frac{3}{2}} + \frac{3}{4} bx^{-\frac{5}{2}}.$$

Substituting, we have:

$$4x^2D_x^2y + 4xD_xy - y = 4x^2\left[\frac{a}{4x^{\frac{3}{2}}} + \frac{3b}{4x^{\frac{5}{2}}}\right] + 4x\left[\frac{a}{2x^{\frac{1}{2}}} - \frac{b}{2x^{\frac{3}{2}}}\right]$$

$$-\left[ax^{\frac{1}{2}} + \frac{b}{x^{\frac{1}{2}}}\right]$$

$$= -\frac{4ax^2}{4x^{\frac{3}{2}}} + \frac{12bx^2}{4x^{\frac{5}{2}}} + \frac{4ax}{2x^{\frac{1}{2}}} - \frac{4bx}{2x^{\frac{3}{2}}} - ax^{\frac{1}{2}} - \frac{b}{x^{\frac{1}{2}}}$$

$$= -ax^{\frac{1}{2}} + \frac{3b}{x^{\frac{1}{2}}} + 2ax^{\frac{1}{2}} - \frac{2b}{x^{\frac{1}{2}}} - ax^{\frac{1}{2}} - \frac{b}{x^{\frac{1}{2}}}$$

$$= 2ax^{\frac{1}{2}} - 2ax^{\frac{1}{2}} + \frac{3b}{x^{\frac{1}{2}}} - \frac{3b}{x^{\frac{1}{2}}} = 0.$$

Find $y'' \frac{d^2y}{dx^2}$ for the expression $xy^3 = 1$.

To find the second derivative, y'', we must first find the first derivative and then differentiate that to obtain the second derivative. We could solve for y and then differentiate to obtain y', but an alternative is implicit differentiation,

$$xy^3 = 1.$$

Differentiating implicitly,

$$3xy^2 \cdot y' + y^3 = 0.$$

$$3xy^2 \cdot y' = -y^3 .$$

$$y' = \frac{-y^3}{3xy^2}$$

$$= -\frac{y}{3x} .$$

Now we take the derivative of y' to find y''.

$$y'' = -\frac{1}{3}\left[\frac{x \cdot y' - y}{x^2}\right].$$

Substituting $y' = -\frac{y}{3x}$ in the expression for y'' and simplifying

$$y'' = -\frac{1}{3}\left[\frac{-x\left(\frac{-y}{3x}\right) - y}{x^2}\right]$$

$$= -\frac{1}{3}\left[\frac{-\frac{y}{3} - y}{x^2}\right]$$

$$= -\frac{1}{3}\left[\frac{-\frac{4}{3}y}{x^2}\right]$$

$$= \frac{4y}{9x^2}.$$

Draw the graph of:

$$y = \frac{x^3 + x^2}{x^2 - 4}$$

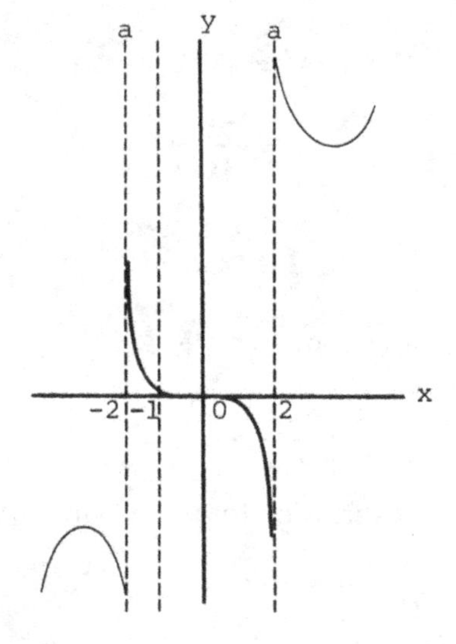

Upon setting the denominator equal to zero, we find that the vertical asymptotes are $x = 2$ and $x = -2$.

Setting $y = 0$, the graph intercepts the x-axis at $x = 0$ and $x = -1$.

The function

$$y = \frac{x^3 + x^2}{x^2 - 4}$$

can be written as

$$y = \frac{\left(1 + \frac{1}{x}\right)}{\left(\frac{1}{x} - \frac{4}{x^3}\right)}.$$

As $x \to \infty, y \to \infty$. This shows that there are no horizontal asymptotes.

To obtain the critical points, we take the first derivative.

$$y' = \frac{dy}{dx} = \frac{x(x^3 - 12x - 8)}{(x^2 - 4)^2}$$

The critical points are $x = 0$ and the solutions of the equation $x^3 - 12x - 8$ (which are approximately, $x = -3.1$, $x = -0.7$ and $x = 3.7$).

Using the second derivative test, we find that $x = -3.1$ and $x = 0$ are relative maxima and $x = -0.7$ and $x = 3.7$ are relative minima.

Draw the graph of:

$$y^2 = \frac{x^2 - 1}{x^2 + 1}$$

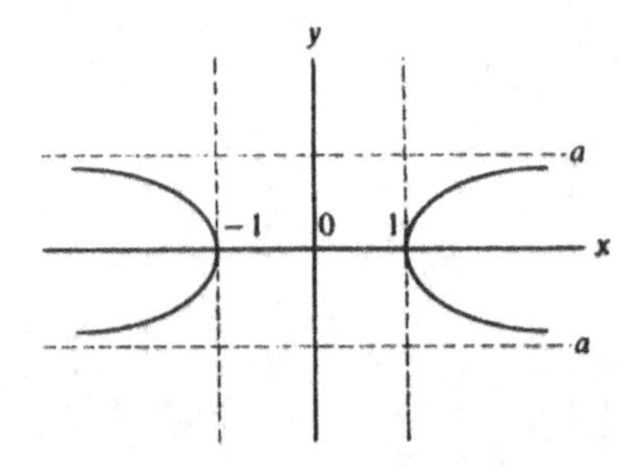

Since we can replace x with (–x) and y with (–y) without changing the value of the equation, this tells us that the graph is symmetric with the x-axis and y-axis.

To determine vertical asymptotes, we set the denominator to zero. There are no real values of x that will satisfy that equation, therefore there are no vertical asymptotes.

$$\text{Note that } y^2 = \frac{x^2 - 1}{x^2 + 1} = \frac{1 - \frac{1}{x^2}}{1 + \frac{1}{x^2}}.$$

As $x \to \infty$, $y^2 \to 1$, which tells us that $y = 1$ and $y = -1$ are horizontal asymptotes.

For values of x in the interval [0,1], y^2 is negative which is impossible. Therefore, the equation is not defined and there is no graph beween 0 and 1, and since the graph is symmetric about the y-axis, there is no graph between –1 and 0.

$$y = \sqrt{\frac{x^2 - 1}{x^2 + 1}}$$

$$y' = \frac{dy}{dx} = \frac{1}{2}\left(\frac{x^2-1}{x^2+1}\right)^{\frac{-1}{2}}\left(\frac{(x^2+1)(2x)-(x^2-1)(2x)}{(x^2+1)^2}\right)$$

$$= \frac{2x}{(x^2+1)^2\sqrt{\frac{x^2-1}{x^2+1}}}$$

$y' = 0$ only if $x = 0$. Since the graph is not defined in the interval $[-1,1]$, the graph has no critical points.

Quiz: The Derivative

1. If $f(x) = \pi^2$, then $f'(1) =$

   (A) 2π

   (B) 0

   (C) π

   (D) 1

   (E) π^2

2. $\lim\limits_{h \to 0} \dfrac{\sin(\pi + h) - \sin \pi}{h}$

   (A) 1

   (B) 0

   (C) -1

   (D) $+\infty$

   (E) $-\infty$

3. If $y = \arccos(\cos^4 x - \sin^4 x)$, then $y'' =$

   (A) 2

   (B) 0

   (C) $-2(\cos x - \sin x)$

   (D) $-2(\sin x + \cos x)$

   (E) -1

4. If $f'(c) = 0$ for $f(x) = 3x^2 - 12x + 9$, where $0 \leq x \leq 4$, then $c =$

   (A) 2

   (B) 3

   (C) 0

   (D) 1

   (E) $\frac{1}{3}$

5. If $g(x) = \frac{-x - f(x)}{f(x)}$, $f(1) = 4$ and $f'(1) = 2$, then $g'(1) =$

   (A) $-\frac{1}{2}$

   (B) $\frac{11}{8}$

   (C) $\frac{3}{16}$

   (D) $\frac{1}{8}$

   (E) $-\frac{1}{8}$

6. If $f(x) = |x|$, then

(A) Domain of f' = Domain of f.

(B) $f'(x) = \frac{|x|}{x}$ for every real number x.

(C) $(f'(x))(f(x)) = f(x)$ for every real number x.

(D) Range of $f' = (-1, 1)$.

(E) The graph of f' is

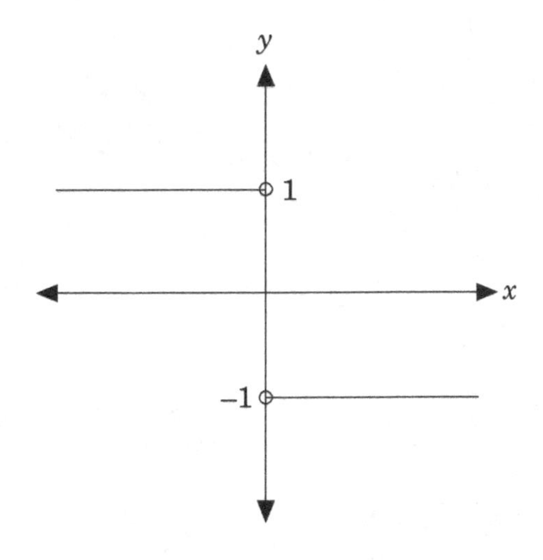

7. If f is differentiable at 0, and $g(x) = [f(x)]^2, f'(0) = f(0) = -1$, then $g'(0) =$

(A) –2

(B) –1

(C) 1

(D) 4

(E) 2

8. If the k-th derivative of $(3x - 2)^3$ is zero, then k is necessarily

   (A) 4

   (B) 3

   (C) ≥ 4

   (D) 2

   (E) 5

9. If arc tan $(x) = \ln(y^2)$, then, in terms of x and y, $\frac{dy}{dx} =$

   (A) $\frac{1}{1 - x^2}$

   (B) $\frac{-1}{1 - x^2}$

   (C) $\frac{y}{1 - x^2}$

   (D) $\frac{y}{1 + x^2}$

   (E) $\frac{y}{2(1 + x^2)}$

10. If $y = \frac{3}{\sin x + \cos x}$ then $\frac{dy}{dx} =$

    (A) $3 \sin x - 3 \cos x$

    (B) $\frac{6 \sin x}{1 + 2 \sin x \cos x}$

    (C) $\frac{3}{\cos x - \sin x}$

(D) $\dfrac{-3}{(\sin x + \cos x)^2}$

(E) $\dfrac{3(\sin x - \cos x)}{1 + 2\sin x \cos x}$

ANSWER KEY

1. (B)
2. (C)
3. (B)
4. (A)
5. (E)
6. (D)
7. (E)
8. (C)
9. (E)
10. (E)

Applications of the Derivative

6.1 Rolle's Theorem

Let f be continuous on a closed interval [a,b]. Assume f'(x) exists at each point in the open interval (a,b).

If f(a) = f(b) = 0 then there is at least one point (x_0) in (a,b) such that $f'(x_0) = 0$.

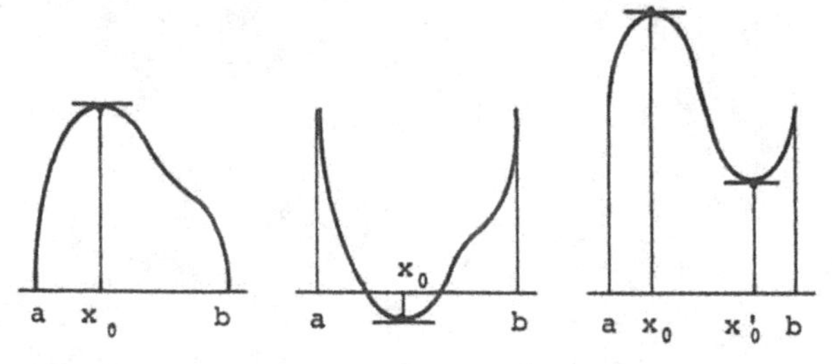

Fig. 6.1 Three functions which satisfy the hypotheses, hence the conclusion, of Rolle's theorem.

6.2 The Mean Value Theorem

If f is continuous on [a,b] and has a derivative at every point in the interval (a,b), then there is at least one number c in (a,b) such that

$$f'(c) = \frac{f(b)-f(a)}{b-a}$$

Notice in Fig. 6.2 that the secant has slope

$$\frac{f(b)-f(a)}{b-a}$$

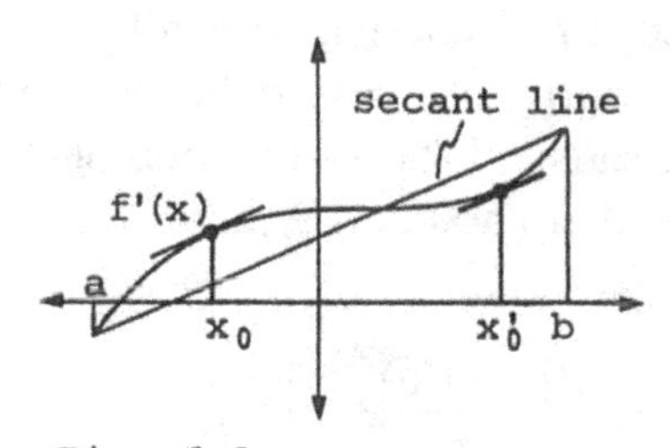

Fig. 6.2

and f'(x) has slope of the tangent to the point (x,f(x)). For some x_0 in (a,b) these slopes are equal.

Problem Solving Examples:

a) State and prove the Mean Value Theorem for the derivative of a real valued function of a single real variable.

b) Give a geometrical interpretation to this result.

A a) Let f be a real valued function of a real variable, x, which is continuous on a closed interval [a,b] and has a derivative in the open interval (a,b). Then the Mean Value Theorem states that there exists a point c in (a,b) such that

$$f(b) - f(a) = f'(c)(b - a) . \qquad (1)$$

To prove this theorem, consider the function

$$\phi(x) = f'(x) - \left(f(a) + \frac{f(b) - f(a)}{b - a}(x - a) \right). \tag{2}$$

As can be seen from the figure, ϕ is the difference of f and the linear function whose graph consists of the line segment passing through the points (a, f(a)) and (b, f(b)). Since f is continous on [a,b], so is ϕ and since f has a derivative at all points in (a, b), so does ϕ. Furthermore, $\phi(a) = \phi(b) = 0$ so that all the conditions of Rolle's Theorem are satisfied for the function $\phi(x)$. Hence the conclusion of that theorem holds; i.e., there is a point $c \in (a, b)$ such that

$$\phi(c) = f'(c) - \frac{f(b) - f(a)}{b - a} = 0$$

or

$$f'(c) = \frac{f(b) - f(a)}{b - a}. \tag{3}$$

Thus, the theorem is proved.

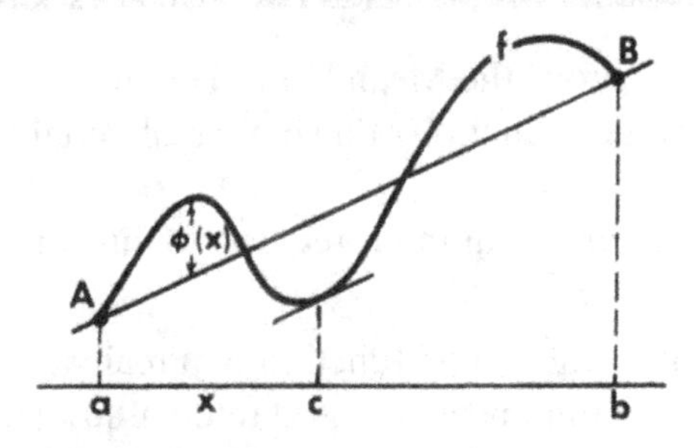

b) The geometrical interpretation of (3) can be seen in the figure. The equation states that there is a point c whose tangent line has the same slope as (i.e., is parallel to) the line connecting A and B.

If $f(x) = 3x^2 - x + 1$, find the point x_0 at which $f'(x)$ assumes its mean value in the interval [2,4].

Recall the Mean Value Theorem. Given a function f(x) which is continuous in [a,b] and differentiable in (a, b), there exists a point x_0 where $a < x_0 < b$ such that:

$$\frac{f(b) - f(a)}{b - a} = f'(x_0),$$

Where x_0 is the mean point in the interval.

In our problem, $3x^2 - x + 1$ is continuous, and the derivative exists in the interval (2,4). We have:

$$\frac{f(4) - f(2)}{4 - 2} = \frac{[3(4)^2 - 4 + 1] - [3(2)^2 - 2 + 1]}{4 - 2}$$

$$= f'(x_0),$$

or

$$\frac{45 - 11}{2} = 17 = f'(x_0) = 6x_0 - 1.$$

$$6x_0 = 18$$

$$x_0 = 3.$$

$x_0 = 3$ is the point where $f'(x)$ assumes its mean value.

6.2.1 Consequences of the Mean Value Theorem

A) If f is defined on an interval (a,b) and if $f'(x) = 0$ for each point in the interval, then $f(x)$ is constant over the interval. Fig. 6.3.

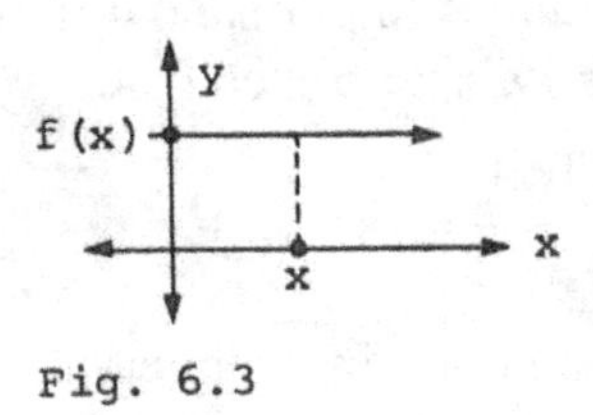

Fig. 6.3

B) Let f and g be differentiable on an interval (a,b). If, for each point x in the interval, $f'(x)$ and $g'(x)$ are equal, then there is a constant, c, such that

$$f(x) + c = g(x) \quad \text{for all } x.$$

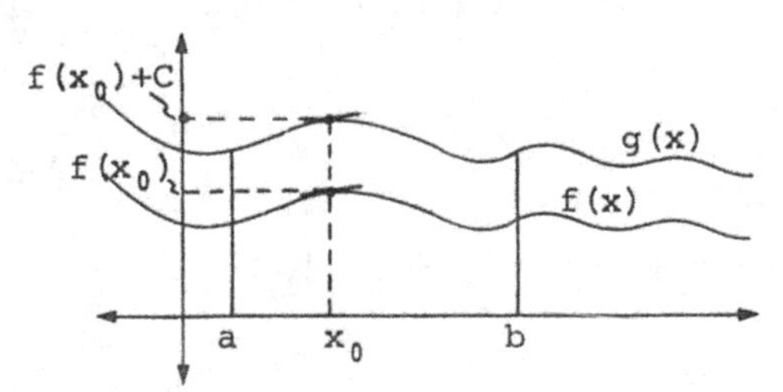

Fig. 6.4 $f(x)+C=g(x)$ for all x

C) The Extended Mean Value Theorem. Assume that the function f and its derivative f' are continuous on [a,b] and that f'' exists at each point x in (a,b) , then there exists at least one point x_0, $a < x_0 < b$, such that

$$f(b) = f(a) + (b-a)f'(a) + \tfrac{1}{2}(b-a)^2 f''(x_0).$$

Problem Solving Examples:

Show: $e^x \geq 1 + x$ for all real numbers x.

Divide the problem into 3 cases; $x = 0, x > 0, x < 0$.

Case 1. $x = 0$

For $x = 0$, we have $e^0 \geq 1 + 0$, or $1 = 1$.

Case 2. $x > 0$

For this case, we apply the Mean Value Theorem. We let

$f(x) = e^x$

and the interval will be $[0, x]$. Applying the theorem, we have

$$\frac{f(x) - f(0)}{x - 0} = \frac{e^x - e^o}{x - 0} = f'(x_0) = e^{x_o} ,$$

where $0 < x_o < x$.

Simplifying, we have:

$e^x = e^{x_o} \cdot x + e^o = xe^{x_o} + 1$

since $x_o > 0$, $e^{x_o} > 1$. Therefore

$e^x > x + 1$

Case 3. $x< 0$

Solution is similar to Case 2 and will be left to the reader as an exercise.

Combining the three results, we have the desired inequality.

6.3 L'Hôpital's Rule

An application of the Mean Value Theorem is in the evaluation of

$$\lim_{x \to a} \frac{f(x)}{g(x)} \text{ where } f(a)=0 \text{ and } g(a)=0.$$

L'Hôpital's Rule states that if the $\lim_{x \to a} \frac{f(x)}{g(x)}$ is an indeterminate form (i.e., $\frac{0}{0}$ or $\frac{\infty}{\infty}$), then we can differentiate the numerator and the denominator separately and arrive at an expression that has the same limit as the original problem.

Thus, $\lim_{x \to a} \frac{f(x)}{g(x)} = \lim_{x \to a} \frac{f'(x)}{g'(x)}$

In general, if f(x) and g(x) have properties

1) $f(a) = g(a) = 0$

2) $f^{(k)}(a) = g^{(k)}(a) = 0$ for $k=1,2,\ldots n$

but 3) $f^{(n+1)}(a)$ or $g^{(n+1)}(a)$ is not equal to zero, then

$$\lim_{x \to a} \frac{f(x)}{g(x)} = \frac{f^{(n+1)}(x)}{g^{(n+1)}(x)}$$

Problem Solving Examples:

Evaluate $\lim_{x \to 2} \dfrac{(2x^2 - 4x)}{x - 2}$.

The function takes the form $\dfrac{0}{0}$ and therefore we can apply L'Hôpital's rule to obtain:

$$\lim_{x \to 2} \frac{4x - 4}{1} = 4$$

We can also solve the problem in a different way by noting that the numerator can be factored.

$$\lim_{\to 2} \frac{2x^2 - 4x}{x - 2} = \lim_{x \to 2} \frac{2x(x - 2)}{x - 2}$$

$$= \lim_{x \to 2} 2x$$

$$= 4.$$

Find $\lim_{x \to 3} \dfrac{x^2 - x - 6}{x - 3}$

This limit may be found by writing

$$\frac{x^2 - x - 6}{x - 3} = \frac{(x + 2)(x - 3)}{x - 3} = x + 2.$$

Hence $\lim_{x \to 3} (x + 2) = 5.$

Since 0/0 (indeterminate) is obtained by substitution in the original function, the limit may also be obtained by L'Hôpital's rule by differentiating seperately numerator and denominator. Thus

$$\lim_{x \to 3} \frac{x^2 - x - 6}{x - 3} = \lim_{x \to 3} \frac{2x - 1}{1} = 5.$$

The application of L'Hôpital's rule is the more systematic approach and should generally be tried first, if another method is not immediately apparent.

6.4 Tangents and Normals

6.4.1 Tangents

A line which is tangent to a curve at a point "a", must have the same slope as the curve. That is, the slope of the tangent is simply

$$m = \lim_{h \to 0} \frac{f(a+h) - f(a)}{h}$$

Therefore, if we find the derivative of a curve and evaluate for a specific point, we obtain the slope of the curve and the tangent line to the curve at that point.

A curve is said to have a vertical tangent at a point $(a, f(a))$ if f is continuous at "a" and $\lim_{x \to a} |f'(x)| = \infty$.

Problem Solving Examples:

Using the Δ-method, find the points on the curve:

$$y = \frac{x}{3} + \frac{3}{x}$$, at which the tangent line is horizontal.

When the slope of a curve equals zero, the curve has a horizontal tangent. We can find the points at which the tangent line is horizontal by calculating the slope $\frac{dy}{dx}$, setting it equal to zero and solving for x. By the Δ-method,

$$\frac{\Delta y}{\Delta x} = \frac{f(x + \Delta x) - f(x)}{\Delta x}$$

$$\frac{\Delta y}{\Delta x} = \frac{\frac{(x + \Delta x)}{3} + \frac{3}{(x + \Delta x)} - \left(\frac{x}{3} - \frac{3}{x}\right)}{\Delta x}$$

$$= \frac{\frac{x}{3} + \frac{\Delta x}{3} + \frac{3}{x + \Delta x} - \frac{x}{3} - \frac{3}{x}}{\Delta x}$$

$$= \frac{\frac{\Delta x}{3} + \frac{3}{x + \Delta x} - \frac{3}{x}}{\Delta x}.$$

$$\frac{\Delta y}{\Delta x} = \frac{\dfrac{x^2\Delta x + x(\Delta x)^2 - 9\Delta x}{3(x + \Delta x)x}}{\Delta x}$$

$$= \frac{x^2 + x\Delta x - 9}{3x^2 + 3x\Delta x}.$$

$$\frac{dy}{dx} = \lim_{\Delta x \to 0} \frac{\Delta y}{\Delta x} = \lim_{\Delta x \to 0} \frac{x^2 + x\Delta x - 9}{3x^2 + 3x\Delta x} = \frac{x^2 - 9}{3x^2}.$$

We can set this value, which is the slope, equal to zero and solve for x.

$$\frac{x^2 - 9}{3x^2} = 0$$

$$\frac{(x - 3)(x + 3)}{x(3x)} = 0$$

$$x = 3, \; x = -3.$$

(Remember that x cannot be zero, for that would give an infinite slope.)

Substituting these values for x back into $y = \frac{x}{3} + \frac{3}{x}$, we can obtain the y coordinates. We find (3, 2) and (–3, –2) to be the required points.

6.4.2 Normals

A line normal to a curve at a point must have a slope perpendicular to the slope of the tangent line. If $f'(x) \neq 0$ then the equation for the normal line at a point (x_0, y_0) is

$$y - y_0 = \frac{-1}{f'(x_0)}(x - x_0).$$

Problem Solving Examples:

Find the equations of the tangent line and the normal to the curve: $y = x^2 - x + 3$, at the point $(2, 5)$.

A

Since the equation of a straight line passing through a given point can be expressed in the form: $y - y_1 = m(x - x_1)$, this is appropriate for finding the equations of the tangent and normal. Here $x_1 = 2$ and $y_1 = 5$. The slope, m, of the tangent line is found by taking the derivative, $\frac{dy}{dx}$, of the curve: $y = x^2 - x + 3$.

$$\frac{dy}{dx} = 2x - 1.$$

At $(2, 5)$, $\frac{dy}{dx} = 2(2) - 1 = 3$, therefore the slope, m, of the tangent line is 3. Substituting x_1, y_1 and m into the equation $y - y_1 = m(x - x_1)$ we obtain:

$$y - 5 = 3(x - 2),$$

as the equation of the tangent line, or

$$3x - y - 1 = 0.$$

Since the slope of the normal is given by: $m' = -\frac{1}{m}$, and since $m = 3$, the slope of the normal is $-\frac{1}{3}$. Substituting $x_1 = 2$, $y_1 = 5$

and the slope of the normal, $m' = -\frac{1}{3}$, into the equation: $y - y_1 = m'(x - x_1)$, we obtain:

$$y - 5 = -\frac{1}{3}(x - 2),$$

or,

$$x + 3y - 17 = 0.$$

This is the equation of the normal.

6.5 Minimum and Maximum Values

If a function f is defined on an interval I, then

a) f is increasing on I if $f(x_1) < f(x_2)$ whenever x_1, x_2 are in I and $x_1 < x_2$.

b) f is decreasing on I if $f(x_1) > f(x_2)$ whenever $x_1 < x_2$ in I.

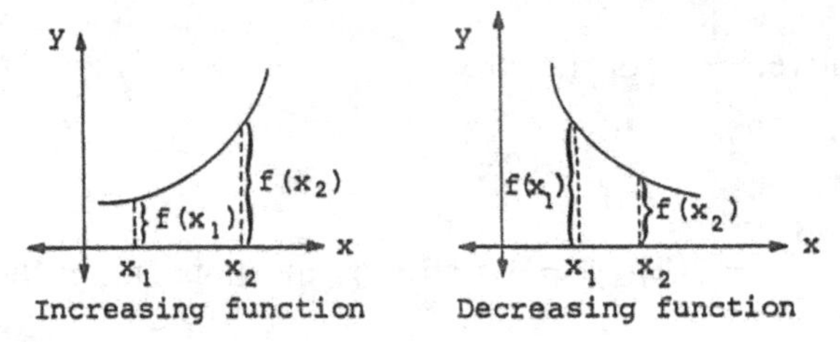

c) f is constant if $f(x_1) = f(x_2)$ for every x_1, x_2 in I.

Suppose f is defined on an open interval I and c is a number in I then,

a) f(c) is a local maximum value if $f(x) \leq f(c)$ for all x in I.

b) f(c) is a local minimum value if $f(x) \geq f(c)$ for all x in I.

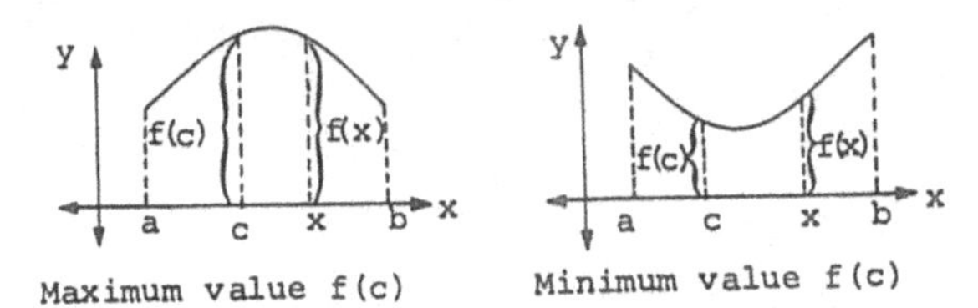

Maximum value f(c) Minimum value f(c)

Fig. 6.5

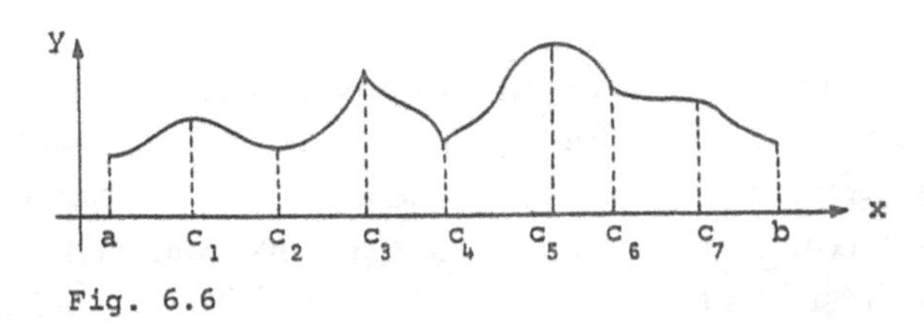

Fig. 6.6

In Fig. 6.6 in the interval [a,b], the local maxima occur at c_1, c_3, c_5 with an absolute maximum at c_5 and local minima occur at c_2 , c_4 .

To find Absolute Extrema for functions, first calculate f(c) for each critical number c, then calculate f(a) and f(b). The absolute extrema of f on [a,b] will then be the largest and the smallest of these functional values. If f(a) or f(b) is an extremum we call it an endpoint extremum.

Viewing the derivative as the slope of a curve, there may be points (or critical values) where the curve has a zero derivative. At these values the tangent to the curve is horizontal.

Conversely, if the derivative at a point exists and is not zero, then the point is not a local extrema.

Problem Solving Examples:

Find the maxima and minima of the function $f(x) = x^4$.

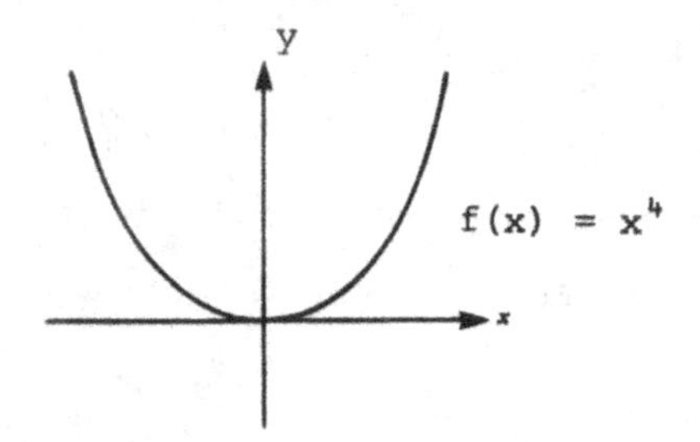

To determine maxima and minima we find $f'(x)$, set it equal to 0, and solve for x to obtain the critical points. We find: $f'(x) = 4x^3 = 0$, therefore $x = 0$ is the critical value. We must now determine whether $x = 0$ is a maximum or minimum value. In this example the Second Derivative Test fails because $f''(x) = 12x^2$ and $f''(0) = 0$. We must, therefore, use the First Derivative Test. We examine $f'(x)$ when $x < 0$, and when $x > 0$. We find that for $x < 0$, $f'(x)$ is negative, and for $x > 0$, $f'(x)$ is positive. Therefore there is a minimum at (0, 0). (See figure.)

Locate the maxima and minima of $y = 2x^2 - 8x + 6$.

To obtain the minima and maxima we find $\frac{dy}{dx}$, set it equal to 0 and solve for x. we find:

$$\frac{dy}{dx} = 4x - 8 = 0.$$

Therefore, $x = 2$ is the critical point. We now use the Second Derivative Test to determine whether $x = 2$ is a maximum or a minimum. We find :

$\frac{d^2y}{dx^2} = 4$, (positive). The second derivative is positive, hence $x = 2$ is a minimum.

Now substitute this back into the original equation to get the corresponding ordinate.

$$y = 2x^2 - 8x + 6 = 2 \cdot 2^2 - 8 \cdot 2 + 6 = 8 - 16 + 6$$

$$= -2.$$

Therefore, the minimum is at $x = 2$, $y = -2$.

6.5.1 Solving Maxima and Minima Problems

Step 1. Determine which variable is to be maximized or minimized (i.e., the dependent variable y).

Step 2. Find the independent variable x.

Step 3. Write an equation involving x and y. All other variables can be eliminated by substitution.

Step 4. Differentiate with respect to the independent variable.

Step 5. Set the derivative equal to zero to obtain critical values.

Step 6. Determine maxima and minima.

Problem Solving Examples:

Locate the maxima and minima of

$$y = \frac{x^3}{3} - \frac{5x^2}{2} + 6x + 4.$$

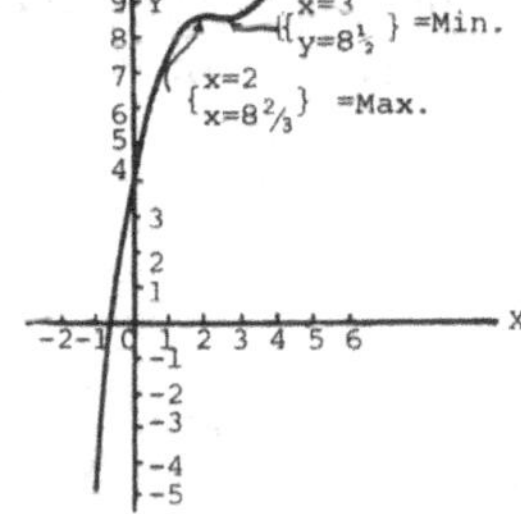

To find the maxima and minima we find $\frac{dy}{dx}$, set it to 0, and solve for x, obtaining the critical points. Doing this we have:

$$\frac{dy}{dx} = x^2 - 5x + 6 = 0, \ (x-2)(x-3) = 0$$

therefore,

$$x = 3 \text{ and } 2.$$

We now use the Second Derivative Test to determine whether the critical values are maximum, minimum or neither. We find:

$$\frac{d^2y}{dx^2} = 2x - 5.$$

For $x = 3$,

$$\frac{d^2y}{dx^2} = 2x - 5 = 2 \cdot 3 - 5 = + \text{ (positive)},$$

which indicates a minimum.

For $x = 2$,

$$\frac{d^2y}{dx^2} = 2x - 5 = 2 \cdot 2 - 5 = - \text{ (negative)},$$

which indicates a maximum.

Therefore, we have a minimum at $x = 3$ and a maximum at $x = 2$. We now wish to find the corresponding ordinates. Going back to the original equation, we have:

For $x = 3$,

$$y = \frac{x^3}{3} - \frac{5x^2}{2} + 6x + 4 = \frac{3^3}{3} - \frac{5 \cdot 3^2}{2} + 6 \cdot 3 + 4$$

$$= 9 - \frac{45}{2} + 18 + 4 = 8\ \frac{1}{2}.$$

For x = 2,

$$y = \frac{x^3}{3} - \frac{5x^2}{2} + 6x + 4 = \frac{2^3}{3} - \frac{5 \cdot 2^2}{2} + 6 \cdot 2 + 4$$

$$= \frac{8}{3} - 10 + 12 + 4 = 8\frac{2}{3}.$$

Therefore, minimum is at x = 3, $y = 8\frac{1}{2}$, and maximum is at x = 2, $y = 8\frac{2}{3}$.

6.6 Curve Sketching and the Derivative Tests

Using the knowledge we have about local extrema and the following properties of the first and second derivatives of a function, we can gain a better understanding of the graphs (and thereby the nature) of a given function.

A function is said to be smooth on an interval (a,b) if both f' and f" exist for all $x \in (a,b)$.

6.6.1 The First Derivative Test

Suppose that c is a critical value of a function, f, in an interval (a,b) , then if f is continuous and differentiable we can say that,

a) if $f'(x) > 0$ for all $a < x < c$
and $f'(x) < 0$ for all $c < x < b$,
then f(c) is a local maximum.

b) if $f'(x) < 0$ for $a < x < c$ and
$f'(x) > 0$ for $c < x < b$, then
$f(c)$ is a local minimum.

c) if $f'(x) > 0$ or if $f'(x) < 0$ for all $x \in (a,b)$ then $f(c)$ is not a local extrema.

Problem Solving Examples:

Find the maxima and minima of $f(x) = 3x^5 - 5x^3$.

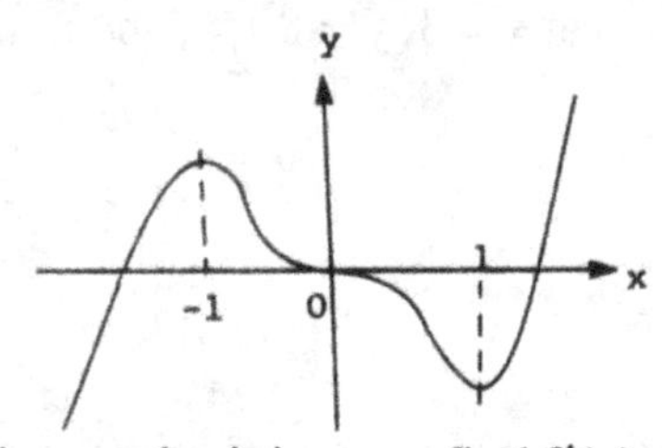

To determine maxima and minima we find $f'(x)$, set it equal to 0 and solve for x, obtaining the critical points. We find:

$$f'(x) = 15x^4 - 15x^2 = 15x^2(x^2 - 1).$$

Therefore, $x = 0, \pm 1$ are the critical points. We must now determine whether the function reaches a maximum, minimum or neither at each of these values. To do this we will use the Second Derivative Test. Computing the second derivative f'', we have:

$$f''(x) = 60x^3 - 30x = 30x(2x^2 - 1).$$

$f''(1) = 30 > 0$. Therefore, $x = 1$ is a relative minimum point and $x = -1$ is a relative maximum since $f''(-1) = -30 < 0$. Now, $f''(0) = 0$. Therefore the Second Derivative Test indicates a point which is neither maximum nor minimum. This is known as a point of inflection. For further study of the behavior of f at 0 we must use the First Derivative Test. We examine $f'(x)$ when $-1 < x < 0$ and when $0 < x < 1$. Let us select a representative value from each interval. We will use

$x = -\dfrac{1}{2}$ and $x = \dfrac{1}{2}$. For $f'\left(-\dfrac{1}{2}\right)$ we obtain a negative value, and

for $f'\left(\dfrac{1}{2}\right)$ we again obtain a negative value. Because there is no

change in sign we conclude that at $x = 0$ there is neither a maximum nor a minimum, as can also be seen from the graph.

6.6.2 Concavity

If a function is differentiable on an open interval containing c, then the graph at this point is

a) concave upward (or convex) if $f''(c) > 0$;

b) concave downward if $f''(c) < 0$.

If a function is concave upward than f' is increasing as x increases. If the function is concave downward, f' is decreasing as x increases.

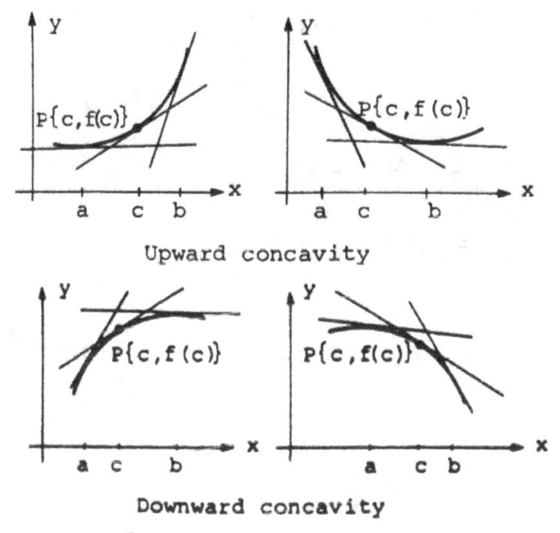

Fig. 6.7

Problem Solving Examples:

Find the intervals of x for which the curve

$$y = 2x^3 - 9x^2 + 12x - 3$$

is concave downward and concave upward.

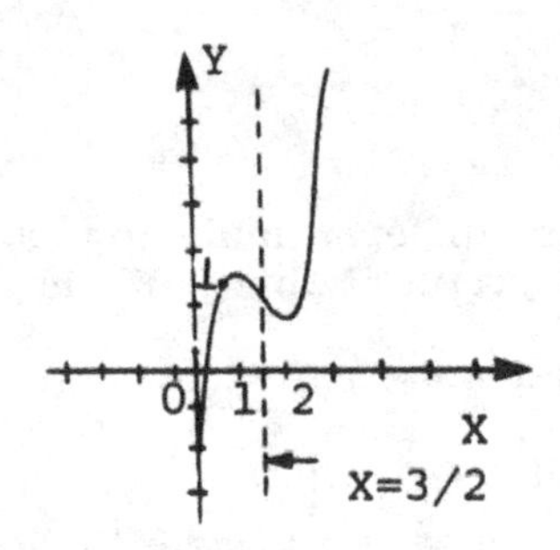

Differentiating twice,

$$\frac{dy}{dx} = 6(x^2 - 3x + 2)$$

and

$$\frac{d^2y}{dx^2} = 6(2x - 3)$$

By setting $\frac{d^2y}{dx^2} = 0$, we have $x = \frac{3}{2}$

y'' is positive or negative according to $x > \frac{3}{2}$ 0r $< \frac{3}{2}$. Hence the graph is concave downward to the left of $x = \frac{3}{2}$ and concave upward to the right of $x = \frac{3}{2}$.

6.6.3 Points of Inflection

Points which satisfy $f''(x) = 0$ may be positions where concavity changes. These points are called the points of inflection. It is the point at which the curve crosses its tangent line.

6.6.4 Graphing a Function Using the Derivative Tests

The following steps will help us gain a rapid understanding of a function's behavior.

A) Look for some basic properties such as oddness, evenness, periodicity, boundedness, etc.

B) Locate all the zeros by setting $f(x) = 0$.

C) Determine any singularities, $f(x) = \infty$.

D) Set $f'(x)$ equal to zero to find the critical values.

E) Find the points of inflection by setting $f''(x) = 0$.

F) Determine where the curve is concave, $f''(x) < 0$, and where it is convex $f''(x) > 0$.

G) Determine the limiting properties and approximations for large and small $|x|$.

H) Prepare a table of values x, $f(x)$, $f'(x)$ which includes the critical values and the points of inflection.

I) Plot the points found in Step H and draw short tangent lines at each point.

J) Draw the curve making use of the knowledge of concavity and continuity.

Problem Solving Examples:

Find the maxima and minima of: $y = \dfrac{4}{x^2 - 4}$, and trace the curve.

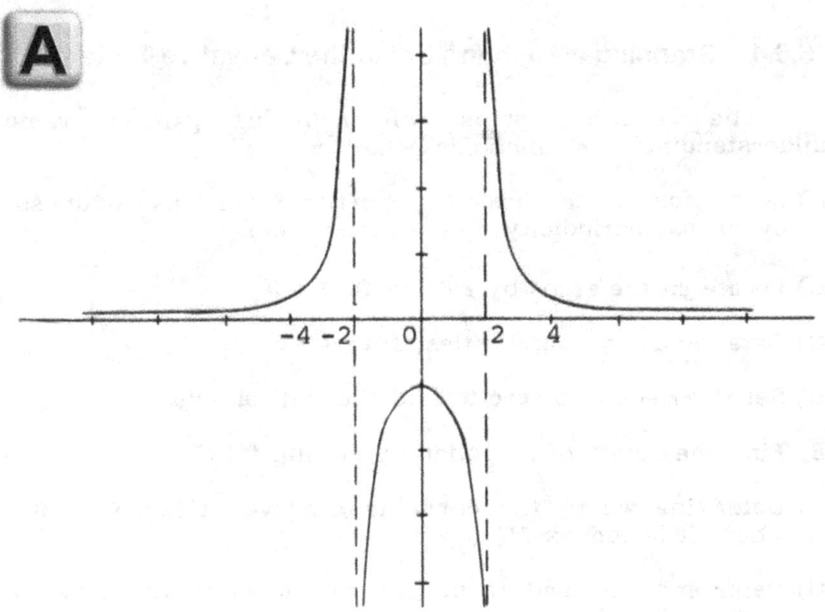

To find the maxima and minima we determine $\dfrac{dy}{dx}$, equate it to 0, and solve for x, obtaining the critical value. We find:

$$\frac{dy}{dx} = \frac{(x^2 - 4)(0) - 4(2x)}{(x^2 - 4)^2} = -\frac{8x}{(x^2 - 4)^2}.$$

$\dfrac{-8x}{(x^2 - 4)^2} = 0$, or $-8x = 0$, therefore $x = 0$. Substituting $x = 0$, into the original equation, $y = -1$, therefore the critical point is $(0, -1)$.

To determine whether a maximum, minimum, or neither occurs at this point, we use the First Derivative Test. We examine $\frac{dy}{dx}$ at a point less than 0, (use –1), and at a point greater than 0 (use 1). If $\frac{dy}{dx}$ changes sign from + to –, a maximum occurs at x = 0, from – to +, a minimum occurs, and if there is no change in sign, neither a maximum nor a minimum occurs. We find that at $x = -1$, $\frac{dy}{dx} = \frac{8}{9}$, a positive value, and at $x = 1$, $\frac{dy}{dx} = -\frac{8}{9}$, a negative value. Therefore, at the point (0, –1), a maximum occurs.

Upon further investigation of the curve: $y = \frac{4}{x^2 - 4}$, we observe that for the values $x = 2, -2$, y is undefined or $y = \pm\infty$. Therefore, the graph of the curve has asymptotes at 2 and –2, as shown in the accompanying graph.

6.7 Rectilinear Motion

When an object moves along a straight line we call the motion rectilinear motion. Distance s, velocity v, and acceleration a, are the chief concerns of the study of motion.

Velocity is the proportion of distance over time.

$$\boxed{v = \frac{s}{t}}$$

$$\text{Average velocity} = \frac{s(t_2) - s(t_1)}{t_2 - t_1}$$

where t_1, t_2 are time instances and $s(t_2)-s(t_1)$ is the displacement of an object.

Instantaneous velocity at time t is defined as

$$v = D\, s(t) = \lim_{h \to 0} \frac{s(t+h)-s(t)}{h}$$

We usually write

$$v(t) = \frac{ds}{dt}.$$

Acceleration, the rate of change of velocity with respect to time is

$$a(t) = \frac{dv}{dt}.$$

It follows clearly that

$$a(t) = v'(t) = s''(t).$$

When motion is due to gravitational effects, $g = 32.2$ ft/sec^2 or $g = 9.81$ m/sec^2 is usually substituted for acceleration.

Speed at time t is defined as $|v(t)|$. The speed indicates how fast an object is moving without specifying the direction of motion.

Problem Solving Examples:

Q A rope attached to a boat is being pulled in at a rate of 10 ft/sec. If the water is 20 ft below the level at which the rope is being drawn in, how fast is the boat approaching the wharf when 36 ft of rope are yet to be pulled in?

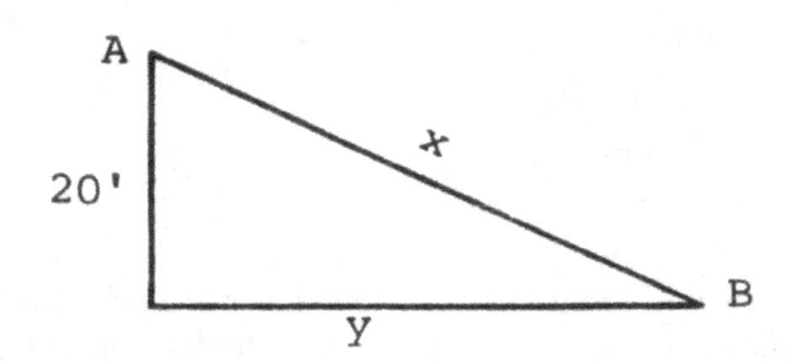

The length AB denotes the rope, and the position of the boat is at B. Since the rope is being drawn in at a rate of 10ft/sec.

$$\frac{dx}{dt} = 10.$$

To find how fast the boat is being towed in when 36 ft. of rope are left,

$$\frac{dy}{dt} \text{ must be found at } x = 36.$$

From the right triangle, $20^2 + y^2 = x^2$ or $y = \sqrt{x^2 - 400}$. Differentiating with respect to t,

$$\frac{dy}{dt} = \frac{dy}{dx} \cdot \frac{dx}{dt} = \frac{1}{2}(x^2 - 400)^{-\frac{1}{2}}(2x)\frac{dx}{dt} = \frac{x\frac{dx}{dt}}{\sqrt{x^2 - 400}}.$$

Substituting the conditions that:

$$\frac{dx}{dt} = -10, \text{ and } x = 36,$$

$$\frac{dy}{dt} = \frac{-360}{\sqrt{896}} = -\frac{45}{\sqrt{14}}.$$

It has now been found that, when there are 36 ft of rope left, the boat is moving in at the rate of:

$$\frac{45}{\sqrt{14}} \text{ ft/sec.}$$

Q A boat is being hauled toward a pier at a height of 20 ft above the water level. The rope is drawn in at a rate of 6ft/sec. Neglecting sag, how fast is the boat approaching the base of the pier when 25 ft of rope remain to be pulled in?

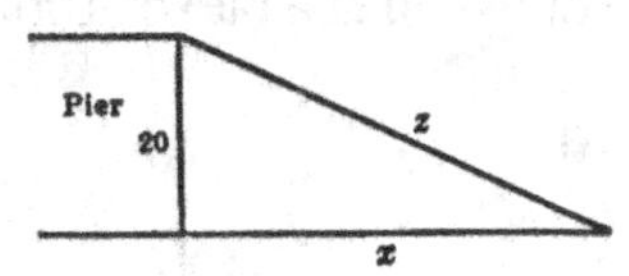

Formulating the given data, we have:

$$\frac{dz}{dt} = 6, z = 25, \text{ and } \frac{dx}{dt} \text{ is to be found.}$$

At any time t we have, from the Pythagorean theorem,

$$20^2 + x^2 = z^2$$

By differentiation, we obtain:

$$x\frac{dx}{dt} = z\frac{dz}{dt}$$

When $z = 25$, $x = \sqrt{25^2 - 20^2} = 15$; therefore

$$15 \frac{dx}{dt} = 25(-6)$$

$$\frac{dx}{dt} = -10 \text{ ft/sec}$$

(The boat approaches the base at 10 ft/sec).

6.8 Rate of Change and Related Rates

6.8.1 Rate of Change

In the last section we saw how functions of time can be expressed as velocity and acceleration. In general, we can speak about the rate of change of any function with respect to an arbitrary parameter (such as time in the previous section).

For linear functions $f(x) = mx+b$, the rate of change is simply the slope m.

For non-linear functions we define the

1) average rate of change between points c and d to be

$$\frac{f(d)-f(c)}{d-c}$$

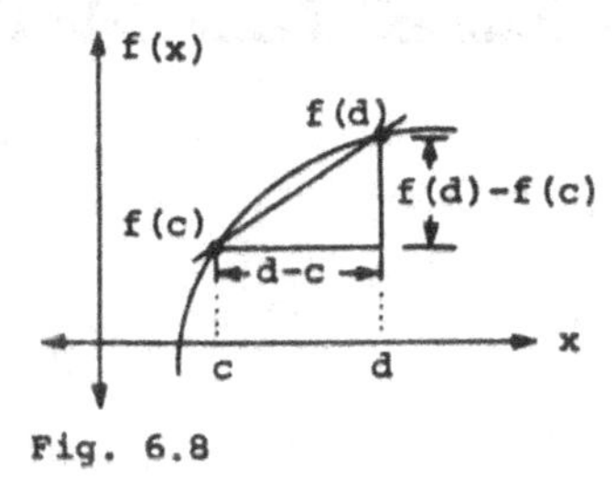

Fig. 6.8

2) instantaneous rate of change of f at the point x to be

$$f'(x) = \lim_{h \to 0} \frac{f(x+h)-f(x)}{h}$$

If the limit does not exist, then the rate of change of f at x is not defined.

The form, common to all related rate problems, is as follows:

a) Two variables, x and y are given. They are functions of time, but the explicit functions are not given.

b) The variables, x and y are related to each other by some equation such as $x^2+y^3-2x-7y^2+2 = 0$.

c) An equation which involves the rate of change $\frac{dx}{dt}$ and $\frac{dy}{dt}$ is obtained by differentiating with respect to t and using the chain rule.

As an illustration, the previous equation leads to

$$2x \frac{dx}{dt} + 3y^2 \frac{dy}{dt} - 2 \frac{dx}{dt} - 14y \frac{dy}{dt} = 0$$

The derivatives $\frac{dx}{dt}$ and $\frac{dy}{dt}$ in this equation are called the related rates.

Problem Solving Examples:

Compute the average rate of change of $y = f(x) = x^2 - 2$ between $x = 3$ and $x = 4$.

Average rate of change is defined as:

$\frac{\Delta y}{\Delta x}$ with $\Delta y = f(x + \Delta x) - f(x)$.

Given: $x = 3, \Delta x = 4 - 3 = 1,$

$y = f(x) = f(3) = 3^2 - 2 = 7$

For $x = 4,$

$y + \Delta y = f(x + \Delta x) = 4^2 - 2 = 14$

$\Delta y = f(x + \Delta x) - f(x) = f(4) - f(3)$

$= (4^2 - 2) - (3^2 - 2) = 14 - 7 = 7$

$\frac{\Delta y}{\Delta x} = \frac{7}{1} = 7$, the average rate of change.

Find the rate of change of y with respect to x at the point x = 5, if $2y = x^2 + 3x - 1.$

Rate of change is defined as

$\lim_{\Delta x \to 0} \frac{\Delta y}{\Delta x}$, with

$\Delta y = f(x + \Delta x) - f(x).$

We have:

$2\Delta y = (x + \Delta x)^2 + 3(x + \Delta x) - 1 - (x^2 + 3x - 1)$

$= x^2 + 2x \cdot \Delta x + (\Delta x)^2 + 3x + 3\Delta x - 1 - x^2 - 3x + 1$

$= 2x \cdot \Delta x + (\Delta x)^2 + 3\Delta x.$

Dividing by Δx,

$$\frac{2\,\Delta y}{\Delta x} = \frac{2x \cdot \Delta x}{\Delta x} + \frac{(\Delta x)^2}{\Delta x} + \frac{3\,\Delta x}{\Delta x}$$

$$= 2x + \Delta x + 3$$

and

$$\frac{\Delta y}{\Delta x} = x + \frac{\Delta x}{2} + \frac{3}{2} .$$

Now,

$$\lim_{\Delta x \to 0} \frac{\Delta y}{\Delta x} = \lim_{\Delta x \to 0} x + \frac{\Delta x}{2} + \frac{3}{2} = x + \frac{3}{2}.$$

For $x = 5$,

$$\lim_{\Delta x \to 0} \frac{\Delta y}{\Delta x} = 5 + \frac{3}{2} = 6\frac{1}{2}$$

This means that the instantaneous rate of change of the function represented by the curve at the point $x = 5$ is $6\frac{1}{2}$.

The function, it is seen, changes $6\frac{1}{2}$ times as fast as the independent variable x at $x = 5$.

The slope of the tangent at $x = 5$ is $6\frac{1}{2}$.

Quiz: Applications of the Derivative

1. The position of a particle moving along a straight line at any time t is given by $s(t) = 2t^3 - 4t^2 + 2t - 1$. What is the acceleration of the particle when $t = 2$?

   (A) 32

   (B) 16

   (C) 4

   (D) 8

   (E) 0

2. $\lim_{x \to 0} \frac{\cos^2 x - 1}{2x \sin x} =$

   (A) -1

   (B) $-\frac{1}{2}$

   (C) 1

   (D) $\frac{1}{2}$

   (E) 0

3. If the graph of f is as in the figure on the following page, where slope of $L_1 = 2$, then $f'(x_0)$ is

   (A) $\frac{1}{2}$

   (B) -2

   (C) 2

(D) $-\frac{1}{2}$

(E) 0

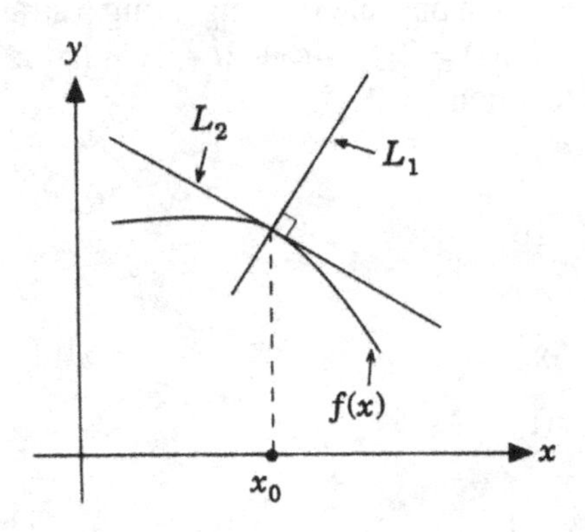

4. Which of the following is NOT true about $y = \cos(-x + \pi)$?

   (A) y has the same period as $\cos(x - \pi)$.

   (B) y has the same period as $\tan\left(2 - \frac{x}{2}\right)$.

   (C) y has only one inflection point in $(-\pi, \pi)$.

   (D) $\frac{d^2y}{dx^2} + y = 0$.

   (E) y has minimum at $x = 0$.

5. For what value of x will the tangent lines to $y = \ln x$ and $y = 2x^2$ be parallel?

   (A) 0

(B) $\frac{1}{4}$

(C) $\frac{1}{2}$

(D) 1

(E) 2

6. Let $f(x) = e^{bx}$, $g(x) = e^{ax}$ and find the value of b such that

$$D_x \left[\frac{f(x)}{g(x)}\right] = \frac{f'(x)}{g'(x)}$$

(A) $\frac{a^2}{a^2 - 1}$

(B) $\frac{a^2}{a + 1}$

(C) $\frac{a + 1}{a^2}$

(D) $\frac{a - 1}{a^2}$

(E) $\frac{a^2}{a - 1}$

7. For which of the following intervals is the graph of $y = x^4 - 2x^3 - 12x^2$ concave down?

(A) $(-2, 1)$

(B) $(-1, 2)$

(C) $(-1, -2)$

(D) $(-\infty, -1)$

(E) $(-1, +\infty)$

8. The graph shown represents $y = f(x)$. Which one of the following is NOT true?

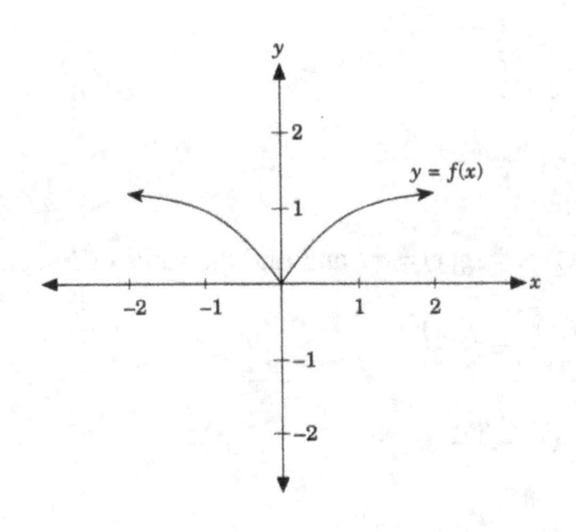

(A) f is continuous on $(-2, 2)$.

(B) $\lim\limits_{x \to 0} f(x) = f(0)$.

(C) f is differentiable on $(-2, 2)$.

(D) $\lim\limits_{x \to 0^-} f(x) = f(0)$.

(E) $f'(x) < 0$ for $x < 0$.

9. The rate of change of the area of an equilateral triangle with respect to its side S at $S = 2$ is approximately:

(A) 0.43

(B) 1.73

(C) 0.87

(D) 7.00

(E) 0.50

10. If $3x^2 - x^2y^3 + 4y = 12$ determines a differentiable function such that $y = f(x)$, then $\frac{dy}{dx} =$

(A) $\dfrac{-3x + 2xy^3}{2}$

(B) $\dfrac{-6x + 2xy^3}{-3x^2y^2 + 4}$

(C) $\dfrac{-3x + 2xy^3 + 6}{2}$

(D) $\dfrac{-6x + 2xy^3 + 3x^2y^2}{4}$

(E) $\dfrac{-6x + 2xy^3}{3x^2y^2 + 4}$

ANSWER KEY

1. (B)
2. (B)
3. (D)
4. (C)
5. (C)
6. (E)
7. (B)
8. (C)
9. (B)
10. (B)

CHAPTER 7

The Definite Integral

7.1 Antiderivatives

Definition:

If $F(x)$ is a function whose derivative $F'(x) = f(x)$, then $F(x)$ is called the antiderivative of $f(x)$.

THEOREM:

If $F(x)$ and $G(x)$ are two antiderivatives of $f(x)$, then $F(x) = G(x) + c$, where c is a constant.

7.1.1 Power Rule for Antidifferentiation

Let "a" be any real number, "r", any rational number not equal to -1 and "c" an arbitrary constant.

$$\text{If } f(x) = ax^r, \text{ then } F(x) = \frac{a}{r+1}x^{r+1} + c.$$

THEOREM:

An antiderivative of a sum is the sum of the antiderivatives.

$$\frac{d}{dx}(F_1+F_2) = \frac{d}{dx}(F_1) + \frac{d}{dx}(F_2) = f_1 + f_2$$

7.2 Area

To find the area under the graph of a function f from a to b, we divide the interval [a,b] into n subintervals, all having the same length (b-a)/n. This is illustated in Figure 7.1.

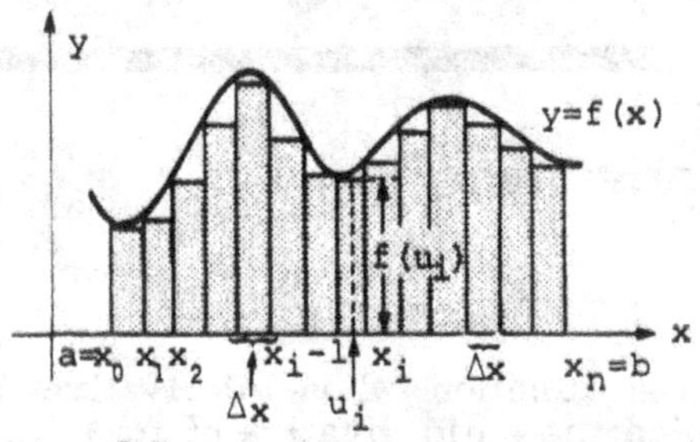

Fig. 7.1

Since f is continuous on each subinterval, f takes on a minimum value at some number u_i in each subinterval.

We can construct a rectangle with one side of length $[x_{i-1}, x_i]$, and the other side of length equal to the minimum distance $f(u_i)$ from the x-axis to the graph of f.

The area of this rectangle is $f(u_i)\Delta x$. The boundary of the region formed by the sum of these rectangles is called the inscribed rectangular polygon.

The area (A) under the graph of f from a to b is

$$A = \lim_{\Delta x \to 0} \sum_{i=1} f(u_i)\Delta x.$$

The area A under the graph may also be obtained by means of circumscribed rectangular polygons.

In the case of the circumscribed rectangular polygons the maximum value of f on the interval $[x_{i-1}, x_i]$, v_i, is used.

Note that the area obtained using circumscribed rectangular polygons should always be larger than that obtained using inscribed rectangular polygons.

Problem Solving Examples:

Determine the area under the curve: $y = f(x) = x^2$ between $x = 2$ and $x = 3$.

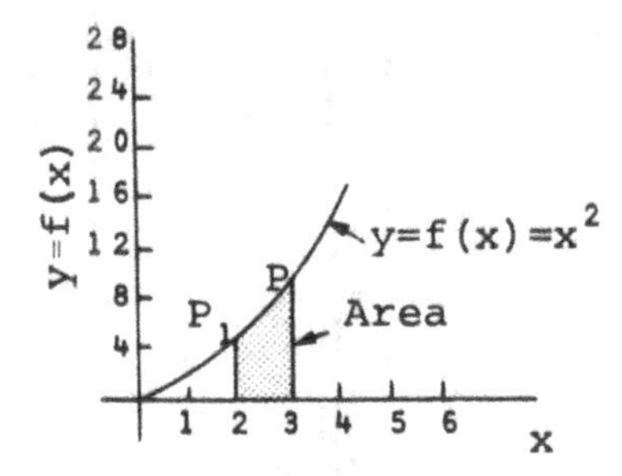

It is given that the area to be evaluated is between $x = 2$ and $x = 3$, therefore, these are the limits of the integral which gives us the required area. Area is equal to the integral of the upper function minus the lower function. From the diagram it is seen that the required area is between $y = x^2$ as the upper function and $y = 0$ (the x-axis) as the lower function. Therefore, we can write:

$$A = \int_2^3 (x^2 - 0)dx$$

$$= \int_2^3 x^2 dx$$

$$= \left.\frac{x^3}{3}\right]_2^3$$

$$A = \frac{3^3}{3} - \frac{2^3}{3} = \frac{19}{3}.$$

Find the area between the curve: $y = x^3$, and the x-axis, from $x = -2$ to $x = 3$.

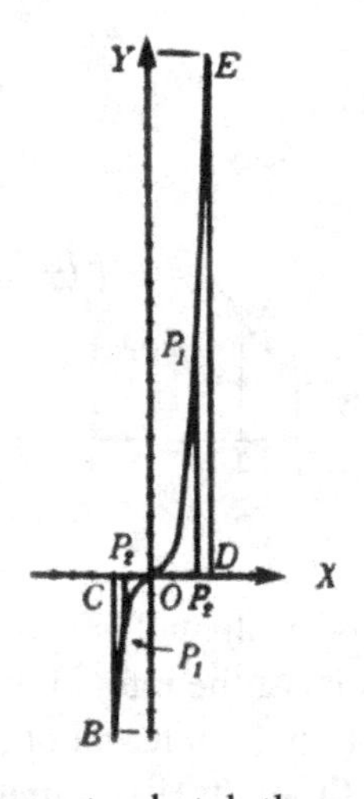

It is generally advantageous to sketch the curve, since parts of the curve may have to be considered separately, particularly when positive and negative limits are given. The desired area is composed of the two parts: BOC and ODE. To find the total area, we can evaluate each area separately and then add. The area is the integral of the upper function minus the lower function. In the first quadrant, the upper function is the curve $y = x^3$, the lower function is $y = 0$, (the x-axis) and the limits are $x = 0$ and $x = 3$. In the third quadrant, the upper function is $y = 0$, the lower function is the curve $y = x^3$, and the limits $x = -2$ and $x = 0$. Hence we can write,

$$A_{total} = \int_0^3 (x^3 - 0)\,dx + \int_{-2}^0 (0 - x^3)\,dx$$

$$= \int_0^3 x^3dx + \int_{-2}^0 -x^3dx$$

$$= \left[\frac{x^4}{4}\right]_0^3 + \left[-\frac{x^4}{4}\right]_{-2}^0$$

$$= \frac{81}{4} + \frac{16}{4}$$

$$= 24\frac{1}{4} \text{ sq. units}$$

Note that refusal to consider this problem in two parts does not give area, but gives "net area" with one area considered positive and the other negative.

7.3 Definition of Definite Integral

A partition P of a closed interval [a,b] is any decomposition of [a,b] into subintervals of the form,

$$[x_0,x_1],\ [x_1,x_2],\ [x_2,x_3],\ldots,[x_{n-1},x_n]$$

where n is a positive integer and x_i are numbers, such that

$$a = x_0 < x_1 < x_2 < \ldots < x_{n-1} < x_n = b.$$

The length of the subinterval is $\Delta x_i = x_i - x_{i-1}$. The largest of the numbers Δx_1, $\Delta x_2 \ldots \Delta x_n$ is called the norm of the partition P and denoted by $||P||$.

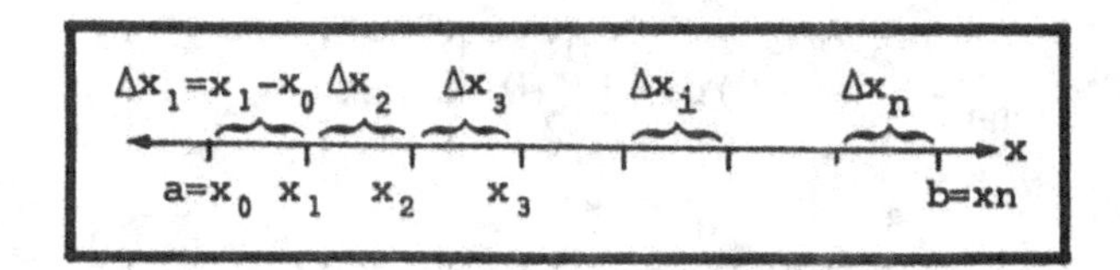

Definition:

Let f be a function that is defined on a closed interval [a,b] and let P be a partition of [a,b]. A Riemann Sum of f for P is any expression R_p of the form,

$$R_p = \sum_{i=1}^{n} f(w_i)\Delta x_i,$$

where w_i is some number in $[x_{i-1}, x_i]$ for $i = 1, 2, \ldots, n$.

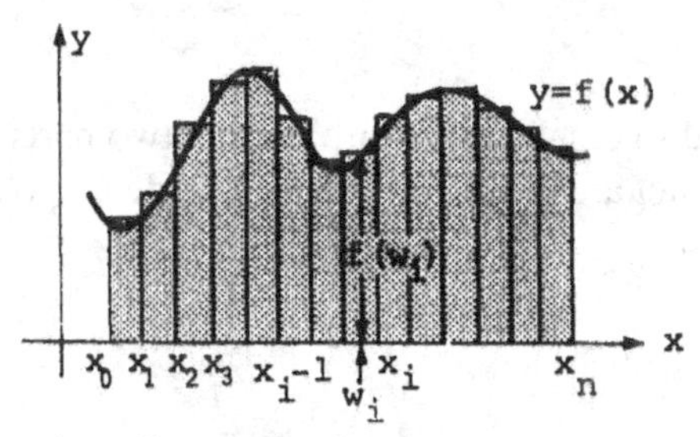

Fig. 7.2

Definition:

Let f be a function that is defined on a closed interval [a,b]. The definite integral of f from a to b, denoted by $\int_a^b f(x)dx$ is given by

$$\int_a^b f(x)dx = \lim_{||P|| \to 0} \sum_i f(w_i)\Delta x_i$$

provided the limit exists.

THEOREM :

If f is continuous on [a,b], then f is integrable on [a,b] that is, the limit $\int_b^a f(x)dx$ exists.

THEOREM :

If f(a) exists, then $\int_a^a f(x)dx = 0$.

Problem Solving Examples:

$\frac{dy}{dx} = (a - bx)^n$. What is $y = F(x)$ when $n = 2$?

$\frac{dy}{dx} = (a - bx)^n$ can be rewritten as $dy = (a - bx)^n dx$. We can now write: $\int dy = \int (a - bx)^n dx$ or, $y = \int (a - bx)^n dx$. To integrate, we consider the formula: $\int u^n du = \frac{u^{n+1}}{n+1} + C$, with $u = (a - bx)$ and $du = -bdx$. Applying the formula, we obtain:

$$y = \int (a - bx)^n dx = -\frac{1}{b} \cdot \frac{(a - bx)^{n+1}}{n+1}$$

$$= -\frac{(a - bx)^{n+1}}{b(n+1)} + C,$$

the integral in the general form.

For $n = 2$,

$$y = \int (a - bx)^2 \cdot dx = -\frac{(a - bx)^3}{3b} + C.$$

Q $\frac{dy}{dx} = (a + bx)^n$. What is $y = F(x)$ when $n = 1$, $n = 2$, and $n = -2$?

In solving, we first find the integral in the general form and then substitute and find the integral for $n = 1$, $n = 2$, and $n = -2$.

$\frac{dy}{dx} = (a + bx)^n$ can be rewitten as

$dy = (a + bx)^n\,dx$. To find the integral, we write:

$\int dy = \int (a + bx)^n\,dx$ or, $y = \int (a + bx)^n\,dx$.

We now consider the formula:

$$\int u^n du = \frac{u^{n+1}}{n+1} + C, \text{ with } u = (a + bx) \text{ and } du = bdx.$$

Applying the formula, we have:

$$y = \int (a + bx)^n \cdot dx$$

$$= \frac{(a + bx)^{n+1}}{b(n+1)} + C,$$

the integral in the general form.

For $n = 1$,

$$\int (a + bx)^1\,dx = \frac{(a + bx)^{1+1}}{b(1+1)} = \frac{(a + bx)^2}{2b} + C.$$

For $n = 2$,

$$\int (a + bx)^2\,dx = \frac{(a + bx)^{2+1}}{b(2+1)} = \frac{(a + bx)^3}{3b} + C.$$

For $n = -2$,

$$\int \frac{dx}{(a+bx)^2} = \int (a+bx)^{-2} \cdot dx = \frac{(a+bx)^{-2+1}}{b(-2+1)}$$

$$= -\frac{1}{b(a+bx)} + C.$$

7.4 Properties of Definite Integral

A) If f is integrable on [a,b], and k is any real number, then kf is integrable on [a,b] and

$$\int_a^b kf(x)dx = k\int_a^b f(x)dx.$$

B) If f and g are integrable on [a,b], then f+g is integrable on [a,b] and

$$\int_a^b [f(x)+g(x)]dx = \int_a^b f(x)dx + \int_a^b g(x)dx.$$

C) If $a<c<b$ and f is integrable on both [a,c] and [c,b], then f is integrable on [a,b] and

$$\int_a^b f(x)dx = \int_a^c f(x)dx + \int_c^b f(x)dx.$$

D) If f is integrable on a closed interval and if a, b, and c are any three numbers in the interval, then

$$\int_a^b f(x)dx = \int_a^c f(x)dx + \int_c^b f(x)dx.$$

E) If f is integrable on [a,b] and if $f(x) \geq 0$ for all x in [a,b], then $\int_a^b f(x)dx \geq 0$.

Problem Solving Examples:

Evaluate the expression: $\int_2^3 \frac{(x+1)dx}{\sqrt{x^2+2x+3}}dx$.

We can rewrite the given integral as:

$$\int_2^3 (x^2 + 2x + 3)^{-\frac{1}{2}} (x + 1)dx,$$

and make use of the formula: $\int u^n du = \frac{u^{n+1}}{n+1}$.

Let $u = x^2 + 2x + 3$. Then $du = (2x + 2)dx$, and $n = -\frac{1}{2}$. Applying the formula, we obtain:

$$\int_2^3 \frac{(x+1)dx}{\sqrt{x^2+2x+3}} = \frac{1}{2}\int_2^3 (x^2 + 2x + 3)^{-\frac{1}{2}} 2(x + 1)dx$$

$$= \frac{1}{2}\left[\frac{(x^2+2x+3)^{\frac{1}{2}}}{\frac{1}{2}}\right]_2^3 = \sqrt{x^2+2x+3}\,\Big]_2^3$$. We now evaluate the definite integral between 3 and 2, obtaining:

$$\sqrt{3^2+(2)(3)+3} - \sqrt{2^2+(2)(2)+3} = \sqrt{18} - \sqrt{11}.$$

Evaluate the expression: $\int_0^2 2x^2\sqrt{x^3+1}\ dx$.

We wish to convert the given integral into a form, to which we can apply the formula for $\int u^n\, du$, with $u = (x^3 + 1)$, $du = 3x^2$ and $n\ \frac{1}{2}$. We obtain:

$$\int_0^2 2x^2\sqrt{x^3+1}\ dx = \frac{2}{3}\int_0^2 (x^3+1)^{\frac{1}{2}}\left(\frac{3}{2}2x^2\, dx\right).$$

Applying the formula for $\int u^n du$, we obtain:

$$\frac{2}{3}\left[\frac{(x^3+1)^{\frac{3}{2}}}{\frac{3}{2}}\right]_0^2 = \frac{4}{9}(x^3+1)^{\frac{3}{2}}\Big]_0^2.$$

Evaluating between 2 and 0, we have:

$$\frac{4}{9}(8+1)^{\frac{3}{2}} - \frac{4}{9}(0+1)^{\frac{3}{2}}$$

$$= \frac{4}{9}(27)(-1)$$

$$= \frac{104}{9}.$$

7.5 The Fundamental Theorem of Calculus

The fundamental theorem of calculus establishes the relationship between the indefinite integrals and differentiation by use of the mean value theorem.

7.5.1 Mean Value Theorem for Integrals

The Mean Value Theorem for the integral has a very simple geometric interpretation.

The Mean Value Theorem says that for a continuous function on the closed interval [a, b], there exists a point x_o, where $a < x_o < b$, such that:

$$f(x_o) = \frac{1}{b-a} \int_a^b f(x)\, dx.$$

If we multiply both sides by (b – a) we have

$$(b-a)\, f(x_o) = \int_a^b f(x)\, dx,$$

which states that the integral from a to b is equal to the area of a rectangle of length (b – a) and height $f(x_o)$. In the diagram, it means that the area in region 1 can be put in region 2, thus forming a rectangle.

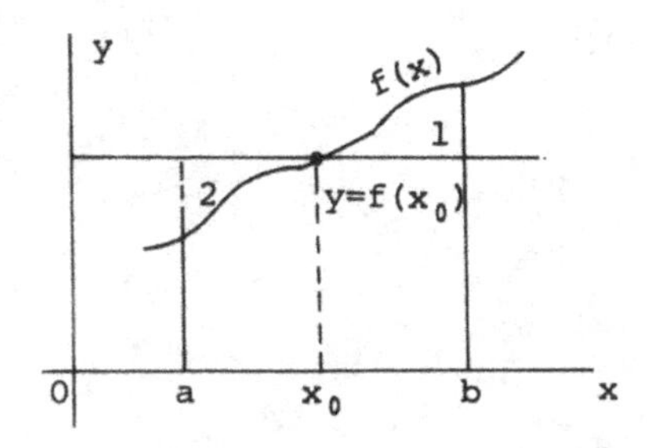

Problem Solving Examples:

What is the mean value or mean ordinate of the positive part of the curve $y = 2x - x^2$?

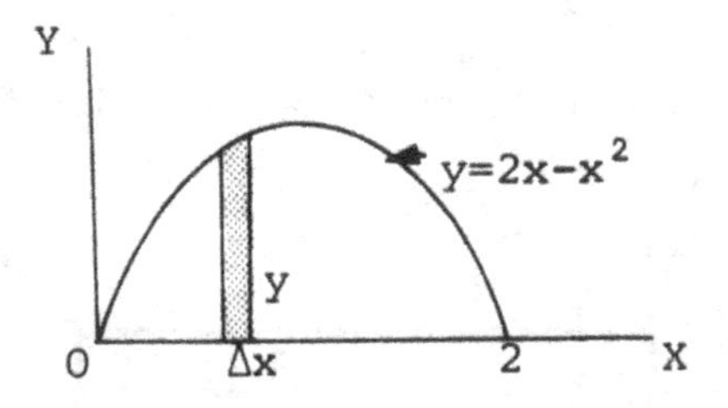

First determine the length of the base to fix the limits of integration for the area by setting y equal to zero, or:

$$0 = 2x - x^2 = x(2 - x).$$

Then, $x_1 = 0$, and $x_2 = 2$.

Now,

$$\bar{y}_x = \frac{1}{x_2 - x_1} \int y \cdot dx$$

$$= \frac{1}{2-0} \int_0^2 (2x - x^2)\, dx$$

$$= \int_0^2 \left(x - \frac{x^2}{2} \right) dx$$

$$= \left| \frac{x^2}{2} - \frac{x^3}{6} \right|_0^2 = \frac{4}{2} - \frac{8}{6} = \frac{2}{3} \; ,$$

the mean ordinate.

7.5.2 Definition of the Fundamental Theorem

Suppose f is continuous on a closed interval [a,b], then

a) If the function G is defined by

$$G(x) = \int_a^x f(t)dt,$$

for all x in [a,b], then G is an antiderivative of f on [a,b].

b) If F is any antiderivative of f, then

$$\int_a^b f(x)dx = F(b)-F(a)$$

Problem Solving Examples:

Find the mean value of the ordinates of the circle $x^2 + y^2 = a^2$ in the first quadrant.

a) With respect to the radius along the x-axis

b) With respect to the arc-length.

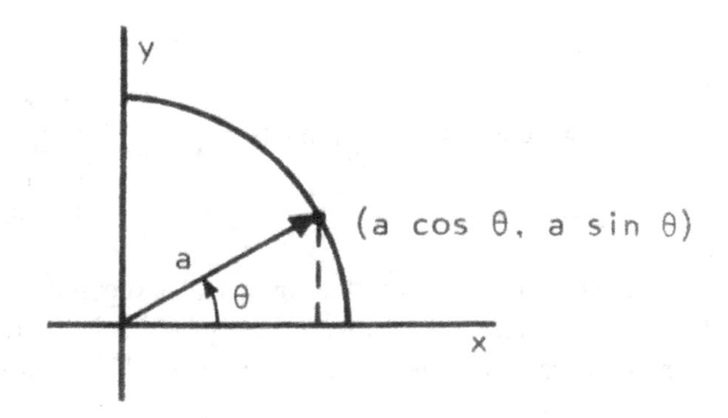

a) Noting that the mean value is defined by:

$$f(x_o) = \frac{1}{b-a} \int_a^b f(x)dx, \text{ where } a < x_o < b,$$

we have:

$$f(x_o) = \frac{1}{a-o} \int_0^a \sqrt{a^2 - x^2}\, dx,$$

where $f(x) = y = \pm\sqrt{a^2 - x^2}$. We take $y = \sqrt{a^2 - x^2}$, because we are in the first quadrant. Also, noting that:

$$\int \sqrt{a^2 - x^2}\, dx = \frac{1}{2} x\sqrt{a^2 - x^2}$$

$$+ \frac{1}{2} a^2 \arcsin \frac{x}{a} + C,$$

we have:

$$f(x_o) = \frac{1}{4}\pi a.$$

(b) The coordinates of any point on the circle can be expressed in terms of θ by the following method. For the point in question, drop a perpendicular line to the x-axis. We have a right triangle with hypotenuse of length a.

$$\cos(\theta) = \frac{x}{a},$$

or

$$x = a\cos(\theta). \qquad \sin(\theta) = \frac{y}{a},$$

or

$$y = a\sin(\theta).$$

Now an element of the arc length is $a\Delta\theta$. (Recall that the length of an arc of a circle is equal to $S = R\theta$ where R is the radius and θ the angle in radians.) Therefore the length of the arc of the circle in the first quadrant is $\frac{1}{2}\pi a$.

By taking the limit, $a\Delta\theta$ becomes $ad\theta$

$$f(\theta) = \frac{1}{\frac{1}{2}\pi a}\int_0^{\frac{\pi}{2}} a\sin\theta(ad\theta)$$

$$= \frac{2a}{\pi}\int_0^{\frac{\pi}{2}} \sin\theta d\theta = \frac{2a}{\pi}[\cos\theta]_0^{\frac{\pi}{2}}$$

$$= \frac{2a}{\pi}(1-0) = \frac{2a}{\pi}.$$

7.6 Indefinite Integral

The indefinite integral of f(x), denoted by $\int f(x)dx$, is the most general integral of f(x), that is

$$\int f(x)dx = F(x) + C.$$

F(x) is any function such that $F'(x) = f(x)$. C is an arbitrary constant.

7.6.1 Integration of Formulas

Table 7.1

1. $\int x^n dx = \frac{x^{n+1}}{n+1} + C, \quad n \neq -1$	7. $\int \sin ax dx = -\frac{1}{a}\cos ax + C$
2. $\int \frac{dx}{x} = \ln \lvert x \rvert + C$	8. $\int \cos ax dx = \frac{1}{a}\sin ax + C$
3. $\int \frac{dx}{x-a} = \ln \lvert x-a \rvert + C$	9. $\int \sec^2 x\, dx = \tan x + C$
4. $\int \frac{dx}{x^2+a^2} = \frac{1}{a}\tan^{-1}\frac{x}{a} + C$	10. $\int e^{ax} dx = \frac{e^{ax}}{a} + C$
5. $\int \frac{xdx}{x^2+a^2} = \frac{1}{2}\ln \lvert x^2+a^2 \rvert + C$	11. $\int \sinh ax dx = \frac{1}{a}\cosh ax + C$
6. $\int \frac{dx}{(a^2-x^2)^{\frac{1}{2}}} = \sin^{-1}\frac{x}{a} + C$	12. $\int \cosh ax dx = \frac{1}{a}\sinh ax + C$

$\ln x \equiv \log_e x$ is called the logarithm of base e where $e \equiv 2.7182818$ ---

Problem Solving Examples:

Integrate the expression: $\int \frac{dx}{1+e^x}$.

A

We wish to convert the given integral into the form $\int \frac{du}{u}$. If we multiply $\frac{1}{1+e^x}$ by $\frac{e^{-x}}{e^{-x}}$ (which is equal to 1) we obtain:

$$\frac{e^{-x}(1)}{e^{-x}(1+e^x)} = \frac{e^{-x}}{e^{-x}+e^0} = \frac{e^{-x}}{e^{-x}+1}.$$

In integrating this, we apply the formula, $\int \frac{du}{u} = \ln |u| + C$, letting

$u = e^{-x} + 1$. Then $du = -e^{-x}dx$. We obtain:

$$\int \frac{e^{-x}}{e^{-x}+1}\, dx = -\int \frac{-e^{-x}\, dx}{e^{-x}+1} = -\ln(1+e^{-x}) + C.$$

7.6.2 Algebraic Simplification

Certain apparently complicated integrals can be made simple by simple algebraic manipulations.

Example: Find $\int \frac{x}{x+1}\, dx$

Write $\frac{x}{x+1} = \frac{x+1-1}{x+1} = 1 - \frac{1}{x+1}$

$$\int \frac{x}{x+1}\, dx = \int dx - \int \frac{dx}{x+1} = x - \ln|x+1| + c$$

Problem Solving Examples:

Integrate: $\int \frac{2x}{x+1} dx$.

To integrate the given expression we manipulate the integrand to obtain the form $\int \frac{du}{u}$. This can be done as follows:

$$\int \frac{2x}{x+1}\, dx = 2\int \frac{x}{x+1}\, dx$$

$$= 2\int \left(\frac{x+1}{x+1} - \frac{1}{x+1}\right) dx$$

$$= 2\int \left(1 - \frac{1}{x+1}\right) dx$$

$$= 2\int dx - 2\int \frac{dx}{x+1}.$$

Now, applying the formula $\int \frac{du}{u} = \ln u$, we obtain:

$$\int \frac{2x}{x+1}\, dx = 2x - 2\ \ln(x+1) + C.$$

7.6.3 Substitution of Variables

Suppose F(x) is expressed as a composite function, $F(x)=f(u(x))$, then $F'(x)=f'(u)\frac{du}{dx}$, and $F'(x)dx = f'(u)du$.

Therefore, $\int F'(x)dx = \int f'(u)du = f(u)+c$

$$= f(u(x)) + c = F(x) + c.$$

THEOREM:

Let f and u be functions satisfying the following conditions:

a) f is continuous on a domain including the closed interval $\{x: a \le x \le b\}$.

b) For each point t in the closed interval $\{t: \alpha \le t \le \beta\}$, the value u(t) is a point in $\{x: a \le x \le b\}$.

c) $u(\alpha) = a$, and $u(\beta) = b$.

d) u is continuous on $\{t: \alpha \le t \le \beta\}$.

The $\int_a^b f(x)dx = \int_\alpha^\beta f(u(t)) \cdot u'(t)dt.$

Example: Evaluate $\int \frac{x}{x^2+a^2} dx$

Let $u = x^2 + a^2$

$du = 2x\, dx$

$\frac{1}{2} du = x dx$

$$\int \frac{x}{x^2+a^2} dx = \frac{1}{2}\int \frac{du}{u} = \frac{1}{2} \text{Ln } |u| + c$$

$$= \frac{1}{2} \text{Ln } |x^2+a^2| + c$$

Problem Solving Examples:

Evaluate the expression: $\int_0^3 x\sqrt{1+x}\, dx$.

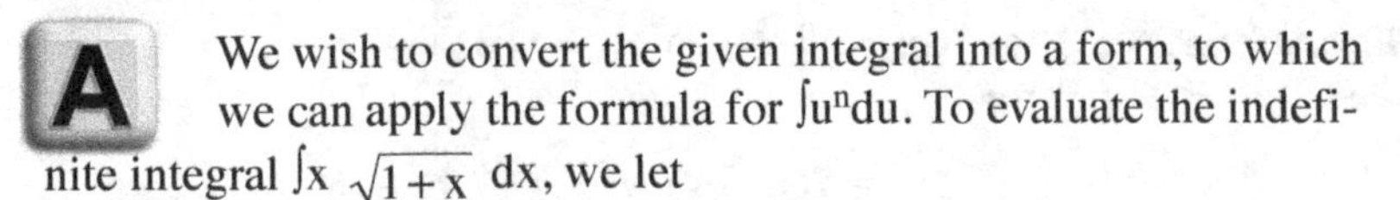

We wish to convert the given integral into a form, to which we can apply the formula for $\int u^n du$. To evaluate the indefinite integral $\int x\sqrt{1+x}\, dx$, we let

$u = \sqrt{1+x}$, $u^2 = 1 + x$, $x = u^2 - 1$, $dx = 2u\ du$.

Substituting, we have:

$$\int x\sqrt{1+x}\ dx = \int (u^2 - 1)\ u\ (2u\ du)$$
$$= 2\int (u^4 - u^2)\ du.$$

We can now apply the formula for $\int u^n du$, and we obtain:

$$\frac{2}{5}u^5 - \frac{2}{3}u^3 + C = \frac{2}{5}(1+x)^{\frac{5}{2}} - \frac{2}{3}(1+x)^{\frac{3}{2}} + C,$$

by substitution. Therefore, the definite integral

$$\int_0^3 x\sqrt{1+x}\ dx = \frac{2}{5}(1+x)^{\frac{5}{2}} - \frac{2}{3}(1+x)^{\frac{3}{2}}\Big]_0^3$$
$$= \frac{2}{5}(4)^{\frac{5}{2}} - \frac{2}{3}(4)^{\frac{3}{2}} - \frac{2}{5}(1)^{\frac{5}{2}} + \frac{2}{3}(1)^{\frac{3}{2}}$$
$$= \frac{64}{5} - \frac{16}{3} - \frac{2}{5} + \frac{2}{3}$$
$$= \frac{116}{15}.$$

7.6.4 Change of Variables

Example: Evaluate $\displaystyle\int_0^1 x(1+x)^{\frac{1}{2}}dx$

Let $u = 1+x$, $du = dx$, $x = u-1$

$$\int_0^1 x(1+x)^{\frac{1}{2}} = \int_1^2 (u-1)u^{\frac{1}{2}}\ du.$$

Notice the change in the limits for $x=0$, $u=1$ and for $x=1$ $u=2$.

$$\int_1^2 (u-1)u^{\frac{1}{2}}\,du = \int_1^2 u^{3/2} - u^{\frac{1}{2}}\,du$$

$$= 2/5\ u^{5/2} - 2/3\ u^{3/2}\Big|_1^2$$

$$= [(2/5)\sqrt{32} - (2/3)\sqrt{8})] - \left(\frac{2}{5} - \frac{2}{3}\right)$$

$$= \frac{4\sqrt{2}}{15} - \frac{4}{15} = \frac{4}{15}(\sqrt{2} - 1).$$

Problem Solving Examples:

Evaluate the expression: $\int_1^2 \frac{x}{(1+2x)^3}\,dx.$

A

This integral is difficult because of the expression 1 + 2x in the denominator. Hence we choose our substitution to eliminate this expression. We let

$$u = 1 + 2x, \text{ then } x = \frac{u-1}{2} \text{ and } dx = \frac{1}{2}\,du.$$

Now

$u = 3$ when $x = 1$

$u = 5$ when $x = 2$,

giving us new limits. Using the substitution, we obtain:

$$\int_1^2 \frac{x}{(1+2x)^3}\,dx = \int_3^5 \left(\frac{\frac{u-1}{2}}{u^3}\right)\left(\frac{1}{2}\right)du$$

$$= \frac{1}{4}\int_3^5 \left(\frac{1}{u^2} - \frac{1}{u^3}\right)du.$$

We can now use the formula for $\int u^n du$ on both terms of the integrand, obtaining:

$$\frac{1}{4}\left[-\frac{1}{u}+\frac{1}{2u^2}\right]_3^5 = \frac{11}{450}.$$

7.6.5 Integration of Parts

This method is based on the formula

$$d(uv) = u\,dv + v\,du.$$

The corresponding integration formula,

$uv = \int u\,dv + \int v\,du$, is applied in the form

$$\boxed{\int u\,dv = uv - \int v\,du}$$

This procedure involves the identification of u and dv and their manipulation into the form of the latter equation. v must be easily determined. If a definite integral is

$$\int_a^b u\frac{dv}{dx}\,dx = uv\Big]_a^b - \int_a^b v\frac{du}{dx}\,dx.$$

Example: Evaluate $\int x \cos x\,dx$

$u = x \quad dv = \cos x\,dx$

$du = dx \quad v = \sin x$

$$\int x\cos x\,dx = x\sin x - \int \sin x\,dx$$

$$= x \sin x - (-\cos x) + c$$

$$= x \sin x + \cos x + c$$

Problem Solving Examples:

Integrate by parts the expression: $\frac{dy}{dx} = x^2 \ln x$.

$$dy = x^2 \ln x \cdot dx.$$

$$y = \int x^2 \ln x \, dx.$$

To integrate by parts we use the equation:

$\int udv = uv - \int vdu$.

Now, let $u = \ln x$.

Then, $du = \frac{1}{x} \cdot dx$.

Let $dv = x^2 \cdot dx$.

Then, $v = \int dv = \int x^2 \, dx = \frac{x^3}{3}$,

by use of the formula for $\int u^n du$. Substituting into the above equation, we have:

$$y = \int \overset{u}{\ln x} \cdot \overset{dv}{x^2 \cdot dx} = \frac{x^3}{3} \ln x - \int \frac{x^3}{3} \cdot \frac{1}{x} \cdot dx.$$

We can now integrate

$$\int \frac{x^3}{3} \cdot \frac{1}{x} \cdot dx,$$

by using the formula for $\int u^n du$, with $u = x$, $du = dx$, and $n = 2$. Doing this, we obtain:

$$y = \frac{x^3}{3} \ln x - \frac{x^3}{9} + C.$$

7.6.6 Trigonometric Integrals

Integrals of the form $\int \sin^n x\, dx$ or $\int \cos^n x dx$ can be evaluated without resorting to integration by parts. This is done in the following manner;

We write $\int \sin^n x\, dx = \int \sin^{n-1} \sin x\, dx$, if n is odd.

Since the integer n-1 is even, we may then use the fact that $\sin^2 x = 1 - \cos^2 x$ to obtain a form which is easier to integrate.

Example: $\int \sin^5 x\, dx = \int \sin^4 x \sin x\, dx$

$$= \int (\sin^2 x)^2 \sin x\, dx$$

but $\sin^2 x = 1 - \cos^2 x$.

Hence, $\int \sin^5 x\, dx = \int (1-\cos^2 x)^2 \sin x\, dx$

$$= \int (1-2\cos^2 x+\cos^4 x)\sin x\, dx$$

Substitute $u = \cos x$, $du = -\sin x\, dx$

$$= -\int (1-2u^2+u^4)du = -u + \frac{2}{3}u^3 - \frac{u^5}{5}$$

$$= -\cos x + \frac{2}{3}\cos^3 x - \frac{1}{5}\cos^5 x + c.$$

A similar technique can be employed for odd powers of cos x.

If the integrand is $\sin^n x$ or $\cos^n x$ and n is even, then the half angle formulas ,

$$\sin^2 x = \frac{1 - \cos 2x}{2} \quad \text{or}$$

$$\cos^2 x = \frac{1+\cos 2x}{2}$$

may be used to simplify the integrand.

Example: $\int \cos^2 x \, dx = \frac{1}{2} \int (1+\cos 2x)dx$

$$= \frac{1}{2} x + \frac{1}{4} \sin 2x + c$$

Problem Solving Examples:

Integrate the expression: $\int \cos^2 x \sin x \, dx$.

In evaluating this integral we use the formula:

$$\int u^n du = \frac{u^{n+1}}{n+1} \quad .$$

Let u = cos x, du = –sin x dx, and n = 2. Applying the formula, we obtain:

$$\int \cos^2 x \sin x \, dx = -\frac{\cos^3 x}{3} + C.$$

Quiz: The Definite Integral

1. $\int \left(x - \frac{1}{x}\right)^2 dx =$

   (A) $\frac{1}{3}\left(x - \frac{1}{x}\right)^3 + C$

   (B) $\frac{1}{3}\left(x - \frac{1}{x}\right)^3\left(1 + \frac{1}{x^2}\right) + C$

   (C) $\frac{1}{3}x^3 - 2x - \frac{1}{x^2} + C$

   (D) $\frac{1}{3}x^3 - 2x - \frac{1}{x} + C$

   (E) $\frac{1}{3}(1 - \ln x)^3 + C$

2. $\int \frac{x - e^x}{xe^x} dx =$

   (A) $-e^{-x} - \frac{1}{x^2} + C$

   (B) $e^{-x} - \ln|x| + C$

   (C) $-e^{-x} - \ln|x| + C$

   (D) $-\frac{1}{e^{2x}} + \ln|x| + C$

   (E) $e^{-x} - \frac{1}{x^2} + C$

3. Let $f(x) = 3x^2$ on $[1, 5]$. For what value of c, $1 < c < 5$, does $f(c) = \dfrac{\int_1^5 f(x)\,dx}{4}$

   (A) 4.7

   (B) 4.4

   (C) 4.1

   (D) 3.8

   (E) 3.2

4. $\int_0^1 (\sec^2 x - \tan^2 x)\,dx =$

   (A) 3

   (B) 5

   (C) 2

   (D) 4

   (E) 1

5. If $0 \le x \le 1$, then $\dfrac{d}{dx}\displaystyle\int_x^0 \dfrac{dt}{2+t} =$

   (A) $\dfrac{1}{x+2}$

   (B) $-\dfrac{1}{x+2}$

   (C) $\ln|2+x|$

   (D) $\ln|2+x| + c$

   (E) $-\ln|2+x| + c$

6. Let $R = \int_0^a \cos(x^2)\,dx$ and $S = \int_0^a \tan x\,dx$

Find $\int_0^a [\cos(x^2) + \tan x]\,dx$.

(A) $2R$

(B) $2S$

(C) $R + 2S$

(D) $S + 2R$

(E) $2R + 2S$

7. $\int_1^e x \ln x\,dx = ?$

(A) e

(B) $\dfrac{e^2 - 1}{2}$

(C) $\dfrac{e^2 + 1}{4}$

(D) $\dfrac{e - 1}{2}$

(E) None of these

8. The graph of f is shown in the figure. Which of the following could be the graph of $\int f(x)\,dx$?

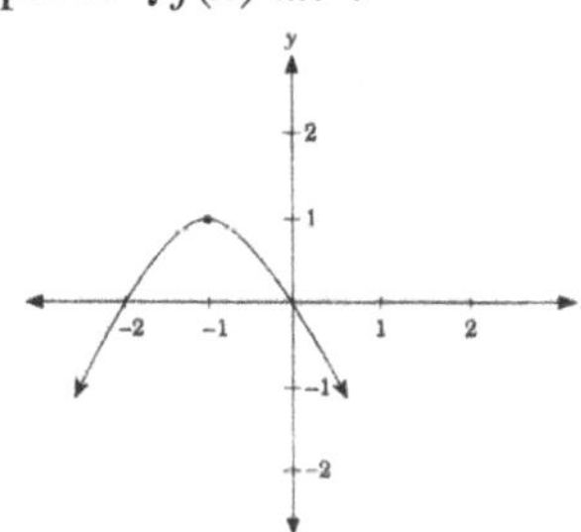

(A)

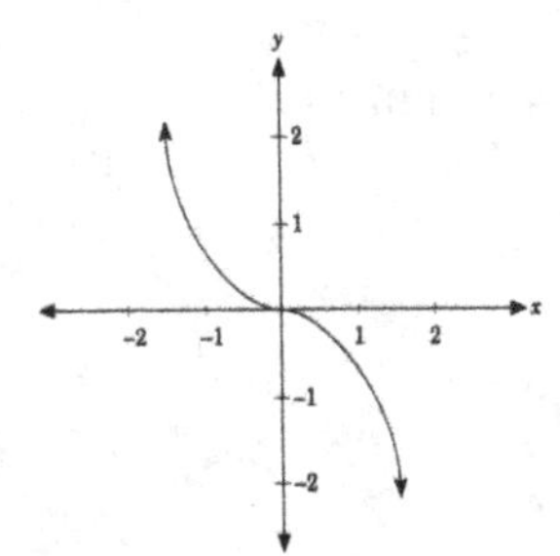

(B)

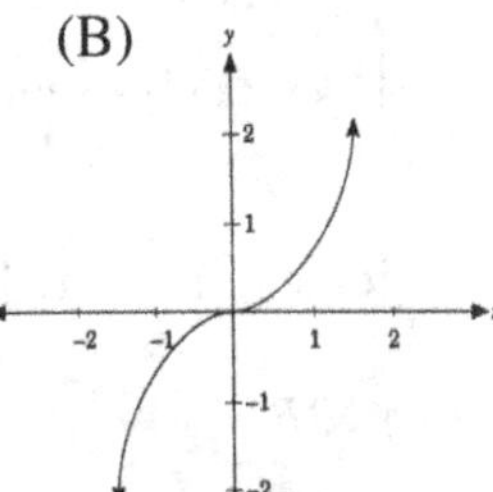

(C)

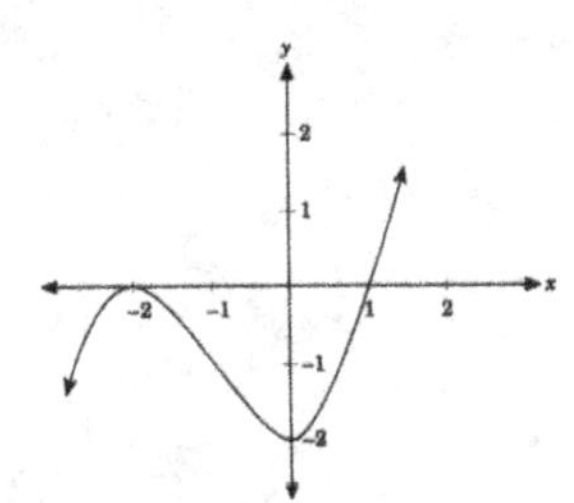

(D)

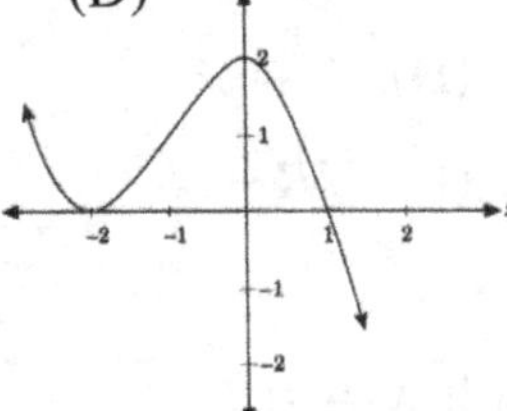

(E)

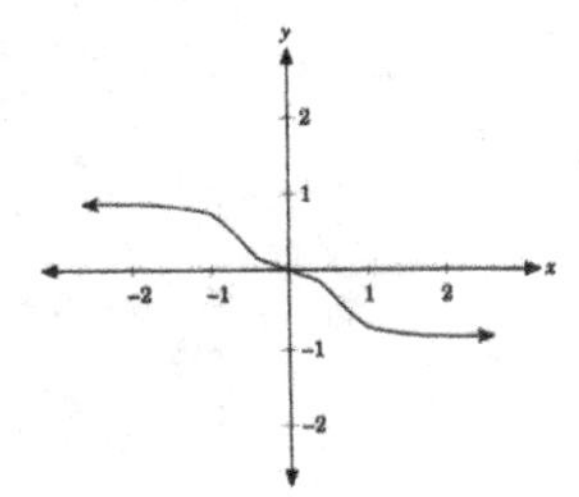

9. If $f(x) = \int (1 - 2x)^3 \, dx$, then the second derivative of $f(x)$ at $x = \frac{1}{2}$ is

(A) –48

(B) –12

(C) 0

(D) 96

(E) None of these

10. $\int x^{-1}\,dx =$

(A) $\dfrac{x^{-2}}{-2} + C$

(B) $x + C$

(C) Undefined

(D) $-x + c$

(E) None of these

ANSWER KEY

1. (D)	6. (A)
2. (C)	7. (C)
3. (E)	8. (D)
4. (E)	9. (C)
5. (B)	10. (E)

CHAPTER 8

Techniques of Integration

8.1 Table of Integrals

1. $\int k dx = kx + C.$

2. $\int x^n dx = \frac{1}{n+1} x^{n+1} + C, \quad n \neq -1.$

3. $\int \frac{dx}{x} = \ln |x| + C.$

4. $\int e^x dx = e^x + C.$

5. $\int p^x dx = \frac{p^x}{\ln p} + C.$

6. $\int \ln x \, dx = x \ln x - x + C.$

7. $\int \cos x \, dx = \sin x + C.$

8. $\int \sin x \, dx = -\cos x + C.$

9. $\int \sec^2 x \, dx = \tan x + C.$

10. $\int \operatorname{cosec}^2 x \, dx = -\cot x + C.$

11. $\int \sec x \tan x \, dx = \sec x + C.$

12. $\int \operatorname{cosec} x \cot x \, dx = -\operatorname{cosec} x + C.$

13. $\int \tan x \, dx = \ln |\sec x| + C.$

14. $\int \cot x \, dx = \ln |\sin x| + C.$

15. $\int \sec x \, dx = \ln |\sec x + \tan x| + C.$

16. $\int \operatorname{cosec} x \, dx = \ln |\operatorname{cosec} x - \cot x| + C.$

17. $\int \frac{dx}{\sqrt{1-x^2}} = \text{arc} \sin x + C.$

18. $\int \frac{dx}{1+x^2} = \text{arc} \tan x + C.$

19. $\int \text{arc} \sin x \, dx = x \, \text{arc} \sin x + \sqrt{1-x^2} + C.$

20. $\int \text{arc} \tan x \, dx = x \, \text{arc} \tan x - \frac{1}{2} \ln (1+x^2) + C.$

21. $\int \frac{dx}{x^2 - 1} = \frac{1}{2} \ln \left| \frac{x-1}{x+1} \right| + C.$

8.2 Integration by Parts

Differential of a product is represented by the formula

$$d(uv) = udv + vdu$$

Integration of both sides of this equation gives

$$uv = \int udv + \int vdu \qquad (1)$$

or

$$\boxed{\int udv = uv - \int vdu} \qquad (2)$$

Equation (2) is the formula for integration by parts.

Example: Evaluate $\int x \ln x \, dx$

Let $\quad u = \ln x \qquad dv = xdx$

$\quad du = 1/x \, dx \qquad v = 1/2 \, x^2$

Thus,

$$\int x \ln x \, dx = (\tfrac{1}{2})x^2 \ln x - \int (\tfrac{1}{2})x^2 \cdot (\tfrac{1}{x})dx$$

$$= (\tfrac{1}{2})x^2 \ln x - \tfrac{1}{2} \int x \, dx$$

$$= (\tfrac{1}{2})x^2 \ln x - (\tfrac{1}{4})x^2 + c$$

Integration by parts may be used to evaluate definite integrals. The formula is:

$$\int_a^b u\,dv = \Big[uv\Big]_a^b - \int_a^b v\,du$$

Problem Solving Examples:

Integrate the expression: $\int x\sqrt{1-3x}\,dx$.

Here we use integration by parts. Then, $\int u\,dv = uv - \int v\,du$.

$x = u$, $\sqrt{1-3x}\,dx = dv$, $dx = du$, and

$$v = \int \sqrt{1-3x}\,dx = -\frac{1}{3}\cdot\frac{2}{3}(1-3x)^{\frac{3}{2}}$$

$$= -\frac{2}{9}(1-3x)^{\frac{3}{2}}$$. This is done by use of the

integration formula:

$\int u^n\,du = \frac{u^{n+1}}{n+1}$, letting $u = (1-3x)$, $du = -3dx$, and $n = \frac{1}{2}$.

Substituting into the above equation, we obtain:

$$\int x\sqrt{1-3x}\,dx = -\frac{2x}{9}(1-3x)^{\frac{3}{2}} + \frac{2}{9}\int(1-3x)^{\frac{3}{2}}\,dx.$$

We now integrate, again using:

$\int u^n du = \frac{u^{n+1}}{n+1}$, with $u = (1-3x)$, $du = -3\,dx$, and $n = \frac{3}{2}$, obtaining:

$$-\frac{2x}{9}(1-3x)^{\frac{3}{2}} \quad -\frac{4}{135}(1-3x)^{\frac{5}{2}} + C.$$

Integrate: $\int x \cdot \cos x \, dx$.

A In this case we use integration by parts, the rule for which states: $\int u dv = uv - \int v du$. Let $u = x$ and $dv = \cos x \, dx$. Then $du = dx$ and $v = \int \cos x \cdot dx = \sin x$. $\int u \cdot dv = uv - \int v \cdot du$ becomes $\int x \cdot \cos x \cdot dx = x \cdot \sin x - \int \sin x \cdot dx$. To integrate $\int \sin x dx$ we use the formula, $\int \sin u du = -\cos u + C$. This gives:

$$\int x \cdot \cos x \cdot dx = x \sin x - (-\cos x) + C$$

$$= x \sin x + \cos x + C.$$

8.3 Partial Fractions

To evaluate rational functions of the form $\int \frac{P(x)}{Q(x)} dx$, where P and Q are polynomials, we apply the following techniques:

1. Factor the denominator, $Q(x)$, into a product of linear and quadratic factors.
   Example: $Q(x) = x^3 + 2x^2 + x + 2$

   $= x^3 + x + 2x^2 + 2 = x(x^2+1) + 2(x^2+1)$

   $= (x^2+1)(x+2)$

   quadratic factor (x^2+1), linear factor ($x+2$)

2. Rewrite $\frac{P}{Q}$ as a sum of simpler rational functions, each of which can be integrated.

If the degree of the numerator $(P(x))$ is larger than the degree of the denominator $(Q(x))$, we divide $P(x)$ by $Q(x)$ to obtain a quotient (polynomial of the form $\frac{P}{Q}$) plus a rational function (remainder divided by the divisor) in which the degree of the numerator is less than the degree of the denominator.

The decomposition of a rational function into the sum of simpler expressions is known as the method of partial fractions. Four ways in which the denominator can be factored are as follows:

1. The denominator $Q(x)$ can be decomposed to give distinct linear factors of the form $\frac{A_1}{x-a_1} + \frac{A_2}{x-a_2} + \ldots + \frac{A_n}{x-a_n}$

2. The denominator $Q(x)$ can be decomposed into linear factors of the form $\frac{A_1}{x-a} + \frac{A_2}{(x-a)^2} + \ldots + \frac{A_k}{(x-a)^k}$, where some of the linear factors are repeated.

   Example: Decomposition of $Q(x) = (x-2)^3$ gives $\frac{A_1}{x-2} + \frac{A_2}{(x-2)^2} + \frac{A_3}{(x-2)^3}$.

3. $Q(x)$ can be factored to give linear and irreducible quadratic factors. Each unrepeated quadratic factor has the form $\frac{Ax+B}{x^2+bx+c}$.

4. $Q(x)$ can be factored to give linear and quadratic factors where some of the quadratic factors can be repeated.

   In this case each repeated quadratic factor can be expressed as follows:

$$\frac{A_1x + B_1}{(x^2+bx+c)} + \frac{A_2x + B_2}{(x^2+bx+c)^2} + \ldots + \frac{A_nx + B_n}{(x^2+bx+c)^n}$$

Problem Solving Examples:

Integrate the expression: $\int \frac{2x^2 + 5x - 1}{x^3 + x^2 - 2x}\, dx.$

To integrate the given expression we use the method of partial fractions. Since

$$x^3 + x^2 - 2x = x(x - 1)(x + 2),$$

the denominator is a product of distinct linear factors, and we try to find A_1, A_2 and A_3 such that

$$\frac{2x^2 + 5x - 1}{x^3 + x^2 - 2x} = \frac{A_1}{x} + \frac{A_2}{x-1} + \frac{A_3}{x+2}$$

$$= \frac{A_1(x - 1)(x + 2) + A_2(x)(x + 2) + A_3(x)(x - 1)}{x^3 + x^2 - 2x}$$

Setting the two numerators equal, we obtain:

$$2x^2 + 5x - 1 = A_1(x - 1)(x + 2) + A_2 x(x + 2) + A_3 x(x - 1).$$

Multiplying out and collecting like powers of x yields:

$$(A_1 + A_2 + A_3)x^2 + (A_1 + 2A_2 - A_3)x - 2A_1 = 2x^2 + 5x - 1.$$

Equating coefficients of like powers of x, we have the equations:

$$A_1 + A_2 + A_3 = 2$$

$$A_1 + 2A_2 - A_3 = 5$$

$$-2A_1 = -1.$$

Solving these equations we find that

$$A_1 = \frac{1}{2}, A_2 = 2 \text{ and } A_3 = -\frac{1}{2}.$$

Therefore we have :

$$\int \frac{2x^2 + 5x - 1}{x^3 + x^2 - 2x} dx = \frac{1}{2}\int \frac{dx}{x} + 2\int \frac{dx}{x-1} - \frac{1}{2}\int \frac{dx}{x+2}.$$

We now integrate, using the formula:

$$\int \frac{du}{u} = \ln |u|,$$

obtaining:

$$\int \frac{2x^2 + 5x - 1}{x^3 + x^2 - 2x} dx = \frac{1}{2} \ln |x| + 2 \ln |x-1| - \frac{1}{2} \ln |x+2| + C.$$

Integrate: $\int \frac{x^3 + 6x^2 + 3x + 6}{x^3 + 2x^2} dx$.

To integrate this expression, we use partial fractions. We first find the factors of the denominator. The denominator, $x^3 + 2x^2$, can be factored as $x^2(x + 2)$. We then have three terms: $\frac{A}{x}$, $\frac{B}{x^2}$, and $\frac{C}{(x+2)}$. Since the denominator is of the same power as the numerator, we can divide the numerator by the denominator, and write the expression as:

$$1 + \frac{4x^2 + 3x + 6}{x^3 + 2x^2}.$$

We now find three numbers: A, B and C, such that:

$$1 + \frac{4x^2 + 3x + 6}{x^3 + 2x^2} = \frac{A}{x} + \frac{B}{x^2} + \frac{C}{(x+2)}.$$

Considering the 1 as a constant, we can ignore it momentarily and solve for the numerical value of A, B and C by multiplying both sides of the equation by $x^3 + 2x^2$. We find:

$$4x^2 + 3x + 6 = Ax(x + 2) + B(x+2) + cx^2.$$

We can combine like powers of x to find

$$4x^2 + 3x + 6 = Ax^2 + 2Ax + Bx + 2B + Cx^2.$$

$$4x^2 + 3x + 6 = (A+C)x^2 + (2A+B)x + 2B.$$

Now we equate like powers of x to find A, B and C. We find

$$A + C = 4.$$
$$2A + B = 3.$$
$$2B = 6.$$

Therefore, $B = 3, A = 0, C = 4.$

We substitute these values, and integrate:

$$\int \frac{x^3 + 6x^2 + 3x + 6}{x^3 + 2x^2}\, dx = \int \left(1 + \frac{0}{x} + \frac{3}{x^2} + \frac{4}{x+2}\right) dx$$
$$= x - \frac{3}{x} + 4\ln(x+2) + C.$$

8.4 Trigonometric Substitution

If the integral contains expressions of the form $\sqrt{a^2-x^2}$, $\sqrt{a^2+x^2}$ or $\sqrt{x^2-a^2}$, where $a > 0$, it is possible to transform the integral into another form by means of trigonometric substitution.

8.4.1 General Rules for Trigonometric Substitutions

1. Make appropriate substitutions.
2. Sketch a right triangle.
3. Label the sides of the triangle by using the substituted information.
4. The length of the third side is obtained by use of the Pythagorean Theorem.
5. Utilize sketch, in order to make further substitutions.

A. If the integral contains the expression of the form $\sqrt{a^2-x^2}$, make the substitution $x = a\sin\theta$.

$$\sqrt{a^2-x^2} = \sqrt{a^2-a^2\sin^2\theta} = \sqrt{a^2(1-\sin^2\theta)} = \sqrt{a^2\cos^2\theta}$$

$$= a\cos\theta$$

In trigonometric substitution the range of θ is restricted. For example, in the sine substitution the range of θ is $-\pi/2 \le \theta \le \pi/2$. The sketch of this substitution is shown in Fig. 8-1.

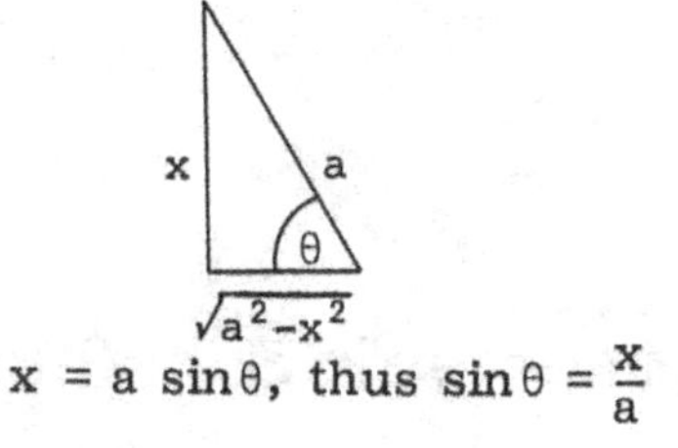

$x = a\sin\theta$, thus $\sin\theta = \dfrac{x}{a}$

Fig. 8-1

B) If the integral contains the expression of the form $\sqrt{x^2-a^2}$, make the substitution $x = a\sec\theta$. The sketch is shown in Fig. 8-2

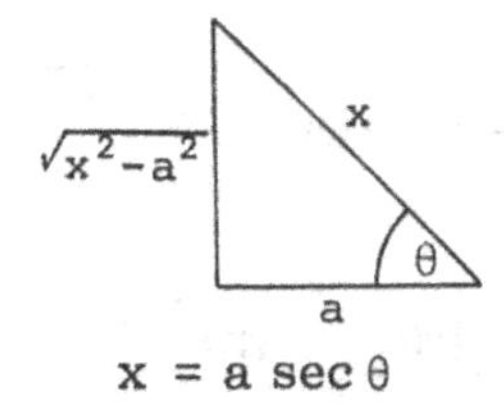

Fig. 8-3

C) If the integral contains the expression of the form $\sqrt{a^2+x^2}$, make the substitution $x = a \tan\theta$.

Example: Evaluate $\int \frac{dx}{\sqrt{4+x^2}}$

Solution: Let

$$x = 2\tan\theta$$

$$dx = 2\sec^2\theta\, d\theta$$

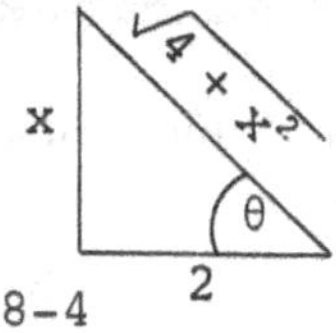

Fig. 8-4

Thus,

$$\int \frac{dx}{\sqrt{4+x^2}} = \int \frac{2\sec^2\theta\, d\theta}{\sqrt{4+(2\tan\theta)^2}}$$

$$= \int \frac{2\sec^2\theta\, d\theta}{\sqrt{4(1+\tan^2\theta)}}$$

$$= \int \frac{2\sec^2\theta\, d\theta}{2\sqrt{\sec^2\theta}} = \int \sec\theta\, d\theta$$

$$= \ln|\sec\theta + \tan\theta| + c$$

To convert from θ back to x we use Fig. 8-4 to find:

$$\sec\theta = \frac{\sqrt{4+x^2}}{2} \text{ and } \tan\theta = \frac{x}{2}$$

Therefore, $\int \frac{dx}{\sqrt{4 + x^2}} = \ln \left| \frac{\sqrt{4 + x^2}}{2} + \frac{x}{2} \right| + c$

8.4.2 Summary of Trigonometric Substitutions

Given expression	Trigonometric substitution
$\sqrt{x^2 - a^2}$	$x = a \sec \theta$
$\sqrt{x^2 + a^2}$	$x = a \tan \theta$
$\sqrt{a^2 - x^2}$	$x = a \sin \theta$

Problem Solving Examples:

Integrate the expression: $\int \frac{dx}{x^2\sqrt{4-x^2}}$.

Let $x = 2 \sin \theta$. Then $dx = 2 \cos \theta \, d\theta$, so that

$$\int \frac{dx}{x^2\sqrt{4-x^2}} = \int \frac{2\cos\theta \, d\theta}{4 \sin^2 \theta\sqrt{4-4 \sin^2 \theta}}$$

$$= \int \frac{2\cos\theta \, d\theta}{4 \sin^2 \theta\sqrt{4(1-\sin^2 \theta)}}$$

Using the identity: $\cos^2 \theta = 1 - \sin^2 \theta$, and the fact that

$\frac{1}{\sin^2 \theta} = \csc^2 \theta$, we obtain:

$$\int \frac{2 \cos \theta \, d\theta}{4 \sin^2 \theta \, (2 \cos \theta)} = \frac{1}{4} \int \csc^2 \theta \, d\theta.$$

Now applying the formula, $\int \csc^2 u \, du = -\cot u + c$, we have:

$$-\frac{1}{4} \cot \theta + C = -\frac{\sqrt{4-x^2}}{4x} + C.$$

Integrate the expression: $\int \frac{x^2\,dx}{\sqrt{a^2 - x^2}}$.

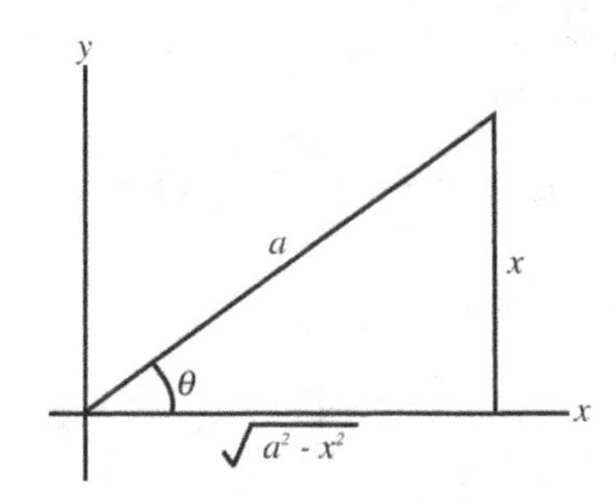

Let $x = a \sin \theta$. $dx = a \cos \theta \, d\theta$. By substitution and the use of the identity:

$\cos^2 \theta = 1 - \sin^2 \theta$, we obtain:

$$a^2 \int \frac{\sin^2 \theta \, a \cos \theta \, d\theta}{a \cos \theta} = a^2 \int \sin^2 \theta \, d\theta = a^2 \int \frac{1 - \cos 2\theta}{2} \, d\theta,$$

using the identity $\sin^2 \theta = \frac{1 - \cos 2\theta}{2}$. Now we evaluate the integral, using: $\int \cos u \, du = \sin u + c$, and then the identity: $\sin 2\theta = 2 \sin \theta \cos \theta$, from which $\frac{\sin 2\theta}{2} = \sin \theta \cos \theta$. Doing this, we obtain:

$$\frac{a^2}{2}(\theta - \frac{1}{2} \sin 2\theta) + c = \frac{a^2}{2}(\theta - \sin \theta \cos \theta) + c.$$

Hence, substituting values from the shown right triangle we obtain:

$$\int \frac{x^2\,dx}{\sqrt{a^2 - x^2}} = \frac{a^2}{2} \left(\arcsin \frac{x}{a} - \frac{x}{a^2} \sqrt{a^2 - x^2}\right) + c.$$

8.5 Quadratic Functions

An integral containing the expression ax^2+bx+c can be simplified by completing the square and making the appropriate substitution.

Thus, $ax^2 + bx + c = a(x^2+\frac{b}{a}x) + c$

$$= a\left(x + \frac{b}{2a}\right)^2 + c - \frac{b^2}{4a}$$

Then substitute $y = x + \frac{b}{2a}$, which changes the expression into an integrable form.

Example: Evaluate $\int \frac{(2x-3)}{x^2+2x+2}\,dx$

Solution: We complete the square, obtaining

$$x^2 + 2x + 2 = (x^2 + 2x+1) + 1$$

$$= (x+1)^2 + 1$$

Let $y = x+1$, $dy = dx$

Thus,

$$\int \frac{(2x-3)dx}{x^2+2x+2} = \int \frac{2y-5}{y^2+1}dy = \int \frac{2ydy}{y^2+1} - 5\int \frac{dy}{y^2+1}$$

$$= \ln(y^2+1) - 5\arctan y + c$$

$$= \ln(x^2+2x+2) - 5\arctan(x+1) + c$$

The technique of completing the square may be used if the quadratic expression appears under the radical sign.

Example: Evaluate $\int \frac{1}{\sqrt{8+2x-x^2}}\,dx$

Solution: We complete the square, obtaining

$$8+2x-x^2 = 8-(x^2-2x) = 8+1-(x^2-2x+1)$$

$$= 9-(x-1)^2$$

Let $u = x-1$; $du = dx$

Thus,

$$\int \frac{1}{\sqrt{8+2x-x^2}} = \int \frac{1}{\sqrt{9-u^2}}\,du$$

$$= \sin^{-1}\frac{u}{3} + c$$

$$= \sin^{-1}\frac{x-1}{3} + c$$

$$= \text{arc sin}\,\frac{x-1}{3} + c$$

Problem Solving Examples:

Integrate the expression: $\int \frac{dx}{\sqrt{x^2+6x}}$

We wish to convert the given integral into a form that will allow us to apply the formula:

$$\int \frac{du}{\sqrt{u^2+a^2}} = \cosh^{-1}\frac{u}{a} + C.$$

We first complete the square under the radical to obtain:

$$\int \frac{dx}{\sqrt{x^2+6x}} = \int \frac{dx}{\sqrt{x^2+6x+9-9}} = \int \frac{dx}{\sqrt{(x+3)^2-9}}.$$

The integral is now in the desired form, with u = (x + 3), du = dx, and a = 3. Applying the formula, we obtain:

$$\int \frac{dx}{\sqrt{(x+3)^2-(3)^2}} = \cosh^{-1}\frac{x+3}{3} + C.$$

Integrate the expression: $\int \frac{dx}{\sqrt{9x^2+4}}$

We wish to convert the given integral into a form, to which we can apply the formula: $\int \frac{du}{\sqrt{u^2+1}} = \sin h^{-1} u + C$. We multiply the expression under the radical by $\frac{1}{4}$ to obtain the desired form. Here, $u = \frac{3x}{2}$, $du = \frac{3}{2}\,dx$. We must also remember to multiply by $\sqrt{\frac{1}{4}} = \frac{1}{2}$ outside the integral sign. Doing this, and using the formula for $\frac{du}{\sqrt{u^2+1}}$, we obtain:

$$\int \frac{dx}{\sqrt{9x^2+4}} = \frac{1}{2}\int \frac{dx}{\sqrt{\left(\frac{3x}{2}\right)^2+1}}$$

$$= \frac{1}{3}\,\frac{\frac{3}{2}dx}{\sqrt{\left(\frac{3x}{2}\right)^2+1}}$$

$$= \frac{1}{3}\sinh^{-1}\frac{3x}{2} + C.$$

CHAPTER 9

Applications of the Integral

9.1 Area

If f and g are two continuous functions on the closed interval [a,b], then the area of the region bounded by the graphs of these two functions and the ordinates x = a and x = b is

$$A = \int_a^b [f(x)-g(x)]dx.$$

where $\quad f(x) \geq 0$ and $f(x) \geq g(x)$

$$a \leq x \leq b$$

This formula applies whether the curves are above or below the x-axis.

The area below f(x) and above the x-axis is represented by $\int_a^b f(x)$. The area between g(x) and the x-axis is represented by $\int g(x)$.

Example: Find the area of the region bounded by the curves $y = x^2$ and $y = \sqrt{x}$.

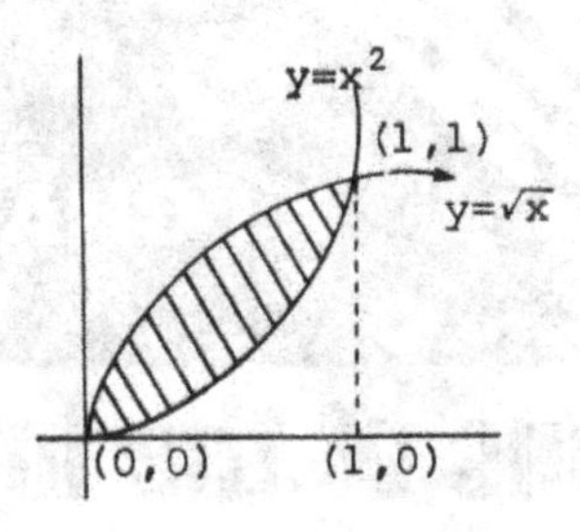

Fig. 9-1

$$\text{Area} = A = \int_0^1 (\sqrt{x} - x^2)\,dx$$

$$= \int_0^1 \sqrt{x}\,dx - \int_0^1 x^2\,dx$$

$$= \left[\frac{2}{3}x^{\frac{3}{2}} - \frac{1}{3}x^3\right]_0^1$$

$$A = \left[\frac{2}{3} - \frac{1}{3}\right] = \frac{1}{3}$$

Problem Solving Examples:

Find the area between the curve: $y = x^3$, and the x-axis, from $x = -2$ to $x = 3$.

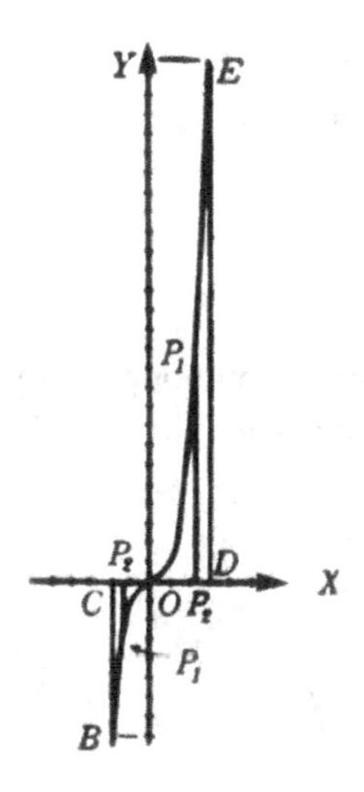

It is generally advantageous to sketch the curve, since parts of the curve may have to be considered separately, particularly when positive and negative limits are given. The desired area is composed of the two parts: BOC and ODE. To find the total area, we can evaluate each area separately and then add. The area is the integral of the upper function minus the lower function. In the first quadrant, the upper function is the curve $y = x^3$, the lower function is $y = 0$, (the x-axis) and the limits are $x = 0$ and $x = 3$. In the third quadrant, the upper function is $y = 0$, the lower function is the curve $y = x^3$, and the limits are $x = -2$ and $x = 0$. Hence, we can write,

$$A_{total} = \int_0^3 (x^3 - 0)dx + \int_{-2}^0 (0 - x^3)\,dx$$

$$= \int_0^3 x^3\,dx + \int_{-2}^0 -x^3\,dx$$

$$= \left[\frac{x^4}{4}\right]_0^3 + \left[-\frac{x^4}{4}\right]_{-2}^0$$

$$= \frac{81}{4} + \frac{16}{4}$$

$$= 24\frac{1}{4} \text{ sq. units.}$$

Note that refusal to consider this problem in two parts does <u>not</u> give area, but gives "net area" with one area considered positive and the other negative.

Find the area of the region bounded by the x-axis, the curve: $y = 6x - x^2$, and the vertical lines: $x = 1$ and $x = 4$.

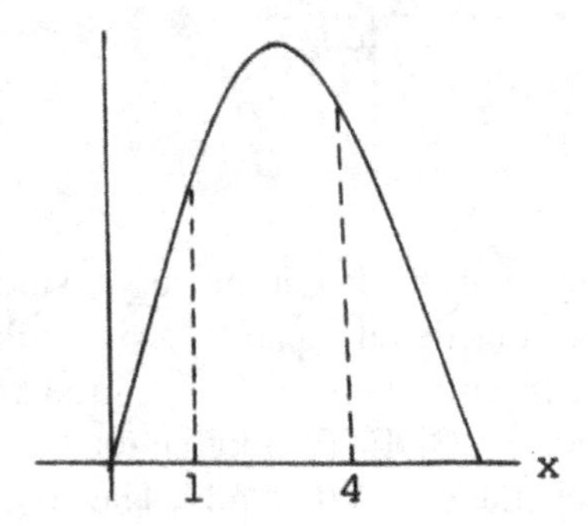

The limits of the integral which give the required area are $x = 1$ and $x = 4$. The function: $y = 6x - x^2$ is above the function $y = 0$ (the x-axis), therefore the area can be found by taking the integral of the upper function minus the lower function, or, $y = 6x - x^2$ minus $y = 0$, from $x = 1$ to $x = 4$. Therefore, we obtain:

$$A = \int_1^4 (6x - x^2) - 0\, dx = \left[3x^2 - \frac{x^3}{3} \right]_1^4$$

$$= \frac{80}{3} - \frac{8}{3} = 24.$$

9.2 Volume of a Solid of Revolution

If a region is revolved about a line, a solid called a solid of revolution is formed. The solid is generated by the region. The axis of revolution is the line about which the revolution takes place.

There are several methods by which we may obtain the volume of a solid of revolution. We shall now discuss three such methods.

1. Disk Method

The volume of the solid generated by the revolution of a region about the x-axis is given by the formula

$$V = \pi \int_a^b [f(x)]^2 \, dx,$$

provided that f is a continuous, nonnegative function on the interval [a,b].

Problem Solving Examples:

Q Find the volume of the solid generated by revolving about the y-axis the region bounded by the parabola: $y^2 = 4x$, the y-axis and the line $y = 2$.

A

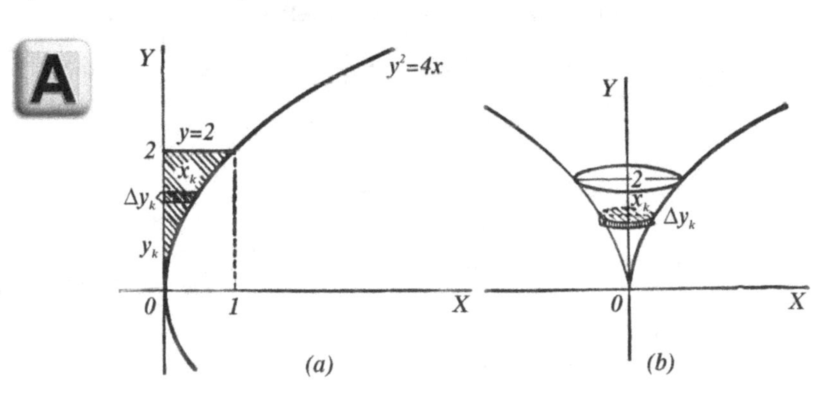

An element of volume is the disk generated by rotating a strip: x by dy, about the y-axis. The volume of the disk is: $dV = \pi x^2\, dy$.

Hence,

$$V = \pi \int_0^2 x^2\, dy$$

$$= \pi \int_0^2 \frac{1}{16} y^4\, dy = \frac{2}{5}\pi.$$

2. Shell Method

This method applies to cylindrical shells exemplified by

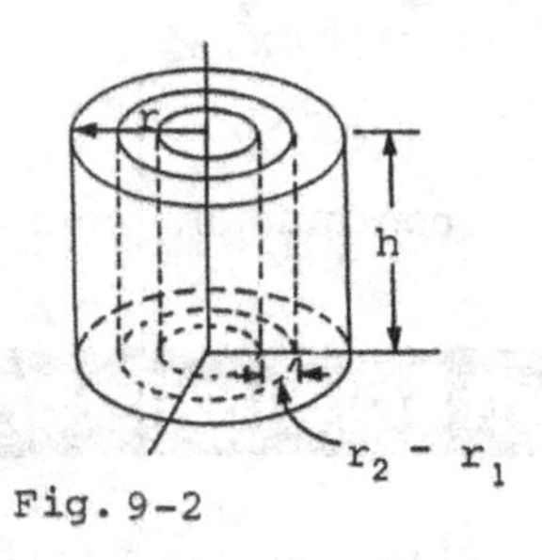

Fig. 9-2

The volume of a cylindrical shell is

$$V = \pi r_2^2 h - \pi r_1^2 h$$

$$= \pi (r_2 + r_1)(r_2 - r_1) h$$

$$= 2\pi \left(\frac{r_2 + r_1}{2} \right) (r_2 - r_1) h$$

where r_1 = inner radius

r_2 = outer radius

h = height.

Let $r = \frac{r_1+r_2}{2}$ and $\Delta r = r_2 - r_1$, then

the volume of a shell becomes

$$V = 2\pi rh\Delta r$$

The thickness of the shell is represented by Δr and the average radius of the shell by r.

Thus,

$$V = 2\pi \int_a^b xf(x)dx$$

is the volume of a solid generated by revolving a region about the y-axis. This is illustrated by Fig. 2-3.

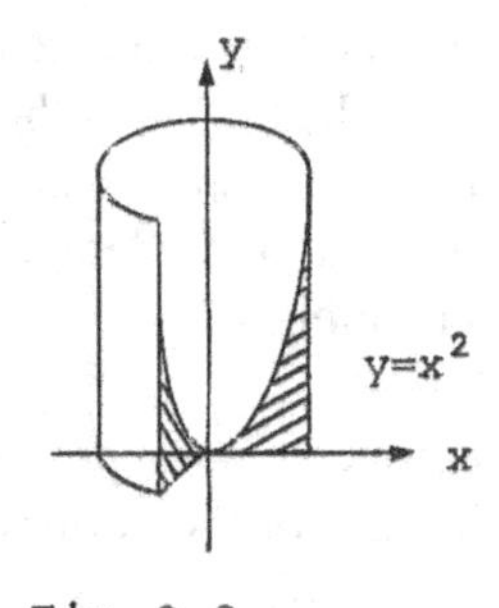

Fig. 9-3

Problem Solving Examples:

Q Find the volume of the solid generated by revolving about the y-axis the region bounded by the parabola:
$y = -x^2 + 6x - 8$, and the x-axis.

A

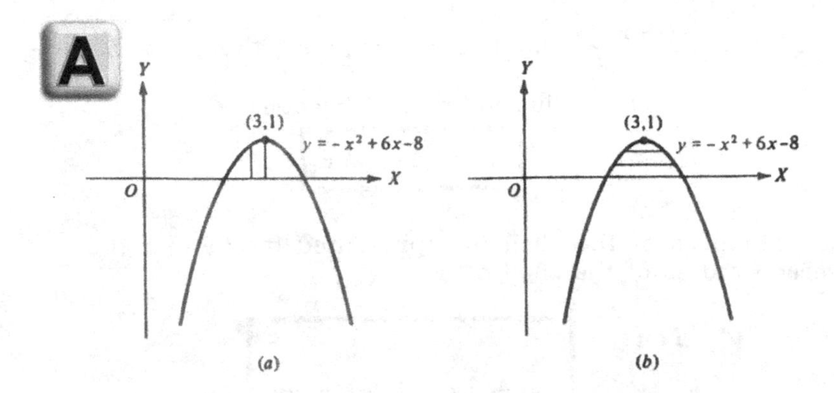

Method 1. We use the method of cylindrical shells. The curve:

$$y = -x^2 + 6x - 8,$$

cuts the x-axis at $x = 2$ and $x = 4$.

The cylindrical shells are generated by the strip formed by the two lines parallel to the y-axis, at distances x and $x + \Delta x$ from the y-axis, $2 \le x \le 4$, as shown in figure (a). When this strip is revolved about the y-axis, it generates a cylindrical shell of average height y^*, $y \le y^* \le y + \Delta y$, thickness Δx, and average radius x^*, $x < x^* \le x + \Delta x$. The volume of this element is:

$$\Delta V = 2\pi x^* y^* \Delta x,$$

where $2\pi x^* y^*$ is the surface area. Expressing y in terms of x and passing to the limits, the sum of the volumes of all such cylindrical shells is the integral:

$$\begin{aligned} V &= 2\pi \int_2^4 x(-x^2 + 6x - 8)\, dx \\ &= 2\pi \int_2^4 (-x^3 + 6x^2 - 8x)\, dx \\ &= 2\pi \left(-\frac{x^4}{4} + 2x^3 - 4x^2 \right) \Bigg|_2^4 \\ &= 2\pi \Big((-64 + 128 - 64) - (-4 + 16 - 16) \Big) \\ &= 8\pi. \end{aligned}$$

Method 2. This can also be thought of as the volume comprising a series of concentric washers with variable outer and inner radii, as sectionally shown in figure (b). The variable radii are as follows: Since

$$y = -x^2 + 6x - 8,$$

we solve for x.

To complete the square, we require a 9, so that

$$x^2 - 6x + 9$$

constitutes a perfect square. Rewriting the equation,

$$x^2 - 6x + 9 - 9 + 8 = -y.$$

$$x^2 - 6x + 9 = 1 - y.$$

$$(x - 3)^2 = 1 - y.$$

Therefore,

$$x = 3 \pm \sqrt{1-y}$$

which shows the washers, y units from the x-axis, have an

inner radius: $x_{in} = 3 - \sqrt{1-y}$, and an

outer radius: $x_o = 3 + \sqrt{1-y}$.

(The particular one on the x-axis has $x_{in} = 2$ and $x_o = 4$.)

The volume of this washer with thickness dy is:

$$dV = \pi\,(x_o^2 - x_{in}^2)\,dy,$$

or $$dV = \pi \left[(x_o + x_{in})(x_o - x_{in}) \right] dy.$$

Substituting the values for x_o and x_{in},

$$dV = \pi \left(\left((3+\sqrt{1-y})+(3-\sqrt{1-y})\right) \cdot \left((3+\sqrt{1-y})-(3-\sqrt{1-y})\right) \right) dy$$

$$= \pi(12\sqrt{1-y})\, dy.$$

Since y varies from 0 to 1, the desired volume is:

$$V = 12\pi \int_0^1 (1-y)^{\frac{1}{2}}\, dy$$

$$= 12\pi \left(-\frac{2}{3}(1-y)^{\frac{3}{2}} \right|_0^1 = 8\pi.$$

3. Parallel Cross Sections

A cross section of a solid is a region formed by the intersection of a solid by a plane. This is illustrated by Fig. 2-4.

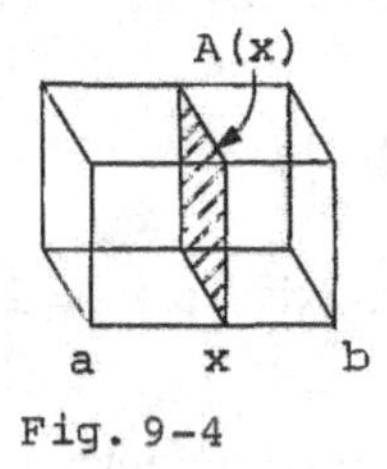

Fig. 9-4

If x is a continuous function on the interval [a,b], then the volume of the cross sectional area A(x) is

$$V = \int_a^b A(x)dx.$$

Problem Solving Examples:

Rotate the curve y = 2x about the line x = 4 and find the volume produced by the rotation of the shaded portion.

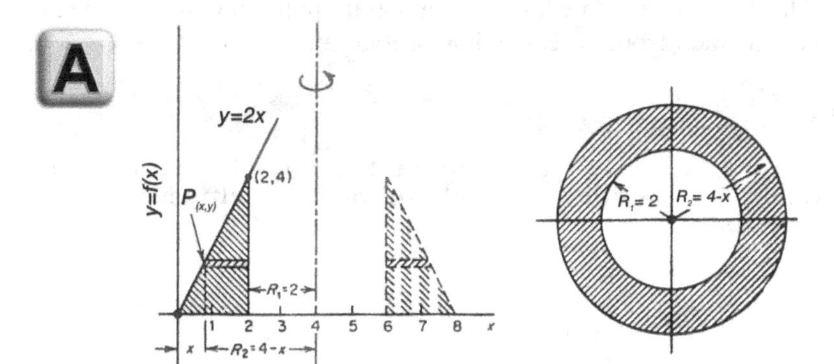

Area of washer = $\pi\left(R_2^2 - R_1^2\right)$

Increment of volume, $dV = \pi\left(R_2^2 - R_1^2\right)\ dy.$

$$V = \int_a^b \pi\left(R_2^2 - R_1^2\right)dy, \text{ where a and b are limits of y.}$$

$$V = \int_0^4 \pi\ \left((4-x)^2 - 2^2\right) dy$$

$$= \int_0^4 \pi\ (16 - 8x + x^2 - 4)\ dy$$

$$= \pi \int_0^4 (12 - 8x + x^2)\ dy.$$

$x = \frac{y}{2}$, from the equation of the curve. Substituting,

$$V = \pi \int_0^4 \left(12 - 4y + \frac{y^2}{4} \right) dy$$

$$= \pi \left(12y - \frac{4y^2}{2} + \frac{y^3}{12} \right)_0^4$$

$$= \pi \left(48 - 32 + - \frac{16}{3} \right) = \frac{64\pi}{3}.$$

This is the washer method. The important point to remember is to obtain the radii from the center line of rotation.

9.3 Work

Force is a physical property that is defined by Newton's law as the mass of an object multiplied by its acceleration:

$$F = ma.$$

where

F = force
m = mass
a = acceleration

An object that is subjected to a constant force which moves it a distance d in the direction of the force is said to have done work.

$$W = Fd$$

Variable forces are forces that are not constant.

Definition:

Let $F(x)$ represent the force at the point x along the x-axis, where F is continuous on the interval $[a,b]$. The work done in moving the object from a to b is

$$W = \int_a^b F(x)dx.$$

Problem Solving Examples:

Find the instantaneous rate of change of force if the equation for work is 1,000 = Fx.

Work	=	force	x	distance
(joules)		(newtons)		(meters)

In this problem,

$$F = \frac{\text{work}}{\text{distance}} = \frac{1{,}000}{x}$$

Find $\frac{dF}{dx}$ by the Δ method:

$$F + \Delta F = \frac{1{,}000}{x + \Delta x}.$$

Subtracting F from both sides,

$$\Delta F = \frac{1{,}000}{x + \Delta x} - F.$$

Substituting the value of F,

$$\Delta F = \frac{1{,}000}{x + \Delta x} - \frac{1{,}000}{x}.$$

Using the common denominator,

$$\Delta F = -\frac{1{,}000(\Delta x)}{(x + \Delta x)x}$$

Dividing the equation by Δx,

$$\frac{\Delta F}{\Delta x} = -\frac{1{,}000}{(x + \Delta x)x} = -\frac{1{,}000}{x^2 + x\Delta x}$$

$$\lim_{\Delta x \to 0} \frac{\Delta F}{\Delta x} = \frac{dF}{dx} = -\frac{1{,}000}{x^2} \text{ newtons/m}^2.$$

A 20-lb. weight is being raised by a 100-ft. rope weighing $\frac{1}{2}$ lb/ft. Determine the work needed to raise the weight 50 ft.

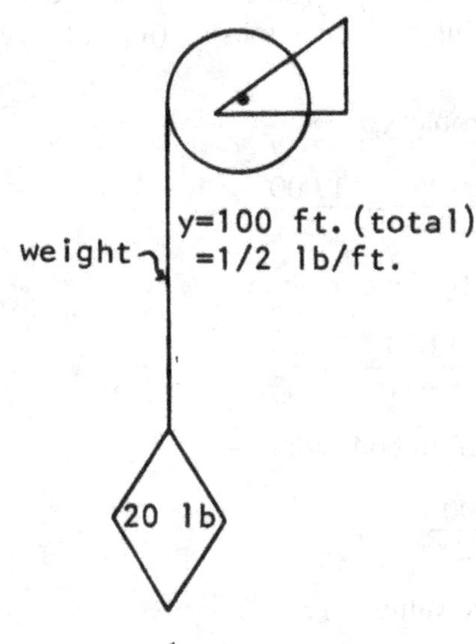

$$F = 20 + (100 - y)\ \frac{1}{2}$$

Let y be the distance moved at any time, hence, the remaining length of the rope is $(100 - y)$ –ft., weighing

$$(100 - y)\text{ ft.} \cdot \frac{1}{2}\text{ lb./ft.} = (100 - y)\,\frac{1}{2}\text{ lb.}$$

Since work is the product of the force applied and the distance moved, a change of work can be expressed in terms of the corresponding change in distance. Thus, for an infinitesimal work we have the following differential equation:

$$dW = Fdy$$

$$dW = 20 + 100 - y \frac{1}{2} \quad dy$$

$$= \left(70 - \frac{y}{2}\right) dy.$$

Integrating both sides,

$$\int dW = \int \left(70 - \frac{y}{2}\right) dy.$$

The weight has to be raised 50 ft., i.e. $0 \le y \le 50$, therefore,

$$W = \int_0^{50} \left(70 - \frac{y}{2}\right) dy$$

$$= 70y - \frac{y^2}{4} \Big|_0^{50} = 2{,}875 \text{ ft.–lb.}$$

9.3.1 Hooke's Law

The force $F(x)$ required to stretch a spring x units beyond its natural length is given by $F(x) = kx$, where k is a constant called the spring constant.

The same formula is used to find the work done in compressing a spring x units from its natural length.

Problem Solving Examples:

Hooke's law states that an elastic body such as a spring stretches an amount proportional to the force applied.

If a spring is stretched $\frac{1}{2}$ inch when a 2-pound force is applied to it, how much work is done in stretching the spring an additional $\frac{1}{2}$ inch?

If the work,

$$W = \int_a^b F(x) \ dx,$$

the problem is to determine the force function F and the interval [a,b]. By Hooke's law, if x denotes the distance the spring has been stretched, then the force function,

$$F(x) = kx.$$

But when $x = \frac{1}{2}$, $F(x) = 2$. Hence $\frac{k}{2} = 2$, or $k = 4$, and it follows that $F(x) = 4x$. The work done in stretching the spring an additional $\frac{1}{2}$ inch is:

$$W = \int_{\frac{1}{2}}^{1} 4x dx = 4 \frac{x^2}{2} \Bigg|_{\frac{1}{2}}^{1} = \frac{3}{2}.$$

If the force is measured in pounds and the distance in feet (inches), then the unit of measurement for work is the foot-pound (inch-pound). Hence for this problem, where the dimensions are in inches,

$$W = \frac{3}{2} \text{ inch-pounds.}$$

9.4 Fluid Pressure

The pressure that a liquid exerts on a plate located at a depth h in a container is

$$P = \rho h$$

where ρ = density of the liquid (weight per unit volume)

h = depth of the liquid.

From this, we may assume that the pressure of the liquid is dependent on the depth, but is independent of the size of the container.

Theorems:

1. The pressure is the same in all directions, at any point x, in a liquid.

2. The total force of a plate which is divided into several parts, is the sum of the forces on each of the parts.

   The total force is defined as

$$F = \int_a^b \rho \, p(h)dh$$

where ρ is density of the liquid;
$p(h)$ is the pressure as a function of depth(h),
a and b are the limits of the region in which the pressure is exerted.

Problem Solving Examples:

Q A vertical floodgate in the form of a rectangle 6 feet long and 4 feet deep has its upper edge 2 feet below the surface of the water, as shown in the diagram. Find the force which it must withstand.

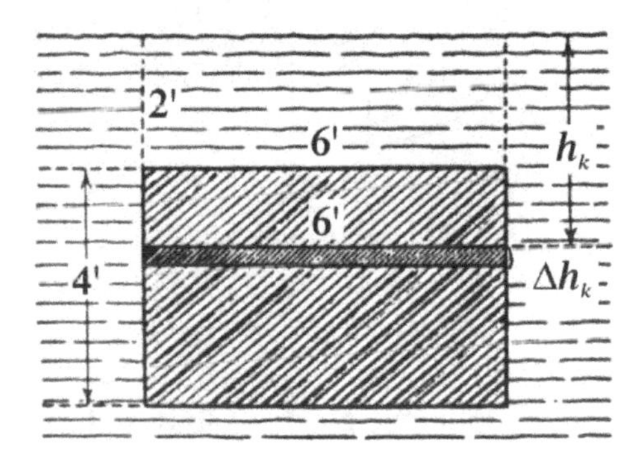

The pressure at a point on the floodgate is a function of its depth. Thus $p = wh$, where p denotes the pressure, w the density of the fluid, and h the depth of the point. The total force on the floodgate is a sum of the infinitesimal forces on the rectangular strips of the gate with lengths parallel to the upper edge. $dF = p dA$, where dF denotes the differential force on an infinitesimal area dA of the floodgate.

A typical infinitesimal rectangular strip of the gate surface has length $l = 6$ and infinitesimal height dh, where h varies from 2 to 6. Hence, $dF = p \cdot 6dh = 6wh\ dh$. The density of water, w, is 62.5 pounds/cu.ft. Then the total force on the gate is:

$$F = \int_2^6 6wh\ dh = \frac{6wh^2}{2}\bigg|_2^6 = 3w(36 - 4)$$

$$= 96w = 96(62.5) = 6{,}000 \text{ pounds.}$$

Compute the total force on a gate closing a circular pipe 4 feet in diameter when the pipe is half full.

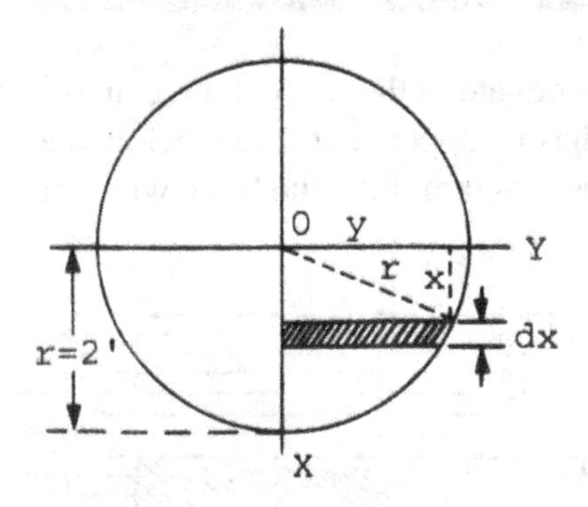

$y \cdot dx$ = a differential of area

xw = fluid pressure

= depth · weight of 1 cu. ft. of fluid,or density, w.

Therefore, $xw \cdot y \cdot dx$ = force on the differential area.

Summing up the differential fluid forces, F = total force = $w\int y \cdot x \cdot dx$. For water, w = 62.5 lb. cu.ft. But

$$y = \sqrt{r^2 - x^2} = \sqrt{4 - x^2}$$

Hence, $F = 62.5 \int_0^2 \sqrt{4 - x^2} \cdot x \cdot dx$ = total force on one-half of the water area.

To integrate, let $u = 4 - x^2$. For x = 0, u = 4. For x = 2, u = 0. This yields new limits. Then,

$$du = -2x \cdot dx,$$

and

$$F = \frac{2}{2} \cdot 62.5 \int_0^2 \sqrt{4 - x^2} \cdot x \cdot dx$$

$$= -\frac{1}{2} \cdot 62.5 \int_4^0 \sqrt{4 - x^2}\ (-2x)\, dx.$$

$$F = -\frac{62.5}{2} \int_4^0 u^{\frac{1}{2}} \cdot du$$

$$= -\frac{62.5}{2} \cdot \left. \frac{u^{\frac{3}{2}}}{\frac{3}{2}} \right|_4^0$$

$$= \frac{62.5}{3} \sqrt{4^3} = \frac{62.5}{3} \cdot 8 = \frac{500}{3} = 167 \text{ lb. for one-half of the}$$

water area. Hence, the force exerted by the water on the gate = 2.167 = 334 lb.

9.5 Area of Surface of Revolution

A surface of revolution is generated when a plane is revolved about a line.

If f' and g' are two continuous functions on the interval [a,b] where g(t) = 0, x = f(t) and y = g(t), then the surface area of a plane revolved about the x-axis is given by the formula

$$S = \int_a^b 2\pi \; g(t) \sqrt{[f'(t)]^2 + [g'(t)]^2}\, dt$$

Since x = f(t) and y = g(t),

$$S = \int_a^b 2\pi y \sqrt{\left(\frac{dx}{dt}\right)^2 + \left(\frac{dy}{dt}\right)^2}\, dt$$

If the plane is revolved about the y-axis, then the surface area is

$$S = \int_a^b 2\pi x \sqrt{\left(\frac{dx}{dt}\right)^2 + \left(\frac{dy}{dt}\right)^2}\, dt$$

These formulas can be simplified to give the following:

$$S = 2\pi \int_a^b y ds$$

for revolution about the x-axis, and

$$S = 2\pi \int_a^b x\,ds$$

for revolution about the y-axis.

In the above equations, ds is given as $ds = \sqrt{1 - [g'(x)]^2}dy$ and $ds = \sqrt{1 + [f'(x)]^2}dx$; respectively.

Problem Solving Examples:

Find the area of the surface formed by revolving about the x-axis the parabola: $y^2 = 2x - 1$, from $y = 0$ to $y = 1$.

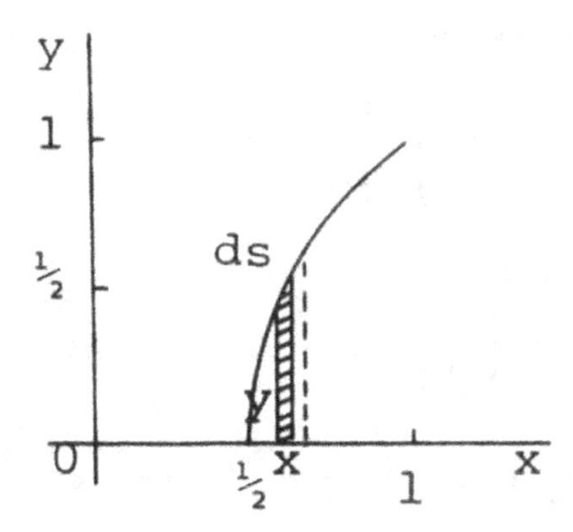

We consider the shaded strip as shown in the figure. Its length is y units and width Δx units. If this strip is rotated about the x-axis, it sweeps a volume that looks like a disk with radius y units, and thickness of Δx units. The circumference $2\pi r = 2\pi y$. Hence its surface area is

$$\Delta S = 2\pi y \, \Delta s$$

Noting that the parabola is made up of such discs for

$$\frac{1}{2} \le x \le 1,$$

the required surface area is the sum of all surface areas of all the disks which compose the entire given solid. Mathematically, if S is the surface area required, then

$$S \cong \sum_{i=1}^{n} 2\pi y_i \, \Delta s_i \,,$$

or

$$S = \lim_{n \to \infty} \sum_{i=1}^{n} 2\pi y_i \, \Delta s_i \,, \quad \frac{1}{2} \le x_i \le 1$$

Since 2y dy = 2 dx, and $\frac{dx}{dy} = y$, the differential of arc length is given by:

$$(ds)^2 = (dx)^2 + (dy)^2$$

$$= \left(1 + \left(\frac{dx}{dy} \right)^2 \right) (dy)^2 \cdot$$

$$ds = \sqrt{1 + \left(\frac{dx}{dy} \right)^2} \cdot dy.$$

Furthermore, $y \ge 0$, so that $|\, y \,| = y$, and hence:

$$s = 2\pi \int_0^1 y\sqrt{1+y^2}\, dy$$

$$= \left(\frac{2\pi}{3}(1+y^2)^{\frac{3}{2}} \right)_0^1$$

$$= \frac{2\pi}{3}(2\sqrt{2} - 1).$$

Find the area of the surface generated by revolving the Hypocycloid: $x^{\frac{2}{3}} + y^{\frac{2}{3}} = a^{\frac{2}{3}}$, about the x-axis.

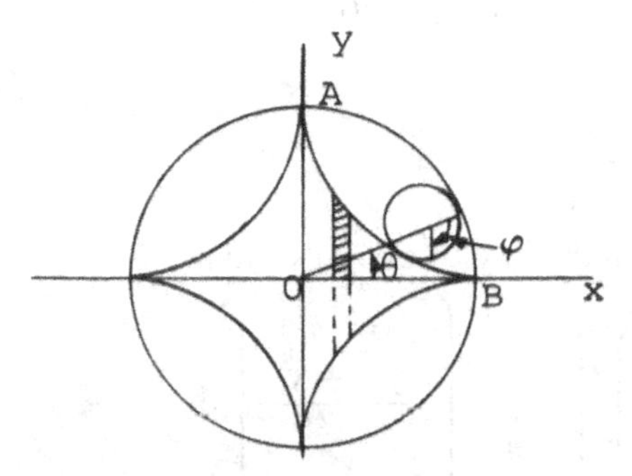

A hypocycloid is the locus of any fixed point on the circumference of a circle (radius = b) that rolls internally, without slipping, on a fixed circle of radius a, where b divides a.

Using the disc method and noting the symmetry of the region about the x-axis, the surface area is given by:

$$s = 2\int_A^B 2\pi y \, ds = 4\pi \int_0^a y \, ds.$$

ds is the arc length, given by:

$$(ds)^2 = (dy)^2 + (dx)^2,$$

Multiplying and dividing the right-hand side of the equation by $(dy)^2$, we have:

$$(ds)^2 = \left(1 + \left(\frac{dx}{dy}\right)^2\right)(dy)^2.$$

Taking the square root, we obtain:

$$ds = \left(1 + \left(\frac{dx}{dy}\right)^2\right)^{\frac{1}{2}} dy.$$

From the given equation: $x^{\frac{2}{3}} + y^{\frac{2}{3}} = a^{\frac{2}{3}}$, we differentiate implicitly:

$$\frac{2}{3} x^{-\frac{1}{3}} dx + \frac{2}{3} y^{-\frac{1}{3}} dy = 0,$$

or

$$\frac{dx}{dy} = -\frac{x^{\frac{1}{3}}}{y^{\frac{1}{3}}}.$$

Substitution in the ds expression yields:

$$ds = \left(1 + \left(\frac{x^{\frac{1}{3}}}{y^{\frac{1}{3}}}\right)^2\right)^{\frac{1}{2}} dy = \left(\frac{y^{\frac{2}{3}} + x^{\frac{2}{3}}}{y^{\frac{2}{3}}}\right)^{\frac{1}{2}} dy$$

$$= \left(\frac{a^{\frac{2}{3}}}{y^{\frac{2}{3}}}\right)^{\frac{1}{2}} dy = \frac{a^{\frac{1}{3}}}{y^{\frac{1}{3}}} dy.$$

Thus,

$$s = 4\pi \int_0^a y \frac{a^{\frac{1}{3}}}{y^{\frac{1}{3}}} dy = 4\pi a^{\frac{1}{3}} \int_0^a y^{\frac{2}{3}} dy$$

$$= 4\pi a^{\frac{1}{3}} \left(\frac{3}{5} y^{\frac{5}{3}}\right|_0^a = \frac{12\pi a^{\frac{1}{3}}}{5} a^{\frac{5}{3}}$$

$$= \frac{12\pi a^2}{5}.$$

9.6 Arc Length

If $f'(x)$ represents the derivative of a function f, and if $f'(x)$ is continuous, then the function is said to be smooth.

We shall now define the arc length of a smooth curve on the interval [a,b].

Definition

The arc length of the graph of f from A(a,f(a)) to B(b,f(b)) is given by the formula

$$S = \int_a^b \sqrt{1+[f'(x)]^2}\, dx$$

if the function is smooth on the interval [a,b].

If g is a continuous function on the closed interval [c,d] and it is defined by x = g(y), then the formula for the arc length is

$$S = \int_c^d \sqrt{1+[g'(y)]^2}\, dy$$

In this case, y is regarded as the independent variable.

Problem Solving Examples:

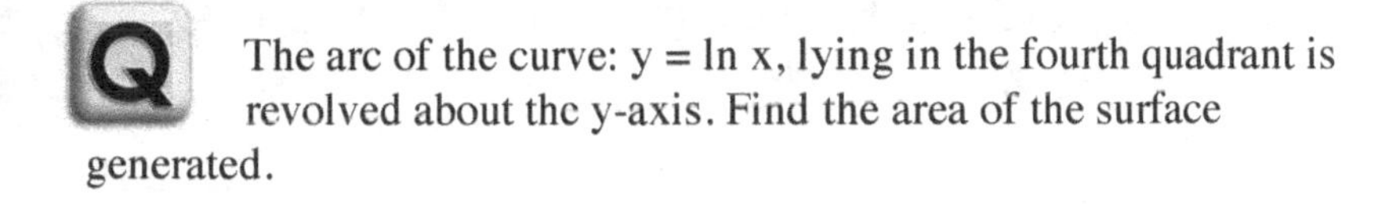

The arc of the curve: y = ln x, lying in the fourth quadrant is revolved about thc y-axis. Find the area of the surface generated.

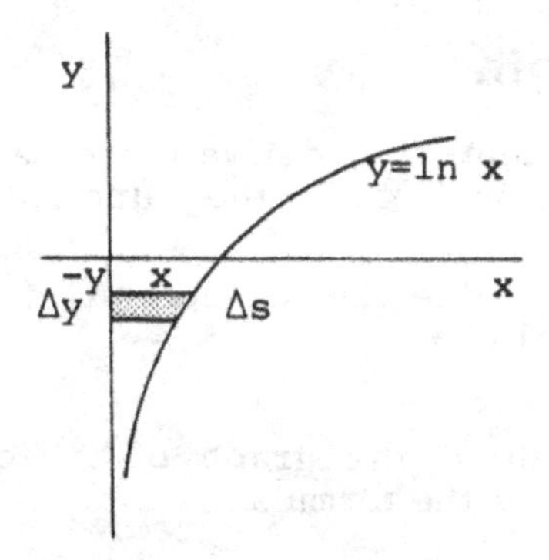

The shaded strip in the figure (x by Δy), when rotated about the y-axis, sweeps a volume approximately equal to that of a truncated cone, base radius x and thickness Δy, with a slanted edge Δs. If we were interested in finding the volume we would merely use Δy. However, for the surface area, Δs is involved, which converges to:

$$(ds)^2 = (dx)^2 + (dy)^2,$$

or,

$$dS = \left(1 = \left(\frac{dx}{dy}\right)^2 \right)^{\frac{1}{2}} dy.$$

The surface area of an elementary disk, under the limit, is:

$$ds = 2\pi x \, ds$$

$$= 2\pi x \left(1 + \left(\frac{dx}{dy}\right)^2 \right)^{\frac{1}{2}} dy.$$

$$y = \ln x \cdot \frac{dy}{dx} = \frac{1}{x} \cdot \frac{dx}{dy} = x.$$

$$dy = \frac{1}{x} dx.$$

Upon substitution,

$$ds = 2\pi x\,(1 + x)^2)^{\frac{1}{2}}\,\frac{1}{x}\,dx.$$

Integration from x = 0 to x = 1 yields the required area in the region bounded by the x- and y-axes.

$$s = 2\pi \int_0^1 \sqrt{1+x^2}\,dx$$

$$= \pi\,\left(x\sqrt{1+x^2} + \ln(x+\sqrt{1+x^2})\right)\Big|_0^1$$

$$= \pi\,\left(\sqrt{2} + \ln(\sqrt{2}+1)\right).$$

(Integration was carried out by the use of integration by parts and then substitution.)

Find the length of the curve: $y = x^2$, from x = 2 to x = 5.

Applying the expression:

$$s = \int_a^b \sqrt{1+\left(\frac{dy}{dx}\right)^2}\,dx,$$

we have $y = x^2 \,.\, y' = 2x.$

Therefore, $s = \int_2^5 \sqrt{1+4x^2}\,dx.$

Integrating by parts and recalling

$$\int udv = uv - \int vdu,$$

let $u = (1 + 4x^2)^{\frac{1}{2}}$, $du = 4x(1 + 4x^2)^{-\frac{1}{2}}$,

$dv = dx, \; v = x.$

$$\int_2^5 (1 + 4x^2)^{\frac{1}{2}} dx$$

$$= (1 + 4x^2)^{\frac{1}{2}} x - \int_2^5 \frac{4x^2}{(1 + 4x^2)^{\frac{1}{2}}} dx. \qquad (1)$$

But $\int_2^5 (1 + 4x^2)^{\frac{1}{2}} \; dx = \int_2^5 \frac{(1 + 4x^2)}{(1 + 4x^2)^{\frac{1}{2}}} dx.$

After expanding the right side, we have:

$$\int_2^5 (1 + 4x^2)^{\frac{1}{2}} dx$$

$$= \int_2^5 \frac{dx}{(1 + 4x^2)^{\frac{1}{2}}} + \int_2^5 \frac{4x^2 dx}{(1 + 4x^2)^{\frac{1}{2}}} . \qquad (2)$$

By adding the equations (1) and (2), the term:

$$\int \frac{4x^2 dx}{(1 + 4x^2)^{\frac{1}{2}}},$$

is eliminated.

At this point we obtain:

$$2 \int_2^5 (1 + 4x^2)^{\frac{1}{2}} \; dx$$

$$= x(1 + 4x^2)^{\frac{1}{2}} + \int_2^5 \frac{dx}{(1 + 4x^2)^{\frac{1}{2}}},$$

or, $\int_2^5 (1 + 4x^2)^{\frac{1}{2}} dx$

$$= \frac{x}{2}(1 + 4x^2)^{\frac{1}{2}} + \int_2^5 \frac{dx}{\left(\frac{1}{4} + x^2\right)^{\frac{1}{2}}}.$$

Note that

$$\int \frac{dx}{\left(\frac{1}{4} + x^2\right)^{\frac{1}{2}}}$$

is of the form:

$$\int \frac{dx}{(a^2 + x^2)^{\frac{1}{2}}} = \ln\left(x + \sqrt{a^2 - x^2}\right)$$

Thus,

$$\int \frac{dx}{\left(\frac{1}{4} + x^2\right)^{\frac{1}{2}}} = \ln\left(x + \sqrt{\frac{1}{4} - x^2}\right).$$

$$\int_2^5 (1 + 4x^2)^{\frac{1}{2}} dx$$

reduces to: $\left[\frac{x}{2}(1 + 4x^2)^{\frac{1}{2}} + \ln\left(x + \sqrt{\frac{1}{4} - x^2}\right)\right]_2^5.$

This expression, evaluated at $x = 5$, is equal to 27.43. At $x = 2$, it is equal to 5.52.

$$\int_2^5 (1 + 4x^2)^{\frac{1}{2}} dx = 27.43 - 5.52$$

$$= 21.91 \text{ units of arc}$$

Quiz: Techniques of Integration and Applications of the Integral

1. The graph of f is shown in the figure. Which of the following could be the graph of $\int f(x)\,dx$?

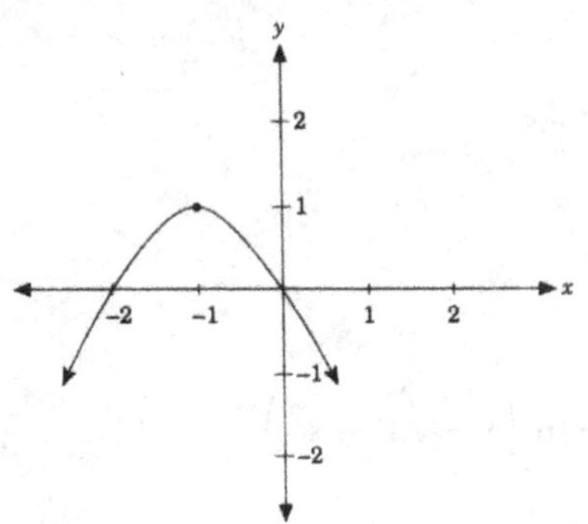

(A)

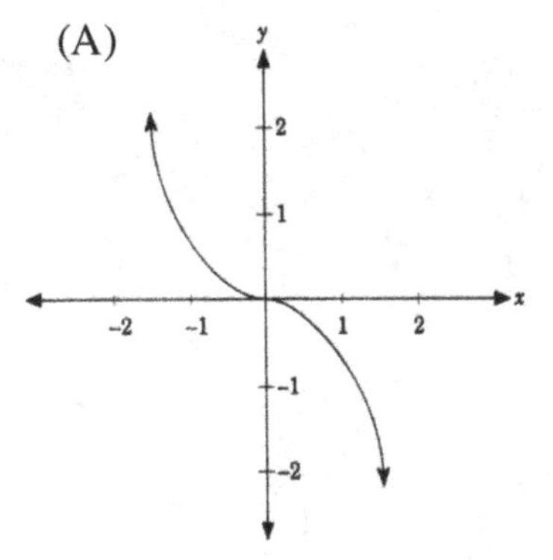

(B)

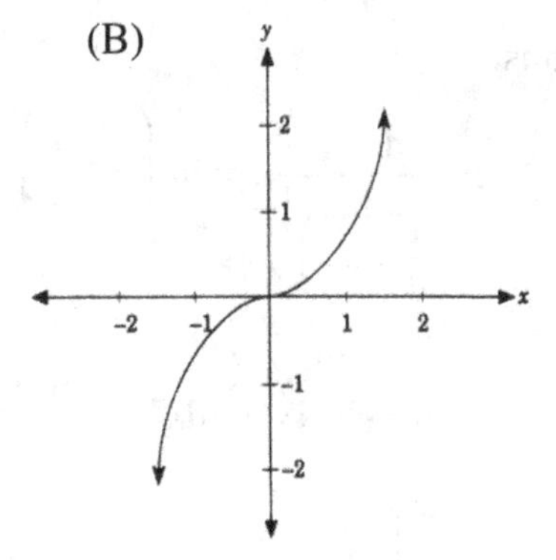

(C)

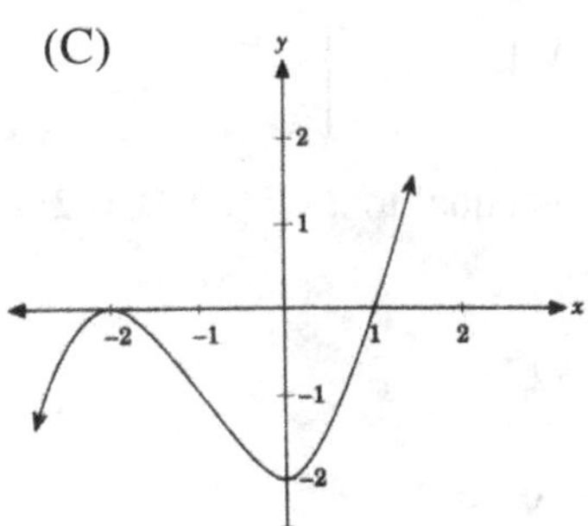

(D)

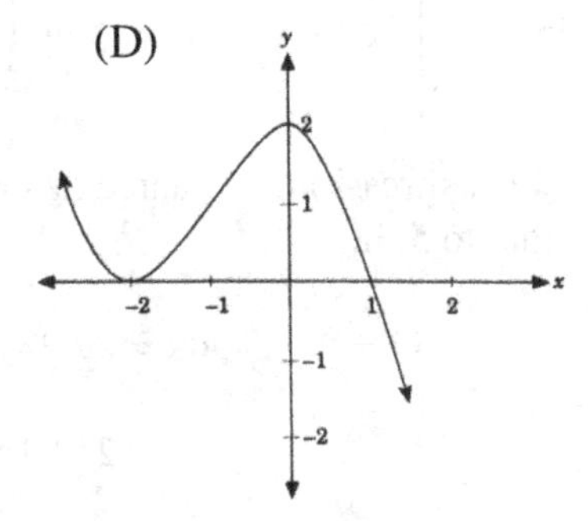

(E)

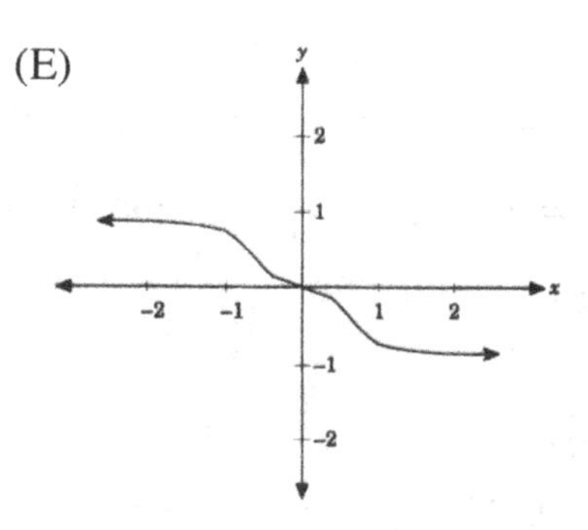

2. Let $F(x) = \int_1^x f(t)\,dt$, and use the graph given of $f(t)$ to find $F^1(1) =$

(A) 0

(B) 1

(C) 2

(D) $\frac{1}{2}$

(E) None of these

3. The area that is enclosed by $y = x^3 + x^2$ and $y = 6x$ for $x \geq 0$ is

(A) $\frac{29}{12}$

(B) 3

(C) $\frac{16}{3}$

(D) 6

(E) $\frac{32}{3}$

4. If the acceleration (ft/sec²) of a moving body is $\sqrt{4t+1}$ and the velocity at $t = 0$, $v(0) = -4\frac{1}{3}$, then the distance traveled between time $t = 0$ and $t = 2$ is

(A) $\frac{119}{30}$

(B) $-\frac{119}{30}$

(C) $-\frac{149}{30}$

(D) $\frac{149}{30}$

(E) -4

5. Given the following graph of the continuous function $f(x)$,

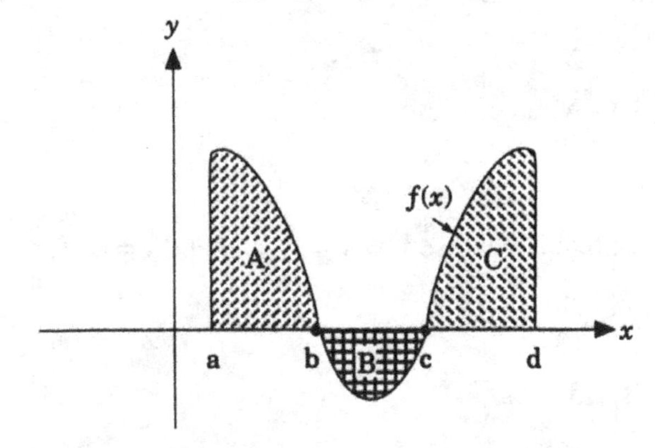

with area of region $A = 3$

area of region $B = 1\frac{1}{2}$

area of region $C = 2$, then $\int_a^d f(x)\,dx =$

(A) $\frac{5}{2}$

(B) $\frac{7}{2}$

(C) $\frac{13}{2}$

(D) 5

(E) $\frac{2}{5}$

6. The area in the first quadrant that is enclosed by the graphs of $x = y$ and $x = 4y$ is

(A) 4

(B) 8

(C) –4

(D) 1

(E) 0

7. $\int \frac{\log (x^3 \cdot 10^x)}{x} dx = ?$ (Note: log stands for $\log_{10}$ and ln stands for $\log_e$)

(A) $\frac{3 \ln 10}{2} (\log x)^2 + x + C$

(B) $\frac{3}{3 \ln 10} (\ln x)^2 + x + C$

(C) $\dfrac{6 \log x}{\ln 10} + x + C$

(D) $\dfrac{3 (\log x)^2}{\ln 10} + x + C$

(E) $3 \ln 10 (\ln x^2) + x + C$

8. Find $\int \arctan x \, dx$ using integration by parts.

(A) $\arctan x + \ln(1 + x^2) + C$

(B) $x \arctan x + C$

(C) $\dfrac{1}{(1 + x^2)} + C$

(D) $x \arctan x - \dfrac{1}{2} \ln(1 + x^2) + C$

(E) $\ln(1 + x^2) + x \arctan x + C$

9. The region in the first quandrant between the x - axis and the graph $y = x^2$ from $x = 0$ to $x = 4$, is rotated about the line $x = 4$. The volume of the resulting solid of revolution is given by

(A) $\int_0^4 2\pi x^2 (4 - x) dx$

(B) $\int_0^4 \pi x^2 (4 - x) dx$

(C) $\int_0^4 2\pi x (4 - x) dx$

(D) $\int_0^4 2\pi x (x - 4)^2 dx$

(E) $\int_0^8 \pi (4 + \sqrt{16 + y})^2 dy$

10. The substitution $x-1=\sin\theta$ transforms the integral $\int_1^2 \frac{dx}{x^2-2x}$ into:

(A) $\int_0^{\frac{\pi}{2}} \frac{d\theta}{\cos\theta}$

(B) $\int_0^{\frac{\pi}{2}} \frac{d\theta}{\sin\theta}$

(C) $-\int_0^{\frac{\pi}{2}} \frac{d\theta}{\cos\theta}$

(D) $-\int_0^{\frac{\pi}{2}} \frac{d\theta}{\sin\theta}$

(E) $-\int_0^{\frac{\pi}{2}} \frac{d\theta}{\cos^2\theta}$

ANSWER KEY

1. (D)
2. (C)
3. (C)
4. (C)
5. (B)
6. (A)
7. (A)
8. (D)
9. (A)
10. (C)

CHAPTER 10

Parametric Equations

10.1 Parametric Equations

A parameter is a quantity whose value determines the value of other quantities. Given the equation $y = t^2+t+3$ and $x = t^2+2t$, where the value of t determines the value of x and y; t is considered a parameter.

Let x and y represent two differentiable functions on the interval [a,b] which are continuous at the endpoints and (x(t),y(t)) represents a point in the plane. If t is in the interval [a,b], then the curve traced out by the point (x(t),y(t)) is given parametrically as

$$x = x(t) \quad \text{and} \quad y = y(t)$$

To sketch the graph of equations given parametrically, we utilize the following steps:

1. Solve the first parametric equation in terms of t.
2. Substitute t into the second equation and simplify it.
3. Tabulate coordinates of the points on the curve.
4. Plot the point in the domain.
5. Sketch graph.

Example: Sketch the graph of the equations given parametrically as

$x = t^2 + 4$ (1)

$y = t^2$. (2)

From (1) we have $\sqrt{x-4} = t$. Hence

$y = (\sqrt{x-4})^2 = x-4$ i.e.

$x - y = 4$

t	x	y
1	5	1
2	8	4
3	13	9
.		
.		
.		
-1	5	1
-2	8	4
-3	13	9

Here $t \in [-\infty, \infty]$.

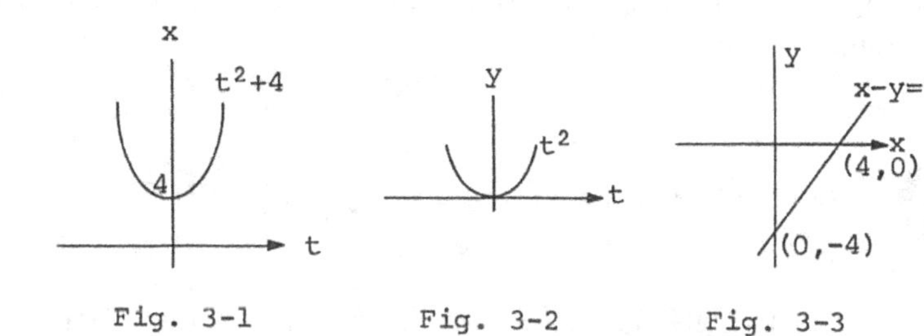

Fig. 3-1 Fig. 3-2 Fig. 3-3

Problem Solving Examples:

Find $\frac{d^2y}{dx^2}$ if $x = t - t^2$ and $y = t - t^3$.

We see here that we can readily obtain $\frac{dx}{dt}$ and $\frac{dy}{dt}$ from the parametric equations. To obtain $\frac{dy}{dx}$, therefore, we apply the chain rule.

$$y' = \frac{dy}{dx} = \frac{\frac{dy}{dt}}{\frac{dx}{dt}} = \frac{1 - 3t^2}{1 - 2t},$$

for which

$$\frac{d^2y}{dx^2} = y'' = \frac{dy'}{dx} = \frac{\frac{dy'}{dt}}{\frac{dx}{dt}} = \frac{\frac{d}{dt}\left[\frac{1 - 3t^2}{1 - 2t}\right]}{(1 - 2t)}$$

Using the theorem for $d\left(\frac{u}{v}\right)$,

$$= \frac{(1 - 2t) \cdot (-6t) - (1 - 3t^2) \cdot (-2)}{(1 - 2t)^3}$$

$$= \frac{2 - 6t + 6t^2}{(1 - 2t)^3}.$$

y'' cannot be directly obtained from y' without using the chain rule, by merely applying the derivative of the quotient, since y' is a function of t rather than of x.

Find $D_x^2 y$, given: $x = 2 \sec \theta - 3$, $y = 4 \tan \theta + 2$.

The two given relations are parametric equations, with parameter θ. For parametric equations, $\frac{dy}{dx}$ is given by:

$$\frac{dy}{dx} = \frac{\frac{dy}{d\theta}}{\frac{dx}{d\theta}}$$

and $\frac{d^2y}{dx^2}$ is given by:

$$\frac{d^2y}{dx^2} = \frac{d}{d\theta}\left(\frac{dy}{dx}\right) \cdot \frac{d\theta}{dx}\ .$$

To find $D_x^2 y$ we determine $\frac{dy}{d\theta}$, $\frac{dx}{d\theta}$, and $\frac{d\theta}{dx}$, and substitute. We find:

$$\frac{dy}{d\theta} = 4 \sec^2 \theta\ ,$$

and

$$\frac{dx}{d\theta} = 2 \sec \theta \tan \theta\ .$$

By substitution,

$$\frac{dy}{dx} = \frac{4 \sec^2 \theta}{2 \sec \theta \tan \theta}\ .$$

$$\frac{d^2y}{dx^2} = \frac{d}{d\theta}\left(\frac{4 \sec^2 \theta}{2 \sec \theta \tan \theta}\right) \cdot \frac{d\theta}{dx}\ ,$$

but $\frac{d\theta}{dx} = \frac{1}{2 \sec \theta \tan \theta}$. Using the quotient rule for differentiation and substituting for $\frac{d\theta}{dx}$ we obtain:

$$\frac{d^2y}{dx^2} = (2 \sec \theta \tan \theta\ )\ (8 \sec \theta \sec \theta \tan \theta\ )$$

$$- \frac{[\ (4 \sec^2 \theta) (2 \sec \theta \sec^2 \theta + 2 \tan \theta \sec \theta \tan \theta)\ ]}{4 \sec^2 \theta \tan^2 \theta}$$

$$\cdot \frac{1}{2 \sec \theta \tan \theta}\ .$$

Multiplying out and factoring $8\sec^3\theta$ from the numerator, we obtain:

$$\frac{d^2y}{dx^2}=\frac{8\sec^3\theta\,(2\tan^2\theta-\sec^2\theta-\tan^2\theta)}{8\sec^3\theta\tan^3\theta}$$

$$=\frac{2\tan^2\theta}{\tan^3\theta}-\frac{\sec^2\theta}{\tan^3\theta}-\frac{\tan^2\theta}{\tan^3\theta}$$

$$=2\cdot\frac{1}{\tan\theta}-\frac{\cos\theta}{\sin^3\theta}-\frac{1}{\tan\theta}.$$

but, $\frac{\cos\theta}{\sin^3\theta}$ can be rewritten as $\frac{\cos\theta}{\sin\theta}\cdot\frac{1}{\sin^2\theta}$, or, $\cot\theta\cdot\csc^2\theta$.

Therefore,

$$\frac{d^2y}{dx^2}=2\cot\theta-\cot\theta-\cot\theta\csc^2\theta$$

$$=\cot\theta-\cot\theta\csc^2\theta.$$

Now we use the identity: $\csc^2\theta=1+\cot^2\theta$, and, substituting, we obtain:

$$\frac{d^2y}{dx^2}=\cot\theta-\cot\theta\,(1+\cot^2\theta)$$

$$=\cot\theta-\cot\theta-\cot^3\theta$$

$$=-\cot^3\theta.$$

Therefore,

$$D_x^2y=-\cot^3\theta.$$

10.2 Derivatives of Parametric Equations

Parametric equations can be used to obtain the slope of a curve (its derivative) by the computation of $\frac{dy}{dx}$, where x and y are given parametrically.

If x = f(t) and y = g(t), then

$$\frac{dy}{dx} = \frac{dy}{dt} \cdot \frac{dt}{dx} = \frac{dy/dt}{dx/dt}$$

is obtained by use of the Chain Rule. The derivative is given in terms of t.

We also use the Chain Rule to obtain the second derivative which is written as:

$$\frac{d^2y}{dx^2} = \frac{d}{dx}\left(\frac{dy}{dx}\right) = \frac{d}{dt}\left(\frac{dy}{dx}\right)\frac{dt}{dx} = \frac{d}{dt}\left(\frac{dy}{dx}\right) \div \frac{dx}{dt}$$

The formula for the third derivative is,

$$\frac{d^3y}{dx^3} = \frac{d}{dx}\left(\frac{d^2y}{dx^2}\right) = \frac{d}{dt}\left(\frac{d^2y}{dx^2}\right) \cdot \frac{dt}{dx} = \frac{\frac{d}{dt}\left(\frac{d^2y}{dx^2}\right)}{dx/dt}$$

Problem Solving Examples:

If $y = u^2$, and $x = \frac{(u-1)}{(u+1)}$, find $\frac{dy}{dx}$.

We can readily obtain $\frac{dy}{du}$ and $\frac{dx}{du}$ from the parametric equations. If we then apply the chain rule, we obtain:

$$\frac{dy}{dx} = \frac{\frac{dy}{du}}{\frac{dx}{du}},$$

Consequently,

$$\frac{dy}{du} = 2u$$

$$\frac{dx}{du} = \frac{2}{(u+1)^2},$$

and

$$\frac{dy}{dx} = \frac{2u}{\frac{2}{(u+1)^2}} = u(u+1)^2 .$$

If $y = t^2 + 2$ and $x = 3t + 4$, find $\frac{dy}{dx}$.

Method 1. We may solve the equation: $x = 3t + 4$, for t and substitute this value of t in the first equation:

$$y = \left(\frac{x-4}{3}\right)^2 + 2 = \frac{x^2}{9} - \frac{8}{9}x + \frac{34}{9},$$

$$\frac{dy}{dx} = \frac{2}{9}x - \frac{8}{9} = \frac{2}{9}(3t+4) - \frac{8}{9} = \frac{2t}{3}.$$

Method 2. Using the chain rule we have:

$$\frac{dy}{dx} = \frac{dy}{dt} \cdot \frac{dt}{dx} \qquad \text{or}$$

$$\frac{dy}{dx} = \frac{\frac{dy}{dt}}{\frac{dx}{dt}} = \frac{\frac{d(t^2+2)}{dt}}{\frac{d(3t+4)}{dt}}$$

$$= \frac{2t}{3} = \frac{2}{3}t\,.$$

10.3 Arc Length

If a graph is given by the parametric equation

$$x = f(t), \quad y = g(t) \quad a \le t \le b,$$

then f and g are continuous on the interval [a,b] and the parameters t_0 and t_1 have two different values.

The formula for arc length is

$$S = \int_a^b \sqrt{[x'(t)]^2 + [y'(t)]^2}\, dt$$

The differential of the arc length(s) is written as:

$$S'(t) = \frac{ds}{dt} = \sqrt{\left(\frac{dx}{dt}\right)^2 + \left(\frac{dy}{dt}\right)^2} \quad dt$$

If the arc is in the form y = f(x) or x = g(y), then the length of the arc is

$$\frac{ds}{dx} = \sqrt{1 + \left(\frac{dy}{dx}\right)^2} \quad \text{or} \quad \frac{ds}{dy} = \sqrt{1 + \left(\frac{dx}{dy}\right)^2}$$

Problem Solving Examples:

Find the length of the arc of the curve $x = t^2$, $y = t^3$, between the points for which t = 0 and t = 2.

The formula for the arc length is:

$$s = \int_a^b \sqrt{1+y'^2}\ dx$$

$$dx = 2t\ dt,\ dy = 3t^2dt,\ \frac{dy}{dx} = y' = \frac{3}{2}t\ .$$

$$1 + y'^2 = 1 + \frac{9}{4}t^2 = \frac{1}{4}(4 + 9t^2)\ .$$

Then

$$s = \int_0^2 \frac{1}{2} \sqrt{4+9t^2}\ (2t\ dt)$$

This can be written as:

$$s = \frac{1}{2}\int_0^2 (4 + 9t^2)^{\frac{1}{2}} \cdot 2t\ dt$$

$$= \frac{2}{3} \cdot \frac{1}{18}\left[4 + 9t^2\right]^{\frac{3}{2}}\Bigg|_0^2$$

$$= \frac{1}{27}\left[40^{\frac{3}{2}} - 4^{\frac{3}{2}}\right]$$

$$= \frac{8}{27}\left[10^{\frac{3}{2}} - 1\right].$$

Find the length of arc of one arch of the cycloid $x = a(\theta - \sin\theta),\ y = a(1 - \cos\theta)$.

We employ the formula:

$$s = \int_a^b ds = \int_a^b \sqrt{\left(\frac{dx}{d\theta}\right)^2 + \left(\frac{dy}{d\theta}\right)^2} \ .$$

$$dx = a(1 - \cos\theta)\, d\theta\ ,\ \ dy = a \sin\theta\, d\theta.$$

$$(ds)^2 = (dx)^2 + (dy)^2$$

$$= [a^2 (1 - \cos\theta)^2 + a^2 \sin^2\theta]\,(d\theta)^2$$

$$= 2a^2 (1 - \cos\theta)\,(d\theta)^2 = 4a^2 \sin^2 \frac{1}{2}\theta\,(d\theta)^2 \ .$$

$$ds = 2a \sin \frac{1}{2}\theta\, d\theta.$$

Hence, $s = \int_0^{2\pi} 2a \sin \frac{1}{2}\theta\, d\theta = -4a \cos \frac{1}{2}\theta \Big|_0^{2\pi} = 8a.$

Polar Coordinates

11.1 Polar Coordinates

Polar coordinates is a method of representing points in a plane by the use of ordered pairs.

The polar coordinate system consists of an origin (pole), a polar axis and a ray of specific angle.

The polar axis is a line that originates at the origin and extends indefinitely in any given direction.

The position of any point in the plane is determined by its distance from the origin and by the angle that the line makes with the polar axis.

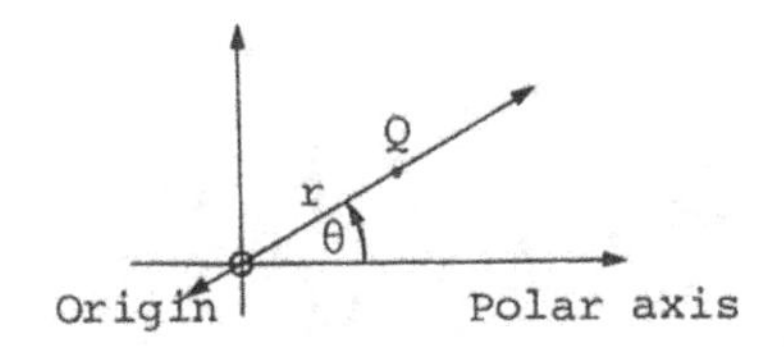

The coordinates of the polar coordinate system are (r, θ).

The angle (θ) is positive if it is generated by a counterclockwise rotation of the polar axis, and is negative if it is generated by a clockwise rotation.

A point in the polar coordinate system can be represented by many $[r, \theta]$ pairs, thus,

1. if $r = 0$, then the resulting point has coordinates $[0, \theta]$, for all values of θ.

2. there is no difference between angles that differ by an integral multiple of 2π. Consequently,

$$[r, \theta] = [r, \theta + 2n\pi] \quad \text{for all integers } n.$$

3. to change the sign of the first coordinate, add π to the second coordinate:

$$[r, \theta + \pi] = [-r, \theta]$$

The relationship between polar coordinates (r, θ) and Cartesian coordinates (x,y) is given by the following equations:

1. if r and θ are given, then

   $x = r\cos\theta$, $y = r\sin\theta$

2. if given x and y,

   $\tan\theta = y/x$, $r = \sqrt{x^2 + y^2}$ or $r^2 = x^2 + y^2$

   $$\sin\theta = \frac{y}{\sqrt{x^2 + y^2}}, \quad \cos\theta = \frac{x}{\sqrt{x^2 + y^2}}$$

Problem Solving Examples:

Draw the graph of $\rho = 2\cos\theta$.

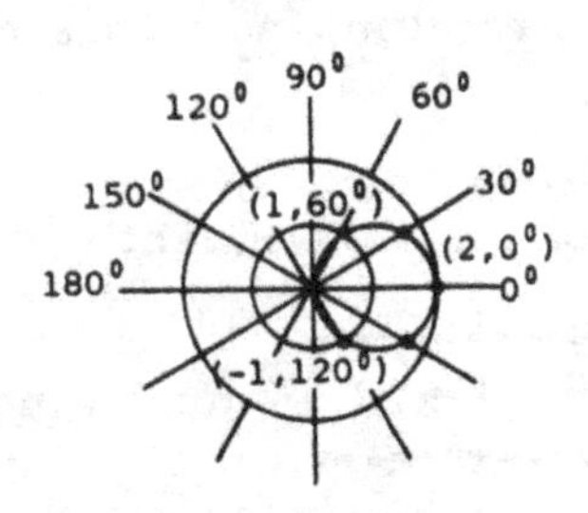

We assign the values to θ and find the corresponding values of ρ, giving the following table:

θ	cosθ	ρ=2cosθ
0^0	1	2
30^0	.87	1.74
60^0	.5	1
90^0	0	0
120^0	–.5	–1
150^0	–.87	–1.74
180^0	–1	–2
Values from 180^0 to 360^0 give the same points. (Check this.)		

We then plot the points (ρ,θ) and draw a smooth curve through them. We get the graph of the figure. The equation which defines the path of P may involve only one of the variables (ρ,θ). In that case the variable which is not mentioned may have any and all values.

Transform the equation $x^2 + y^2 - x + 3y = 3$ to a polar equation.

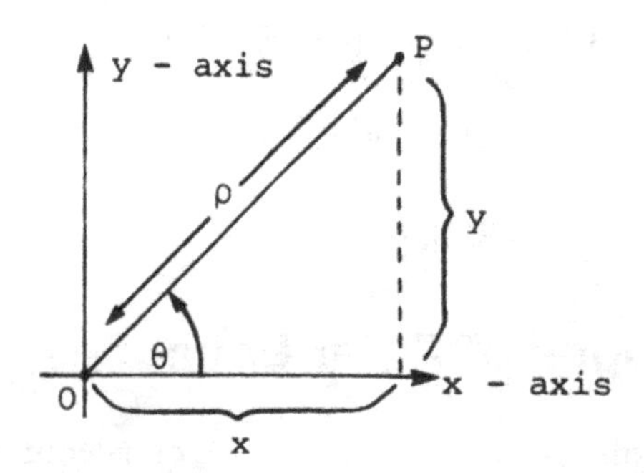

Ordinarily, when we wish to locate a point in a plane, we draw a pair of perpendicular axes and measure specified signed distances from the axes. The points are designated by pairs in the terms of (x,y). These are called rectangular coordinates.

Another way is to designate a point in terms of polar coordinates. (ρ,θ) are the polar coordinates of a point P where ρ is the radius vector of P and θ is the angle that is made with the positive x-axis and the radius vector, OP. (See diagram.)

If P is designated by the coordinates (x,y) in rectangular coordinates and by (ρ,θ) in polar coordinates, then the following relationships hold:

$$\text{Cos } \theta = \frac{\text{adjacent side}}{\text{hypotenuse}} = \frac{x}{\rho} \quad \text{or} \quad x = \rho \text{ Cos } \theta$$

$$\text{Sin } \theta = \frac{\text{opposite side}}{\text{hypotenuse}} = \frac{y}{\rho} \quad \text{or} \quad y = \rho \text{ Sin } \theta$$

Now in this example, we replace x by ρ and Cos θ and y by ρ Sin θ to obtain:

$$(\rho \text{ Cos } \theta)^2 + (\rho \text{ Sin } \theta)^2 - (\rho \text{ Cos } \theta) + 3(\rho \text{ Sin } \theta) = 3$$

$$\rho^2 \cos^2\theta + \rho^2 \sin^2\theta - \rho \cos\theta + 3\rho \sin\theta = 3$$

Factor out ρ^2 and $-\rho$.

$$\rho^2(\cos^2\theta + \sin^2\theta) - \rho(\cos\theta - 3\sin\theta) = 3$$

Apply the identity $\cos^2\theta + \sin^2\theta = 1$. Then

$$\rho^2 - \rho(\cos\theta - 3\sin\theta) = 3.$$

11.2 Graphs of Polar Equations

The graph of an equation in polar coordinates is a set of all points, each of which has at least one pair of polar coordinates (r, θ), which satisfies the given equation.

To plot a graph:

1. Construct a table of values of θ and r.
2. Plot these points.
3. Sketch the curve.

11.2.1 Rules of Symmetry

1. The graph is symmetric with respect to the x-axis, if the substitution of $(r, -\theta)$, for (r, θ) gives the same equation.
2. The graph is symmetric with respect to the y-axis, if the substitution of $(r, \pi - \theta)$ for (r, θ) gives the same equation.
3. The graph is symmetric with respect to the origin, if the substitution of $(-r, \theta)$ or $(r, \theta + \pi)$ for (r, θ) gives the same equation.

The following graphs illustrate equations in polar coordinates.

1. If $r = a \cos\theta$, then (r, θ) describes a circle. The graph is a circle with diameter a. It is symmetric with respect to the x-axis.

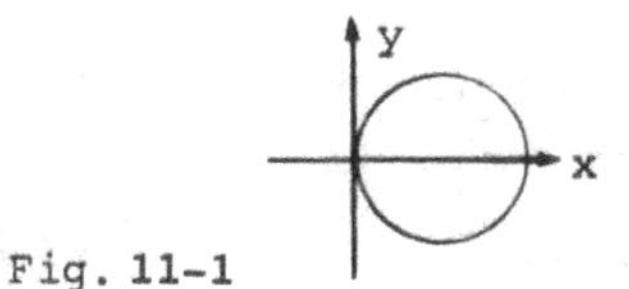

Fig. 11-1

2. $r = a \sin\theta$
The graph is a circle symmetric with respect to the y-axis.

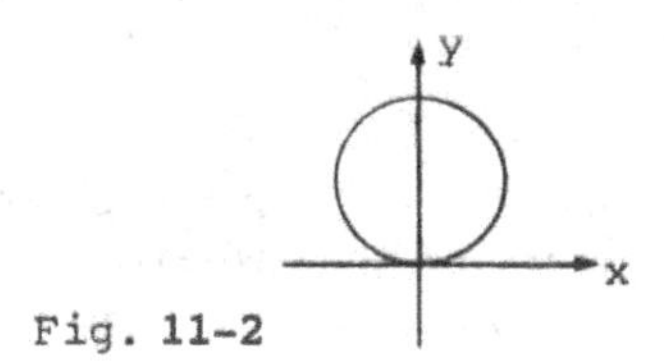

Fig. 11-2

3. $r = a$
The graph is a circle symmetric with respect to the origin.

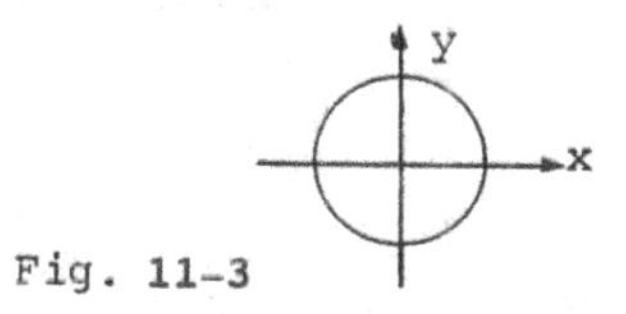

Fig. 11-3

4. $r = a \pm b \cos\theta$, $r = a \pm b \sin\theta$
These graphs are called limancons.

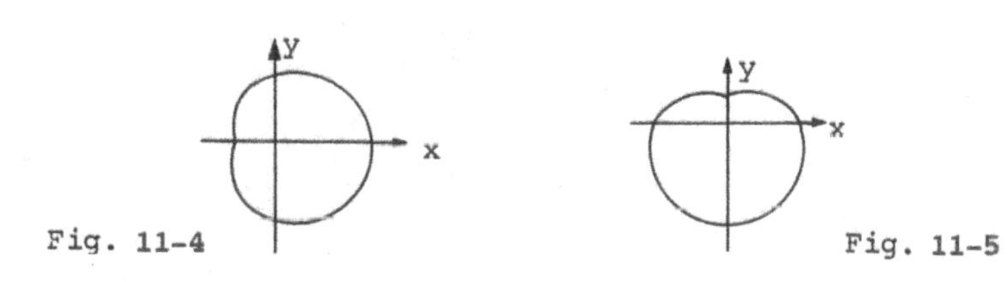

Fig. 11-4 Fig. 11-5

$r = a + b \cos \theta$ where $a > b$

This graph is symmetric with respect to the x-axis.

$r = a - b \sin\theta$ where $a > b$

This graph is symmetric with respect to the y-axis.

5. The special case occurs when a = b. The graph is then called a Cardioid.

$r = a \pm b \cos\theta$

$r = a \pm b \sin\theta$.

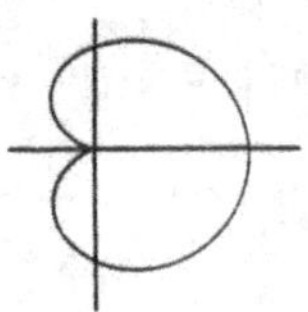

Fig. 11-6

$r = a + b \cos\theta$
where $a = b$

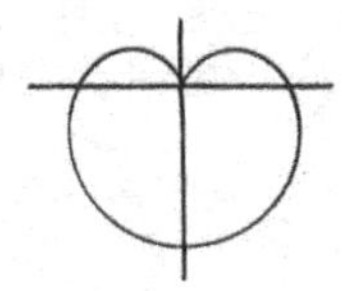

Fig. 11-7

$r = a - b \sin\theta$
where $a = b$.

6. $r = a \cos n\theta$ $r = a \sin n\theta$
These graphs are called rose or petal curves.

The number of petals is equal to n if n is an odd integer and is equal to 2n if n is an even integer. If n is equal to one (n=1), there is one petal and it is circular.

Example: The graph of $r = 2 \sin 2\theta$ is illustrated below:

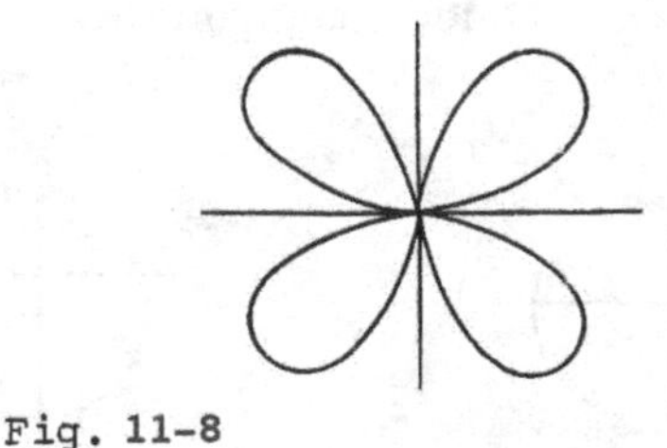

Fig. 11-8

7. $r^2 = a^2 \cos 2\theta \quad r^2 = a^2 \sin 2\theta$
These are called lemniscates. The graphs are illustrated below.

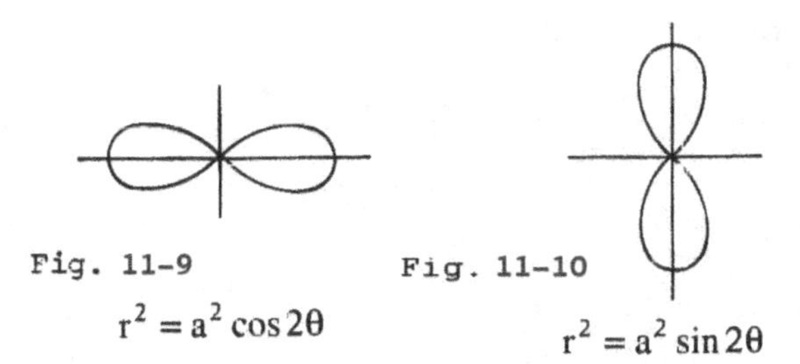

Fig. 11-9 $r^2 = a^2 \cos 2\theta$

Fig. 11-10 $r^2 = a^2 \sin 2\theta$

8. The spiral of Archimedes has an equation of the form $r = k\theta$. The graph is illustrated below.

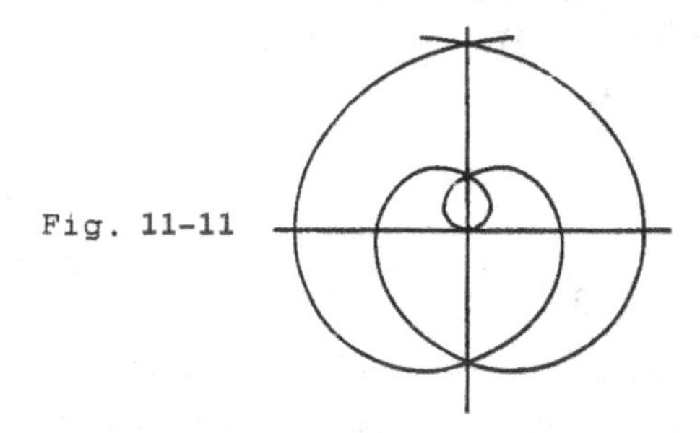

Fig. 11-11

9. The Logarithmic spiral has an equation of the form

$$\log_b r = \log_b a + k\theta, \text{ or } \quad r = ab^{k\theta}$$

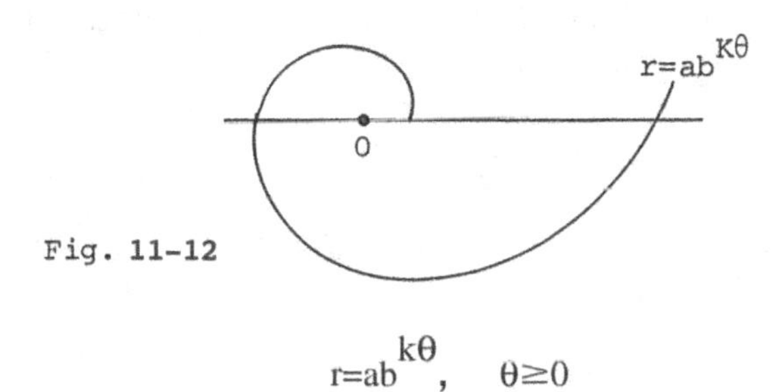

Fig. 11-12

$r=ab^{k\theta}, \quad \theta \geq 0$

Problem Solving Examples:

Find the area inside the curve: $r = \cos\theta$, and outside the curve: $r = 1 - \cos\theta$.

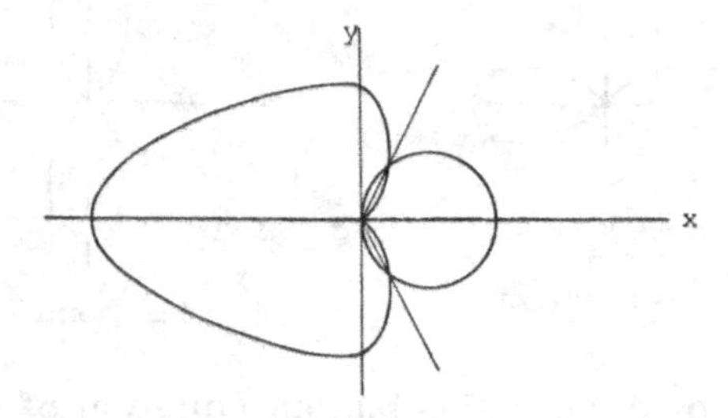

The limits of the integral which give us the required area are the points of intersection of the two curves. To find the points of intersection we set the curves equal and solve for θ.

$$\cos\theta = 1 - \cos\theta$$

$$2\cos\theta = 1$$

$$\cos\theta = \frac{1}{2}$$

$$\theta = \frac{\pi}{3}, -\frac{\pi}{3}.$$

Since the area required is divided evenly above and below the x-axis, we can multiply the value we obtain for the area above the x-axis by 2. The limits for the integral which give us the area above the x-axis are $x = 0$ and $x = \frac{\pi}{3}$. (The value of θ which is $-\frac{\pi}{3}$ would be in the fourth quadrant). Now we can set up the integral. The formula for the area in polar coordinates is

$$A = \int_{\alpha}^{\beta} \frac{1}{2} r^2 d\theta.$$

The total area required in this problem is two times the difference between the areas of the two curves, above the x-axis. Hence,

$$A = 2\int_0^{\frac{\pi}{3}} \frac{1}{2}(\cos\theta)^2\,d\theta - 2\int_0^{\frac{\pi}{3}} \frac{1}{2}(1-\cos\theta)^2\,d\theta$$

$$= \int_0^{\frac{\pi}{3}} [\cos^2\theta - (1-\cos\theta)^2]\,d\theta$$

$$= \int_0^{\frac{\pi}{3}} (\cos^2\theta - 1 + 2\cos\theta - \cos^2\theta)\,d\theta$$

$$= \int_0^{\frac{\pi}{3}} (2\cos\theta - 1)\,d\theta$$

$$= [2\sin\theta - \theta]_0^{\frac{\pi}{3}} = 2\sin\frac{\pi}{3} - \frac{\pi}{3} = \sqrt{3} - \frac{\pi}{3}.$$

What is the area bounded by the curve: $\rho = \sin\theta$, as θ varies from 0 to π?

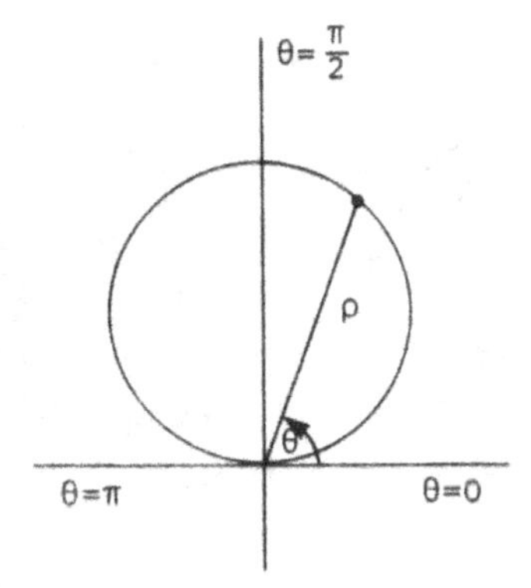

In polar coordinates, the area is given by:

$$A = \int_\alpha^\beta \frac{1}{2}\rho^2\,d\theta$$

Here, the limits are 0 and π. Hence we can write:

$$A = \int_0^{\pi} \frac{1}{2}(\sin\theta)^2\, d\theta$$

$$= \frac{1}{2}\int_0^{\pi} \sin^2\theta\, d\theta$$

Substituting: $\sin^2\theta = \dfrac{1-\cos 2\theta}{2}$, and integrating, we obtain:

$$A = \frac{1}{2}\left[\frac{\theta}{2} - \frac{\sin 2\theta}{2}\right]_0^{\pi}$$

$$= \frac{\pi}{4} \text{ square units.}$$

This is a circle of radius $\frac{1}{2}$ centered at $\left(\frac{1}{2}, \frac{\pi}{2}\right)$. Its area can be checked by $\pi r^2 = \frac{4}{\pi}$.

Derive the equation for the area of a circle.

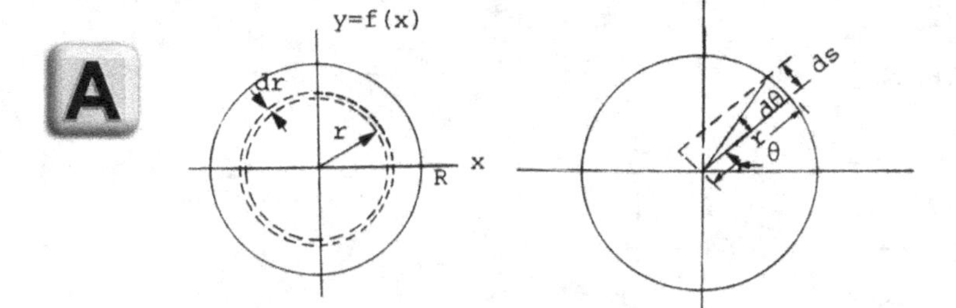

Looking at the diagram, we can say that the ring formed by the two dotted lines is a rectangle curved to close on itself. The rectangle is an infinitesimal part of the entire circle. It has a thickness dr and its length is $2\pi r$. The area of the rectangle is then $2\pi r dr$. The area of the entire circle is then the sum of all these infinitesimal areas, which can be written as the integral of $2\pi r\, dr$ from $r = 0$ to $r = R$. Hence:

$$A = \int_0^R 2\pi r \, dr$$

$$A = 2\pi \int_0^R r \, dr$$

$$= 2\pi \left[\frac{r^2}{2} \right]_0^R = 2\pi \left(\frac{R^2}{2} - \frac{0^2}{2} \right) = \pi R^2 .$$

A different procedure can also be used to solve this problem. A formula can be derived for finding area using polar coordinates. The area of the infinitesimal rectangle shown is length r times width ds, or rds. Therefore the area of the triangle is $\frac{1}{2}$ rds. But, ds = r$d\theta$. The area of the entire circle is the sum of all the infinitesimal triangles, or the integral of $\frac{1}{2}r\,(r d\theta)$ which is

$$\int_\alpha^\beta \frac{1}{2} r^2 d\theta.$$

where α and β are the limits. For the entire circle, θ varies from 0 to

where α and β are the limits. For the entire circle, θ varies from 0 to
2π which are the limits α and β.

$$A = \int_0^{2\pi} \frac{1}{2} r^2 d\theta.$$

But r is a constant. Therefore,

$$A = \frac{r^2}{2} \int_0^{2\pi} d\theta = \frac{r^2}{2}[2\pi - 0]$$

$$= \pi r^2.$$

Find the area of the cardioid: $r = a(1 + \cos\theta)$.

From the diagram we can see that the area is split into two equal parts, therefore the total area can be written as 2 times the area above the polar line A. The formula for area in polar coordinates is:

$$A = \int_{\alpha}^{\beta} \frac{1}{2} r^2 d\theta.$$

For the area above the polar line A, θ goes from 0 to π which are the limits of the integral for area. Therefore we can write:

$$A = 2 \int_0^{\pi} \frac{1}{2}\left(a(1 + \cos \theta)\right)^2 d\theta$$

$$= \int_0^{\pi} a^2 (1 + \cos \theta)^2 d\theta$$

$$= a^2 \int_0^{\pi} (1 + 2 \cos \theta + \cos^2 \theta)\, d\theta.$$

Now substituting: $\cos^2 \theta = \frac{1 + \cos 2\theta}{2}$, and integrating, we obtain:

$$A = a^2 \left(\theta + 2 \sin \theta + \frac{\theta}{2} + \frac{\sin 2\theta}{4} \right)\Bigg|_0^{\pi}$$

$$= 3\frac{\pi a^2}{2} \text{ square units.}$$

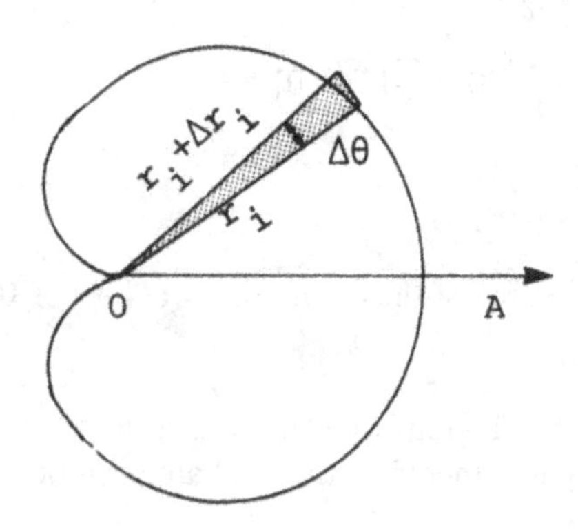

The expression for a four-leaved rose is: $\rho = a \sin 2\theta$. Find its area.

The entire area of the four-leaved rose is divided evenly into each of the four quadrants. Therefore the total area can be expressed as four times the area in the first quadrant. The formula for area expressed in polar coordinates is:

$$A = \int_{\alpha}^{\beta} \frac{1}{2} p^2 d\theta.$$

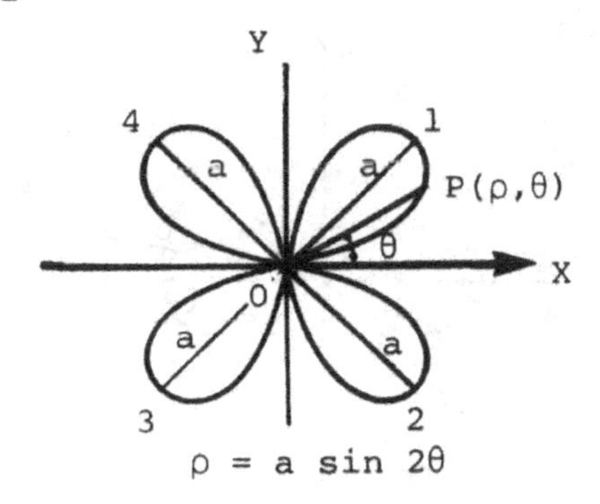

$\rho = a \sin 2\theta$

goes from 0 to $\frac{\pi}{2}$, which are the limits. We can then express the total area as:

$$A = 4 \int_{0}^{\frac{\pi}{2}} \frac{1}{2} (a \sin 2\theta)^2 \, d\theta.$$

$$= 2a^2 \int_{0}^{\frac{\pi}{2}} \sin^2 2\theta \, d\theta.$$

But $\sin^2 2\theta = \frac{1}{2} - \frac{1}{2} \cos 4\theta$. Substituting, we therefore obtain:

$$A = 2a^2 \int_{0}^{\frac{\pi}{2}} \left(\frac{1}{2} - \frac{1}{2} \cos 4\theta \right) d\theta = a^2 \int_{0}^{\frac{\pi}{2}} (1 - \cos 4\theta) \, d\theta$$

$$= a^2\left[\theta - \frac{1}{4}\sin 4\theta\right]_0^{\frac{\pi}{2}} = a^2\left[\frac{\pi}{2} - 0 - 0 + 0\right] = \frac{\pi a^2}{2},$$

the area of the four leaves.

Find the area of the surface generated by revolving the lemniscate: $r^2 = a^2 \cos 2\theta$, about the line: $\theta = \frac{\pi}{2}$.

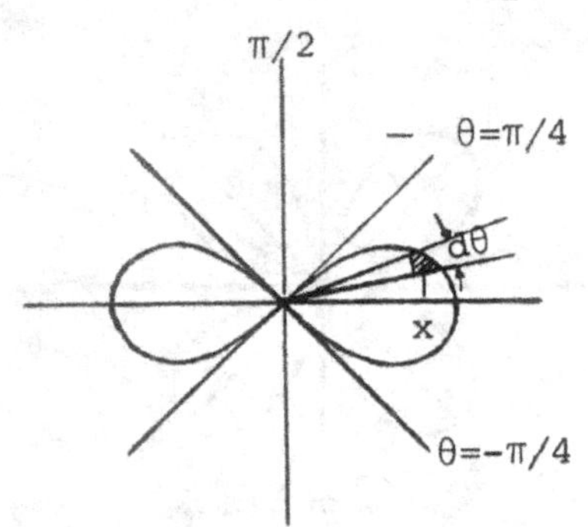

when revolved about the $\frac{\pi}{2}$ -axis, is: $2\pi x ds$, where x is the radius of generation, and ds is the length of the arc subtended by $d\theta$. Since, at any point P on the edge of the lemniscate, $x = r\cos\theta$, $rd\theta = ds$, and due to symmetry,

$$A = 4\pi \int_0^{\frac{\pi}{4}} r\cos\theta\,(r d\theta)$$

$$= 4\pi a^2 \int_0^{\frac{\pi}{4}} \cos\theta \cos 2\theta\, d\theta = 4\pi a^2 \int_0^{\frac{\pi}{4}} \cos\theta\,(1-2\sin^2\theta)\, d\theta$$

$$= 4\pi a^2 \left[\sin\theta - \frac{2\sin^3\theta}{3}\right]_0^{\frac{\pi}{4}}$$

$$= 4\pi a^2 \left(\frac{1}{\sqrt{2}} - \frac{2}{3}\frac{1}{2\sqrt{2}}\right)$$

$$= 4\pi a^2 \ \frac{2}{3}\left(\frac{1}{\sqrt{2}}\right) = 4\frac{\pi a^2\sqrt{2}}{3}$$

Note, that multiplying both sides of the equation by r^2, we obtain:

$$r^4 = r^2a^2 \cos 2\theta = a^2 (r^2 \cos^2 \theta - r^2 \sin^2 \theta),$$

$$= a^2 x^2 - y^2,$$

On the other hand, $r^2 = x^2 + y^2$,

Thus,

$$(x^2 + y^2)^2 = a^2 (x^2 - y^2).$$

Therefore, insertion of $x = r \cos \theta$ does not affect the calculation.

11.3 Polar Equation of Lines, Circles, and Conics

11.3.1 Lines

The equation of a line in Cartesian coordinates is

$\alpha x + \beta y + \gamma = 0$, where α, β and γ are all constants.

To transform this equation to polar coordinates we substitute $x = r \cos\theta$ and $y = r \sin\theta$. The resulting equation is

$$-\frac{\gamma}{r} = \alpha \cos\theta + \beta \sin\theta$$

Problem Solving Examples:

Convert the equation $r = \tan\theta + \cot\theta$ to an equation in Cartesian coordinates.

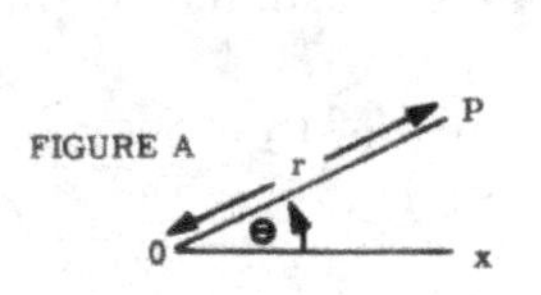

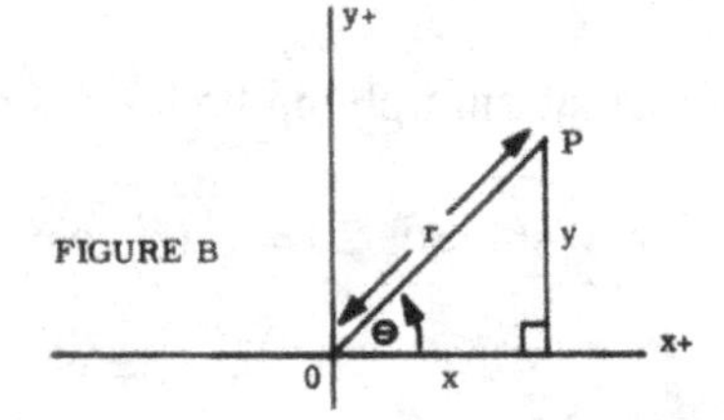

The given equation is expressed in polar coordinates (r, θ) where r is the radius vector, OP, and θ is the angle that r makes with the polar axis, OX. O is the fixed point called the pole. See figure A.

Since

$$\tan\theta \neq -\cot\theta, \text{ then } r \neq 0, \text{ and the graph of } r = \tan\theta + \cot\theta$$

does not pass through the pole. If r were equal to zero, then the curve would pass through (0, 0). Therefore in the transformation of this equation to Cartesian coordinates, we must remember that $(x, y) \neq (0, 0)$. Now we must convert all expressions of r and θ into rectangular coordinates (x, y). If P is designated by the coordinates (x, y) in rectangular coordinates and by (r, θ) in polar coordinates, then the following relationships hold true: (See figure B)

$$\tan\theta = \frac{\text{opposite side}}{\text{adjacent side}} = \frac{y}{x}$$

$$\cot\theta = \frac{\text{adjacent side}}{\text{opposite side}} = \frac{x}{y}$$

By the Pythagorean Identity $x^2 + y^2 = r^2$

Solve for r: $r = \sqrt{x^2 + y^2}$

Substitute these values for r, tan θ, and cot θ.

$$\sqrt{x^2+y^2} = \frac{y}{x} + \frac{x}{y}$$

$$xy\sqrt{x^2+y^2} = x^2 + y^2$$

Divide by $\sqrt{x^2+y^2}$

$$xy = \frac{x^2+y^2}{\sqrt{x^2+y^2}}$$

Rationalize the denominator by multiplying the numerator and denominator by $\sqrt{x^2+y^2}$

$$xy = \frac{x^2+y^2}{\sqrt{x^2+y^2}}\,\frac{\sqrt{x^2+y^2}}{\sqrt{x^2+y^2}}$$

$$xy = \frac{x^2+y^2}{x^2+y^2}\sqrt{x^2+y^2}$$

$$xy = \sqrt{x^2+y^2}$$

Squaring both sides, we obtain:

$$x^2y^2 = x^2 + y^2$$

where $x \neq 0$ and $y \neq 0$.

11.3.2 Circles

The equation of a circle in polar coordinates is

$$r^2 - 2cr\cos(\theta - \alpha) + c^2 = a^2,$$

with radius a and center at (c, α)

Problem Solving Examples:

Graph the following inequalities: $(x + 1)^2 + y^2 < 1$, and $(x + 1)^2 + y^2 > 1$.

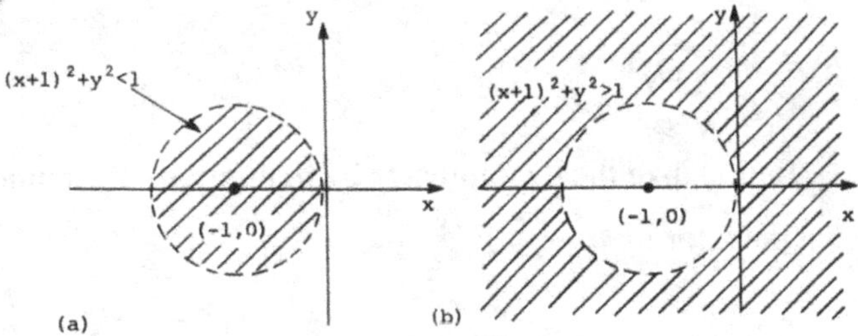

The graph of $(x + 1)^2 + y^2 = 1$ is a circle. By testing a point inside the circle and a point outside the circle, we will determine which inequality holds.

fig. (a) is the graph of:

$(x + 1)^2 + (y^2) < 1,$ and

fig. (b) is the graph of:

$(x + 1)^2 + (y^2) > 1.$

11.3.3 Conics

If the graph of a point moves so that the ratio of its distance from a fixed point to its distance from a fixed line remains constant, then the following theorems are true:

1. If the ratio is equal to one, then the curve is a parabola.

2. If the ratio is between 0 and one, then the curve is an ellipse.

3. If the ratio is greater than one, then the curve is a hyperbola.

Thus, the equation of a conic in polar coordinates is given by the formula

$$r = \frac{be}{1 \pm e\cos\theta}, \qquad r = \frac{be}{1 \pm e\sin\theta}$$

The conic is a parabola if $e = 1$, an ellipse if $0 < e < 1$ or a hyperbola if $e > 1$.

Problem Solving Examples:

Find the equation of the tangent line to the ellipse: $4x^2 + 9y^2 = 40$, at the point $(1, 2)$.

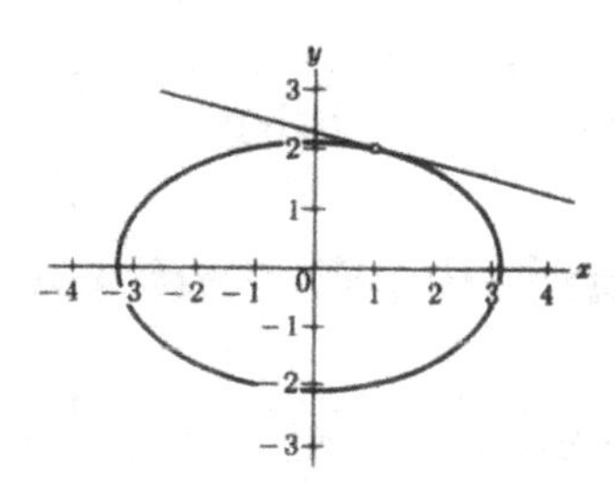

Since y is not given explicitly, the slope of the tangent to the ellipse at any point is best found by treating it as an implicit function. Differentiating, we have:

$$8x + 18yy' = 0,$$

from which

$$y' = \frac{-4x}{9y} .$$

Evaluating this derivative at the point (1, 2), we have

$$y' = \frac{-4}{18} = \frac{-2}{9} .$$

Thus the slope of the desired tangent line is $\frac{-2}{9}$.

The equation of a straight line at a given point can be expressed in the form $y - y_1 = m(x - x_1)$. Here $x_1 = 1$ and $y_1 = 2$, and the slope $m = \frac{-2}{9}$. Substituting, we obtain:

$$y - 2 = \frac{-2}{9}(x - 1).$$

$$9y - 18 = -2x + 2.$$

$2x + 9y - 20 = 0$, which

is the equation of the tangent line.

The slope could also have been found by solving the equation of the curve for y, and then differentiating.

11.4 Areas in Polar Coordinates

The area (A) of a region bounded by the curve $r = f(\theta)$, and by the lines $\theta = a$ and $\theta = b$ is given by the formula:

$$A = \frac{1}{2}\int_a^b r^2 d\theta = \frac{1}{2}\int_a^b [f(\theta)]^2 \, d\theta$$

When finding the area,

1. Sketch graph of the polar equations given.

2. Shade region for which area is sought.

3. Determine limits.

4. Solve using the equation for area.

Example: Find the area outside the circle $r = 2a \cos \theta$ and inside the cardioid $r = a(1+\cos\theta)$.

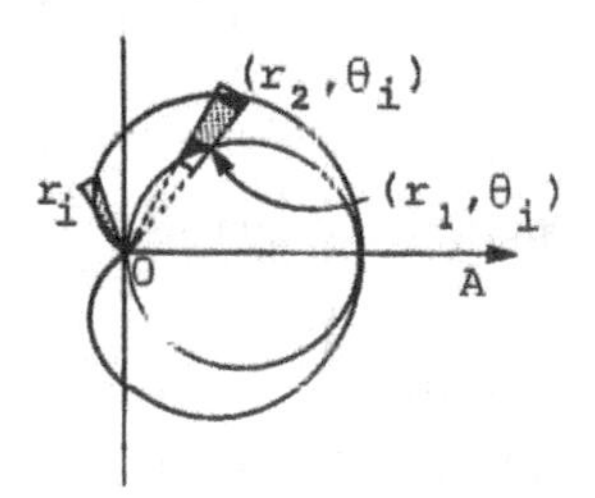

Fig. 4-13

Solution:

The required area is split evenly into two equal parts, above and below the x-axis. Therefore, to obtain the total area we can multiply the value for the area above the x-axis by 2. The area above the x-axis may be split into two parts and expressed as the sum of the area in the first quadrant (area of the cardioid minus the area of the circle, where θ goes from 0 to $\pi/2$) and the area of the second quadrant (the area of the cardioid alone where θ goes from $\pi/2$ to π).

Thus,

$$\text{Area} = A = \frac{1}{2}\int_a^b r^2 d\theta = \frac{1}{2}\int_a^b [f(\theta)]^2 d\theta$$

$$A_{\text{Total}} = 2\int_0^{\frac{\pi}{2}} \frac{1}{2}[a(1+\cos\theta)]^2 \, d\theta - 2\int_0^{\frac{\pi}{2}} \frac{1}{2}(2a\cos\theta)^2 \, d\theta$$

$$+ 2 \int_{\frac{\pi}{2}}^{\pi} \tfrac{1}{2}[a(1+\cos\theta)]^2 d\theta$$

$$A_{Total} = a^2 \int_0^{\frac{\pi}{2}} (1+2\cos\theta+\cos^2\theta)d\theta - a^2 \int_0^{\frac{\pi}{2}} 4\cos^2\theta d\theta$$

$$+ a^2 \int_{\frac{\pi}{2}}^{\pi} (1+2\cos\theta+\cos^2\theta)d\theta$$

$$A_{Total} = a^2 \int_0^{\frac{\pi}{2}} (1+2\cos\theta-3\cos^2\theta)d\theta$$

$$+a^2 \int_{\frac{\pi}{2}}^{\pi} (1+2\cos\theta+\cos^2\theta)d\theta$$

Substitute $\cos^2\theta = \dfrac{1+\cos 2\theta}{2}$ and integrate to obtain

$$A_{Total} = a^2 \left[2\sin\theta - \frac{\theta}{2} - \frac{3\sin 2\theta}{4}\right]_0^{\frac{\pi}{2}}$$

$$+ a^2 \left[2\sin\theta + \frac{3\theta}{2} + \frac{\sin 2\theta}{4}\right]_{\frac{\pi}{2}}^{\pi}$$

$$A_{Total} = a^2(2-\pi/4) + a^2\left(\frac{3\pi}{4} - 2\right) = \frac{\pi a^2}{2}.$$

Problem Solving Examples:

Find the area inside the curve: $r = \cos\theta$, and outside the curve: $r = 1 - \cos\theta$.

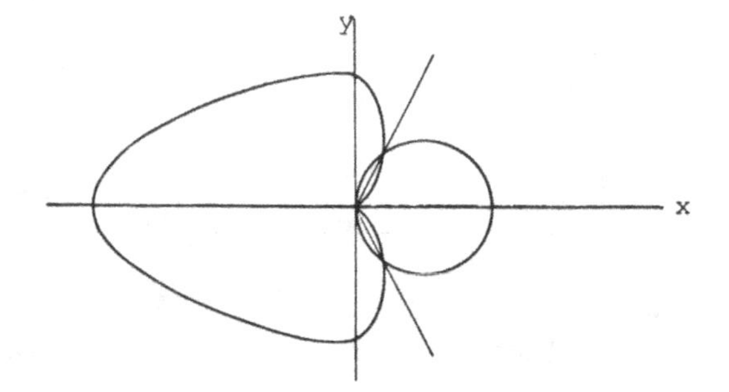

The limits of the integral which give us the required area are the points of intersection of the two curves. To find the points of intersection we set the curves equal and solve for θ.

$$\cos\theta = 1 - \cos\theta$$

$$2\cos\theta = 1$$

$$\cos\theta = \frac{1}{2}$$

$$\theta = \frac{\pi}{3}, -\frac{\pi}{3}.$$

Since the area required is divided evenly above and below the x-axis, we can multiply the value we obtain for the area above the x-axis by 2. The limits for the integral which give us the area above the x-axis are $x = 0$ and $x = \frac{\pi}{3}$. (The value of θ which is $-\frac{\pi}{3}$ would be in the fourth quadrant). Now we can set up the integral. The formula for the area in polar coordinates is

$$A = \int_{\alpha}^{\beta} \frac{1}{2} r^2 d\theta.$$

The total area required in this problem is two times the difference between the areas of the two curves, above the x-axis. Hence,

$$A = 2\int_0^{\frac{\pi}{3}} \frac{1}{2}(\cos\theta)^2\, d\theta - 2\int_0^{\frac{\pi}{3}} \frac{1}{2}(1 - \cos\theta)^2\, d\theta$$

$$= \int_0^{\frac{\pi}{3}} [\cos^2\theta - (1 - \cos\theta)^2]\, d\theta$$

$$= \int_0^{\frac{\pi}{3}} (\cos^2\theta - 1 + 2\cos\theta - \cos^2\theta)\, d\theta$$

$$= \int_0^{\frac{\pi}{3}} (2\cos\theta - 1)\, d\theta$$

$$= [2\sin\theta - \theta]_0^{\frac{\pi}{3}} = 2\sin\frac{\pi}{3} - \frac{\pi}{3} = \sqrt{3} - \frac{\pi}{3}.$$

Find the total area enclosed by the curve: $r^2 = \cos\theta$.

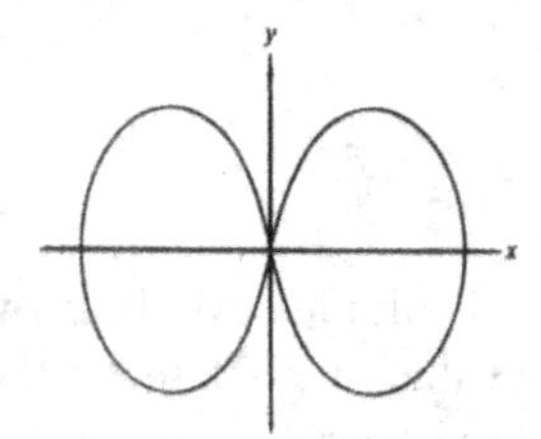

The total area of this curve can be expressed as 4 times the area of the curve in the first quadrant. In the first quadrant, θ goes from zero to $\frac{\pi}{2}$, which are the limits of the integral that gives the required area.

The formula for the area of a curve in polar coordinates is:

$$A = \int_\alpha^\beta \frac{1}{2}\alpha^2\, d\theta.$$

The given curve is $r^2 = \cos\theta$. The integral can therefore be written as follows:

$$A = 4\int_0^{\frac{\pi}{2}} \frac{1}{2}\cos\theta\, d\theta = 2\Big[\sin\theta\Big]_0^{\frac{\pi}{2}} = 2,$$

for the entire area.

Analytic Geometry

12.1 Three-Dimensional Coordinate System

A set of ordered triples of real numbers is called the three-dimensional number space (R_3).

An ordered triple is composed of three numbers called coordinates.

Two ordered triples, (x_0, y_0, z_0) and $(x_1, y_1, z_1,)$, are equal if and only if $x_0 = x_1$, $y_0 = y_1$ and $z_0 = z_1$.

The coordinate axes are composed of three mutually perpendicular lines that intersect at the point (0, 0, 0), called the origin. The axes form a right-hand system if the positive direction of the z- and y-axes lies in the plane of the paper, and if the positive direction of the x-axis is projected out of the plane of the paper. This is illustrated by Figure 12-1.

Fig. 12-1

If the y- and x-axes are interchanged, then the axes form a left-hand system.

A coordinate plane is a plane containing two of the coordinate axes. A Cartesian or rectangle coordinate system is a system in which each point in R_3 space has only one ordered number triple, and each ordered number triple designates only one point in R_3 space.

Problem Solving Examples:

Find the relative dimensions of the right circular cone of maximum volume inscribed in a sphere of radius a.

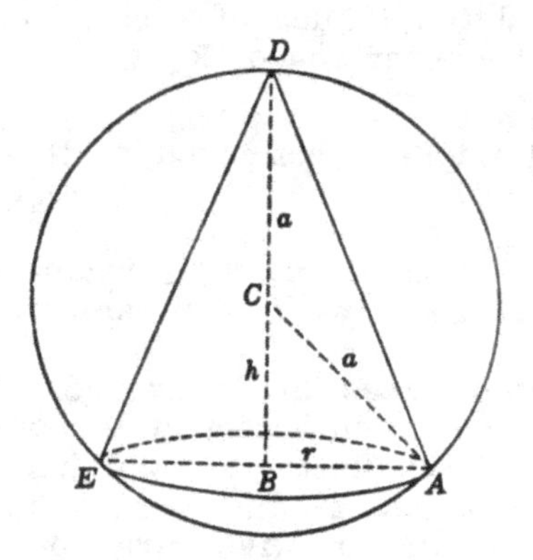

From the drawing we have

a = radius of sphere
r = radius of cone
(h + a) = height of cone, and

$$h^2 + r^2 = a^2 .$$

We also have the volume of the cone =

$$V = \frac{1}{3}\pi r^2 (h + a) .$$

Let r be the independent variable and consider h to be a function of r. Differentiating both equations above with respect to r, we obtain

$$2hh' + 2r = 0,$$

$$\frac{1}{3}\pi r^2 h' + \frac{1}{3}\pi h \cdot 2r + \frac{1}{3}\pi a \cdot 2r = 0,$$

where $$h' = \frac{dh}{dr},$$

and where $\frac{dV}{dr}$ has been set equal to 0 to find the maximum volume.

Eliminating h′ from these two equations, we find: $r^2 = 2h(a + h)$.

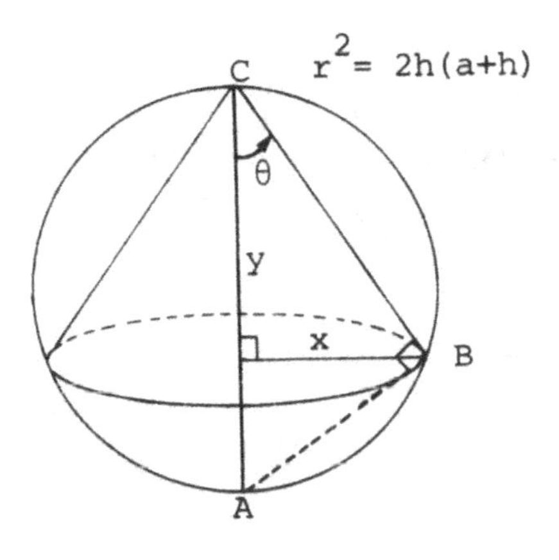

A cylindrical can is to be made to contain 1 quart. Find the relative dimensions so that the least amount of material is required.

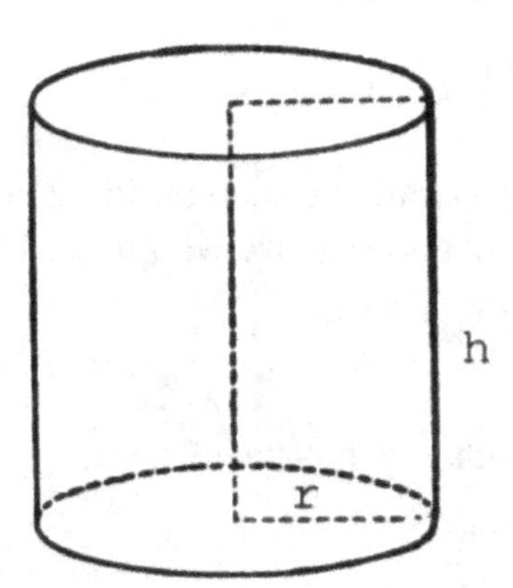

Let r and h denote the radius of base and the altitude of the cylinder. Then

$$\text{Volume} = \pi r^2 h = 1, \quad \text{and}$$

$$2\pi rh + 2\pi r^2 = S \text{ (total surface).}$$

Let r be regarded as the independent variable and consider h a function of r. Differentiate the above equations with respect to r;

denoting $\frac{dh}{dr} = D_r h$ by h', we obtain:

$$\pi r^2 h' + 2\pi rh = 0$$

or

$$rh' + 2h = 0,$$

and

$$2\pi rh' + 2\pi h + 4\pi r.$$

We set this equal to 0 because S is to be a minimum. Doing this we have,

$$2\pi rh' + 2\pi h + 4\pi r = 0$$

or

$$rh' + h + 2r = 0.$$

We now solve:

$$rh' + 2h = 0$$

and,

$$rh' + h + 2r = 0$$

simultaneously for h. We eliminate h′ by subtraction, obtaining $h - 2r = 0$ or $h = 2r$. Hence for minimum total surface S, we must have: altitude = diameter of base.

12.1.1 Three-Dimensional Distance Formula

The distance (d) between the two points $P_1(x_1, y_1, z_1)$ and $P_2(x_2, y_2, z_2)$ is given by the formula,

$$d = \sqrt{(x_2 - x_1)^2 + (y_2 - y_1)^2 + (z_2 - z_1)^2}\,.$$

Problem Solving Examples:

Find the distance of the point (x, y, z,) from the origin 0.

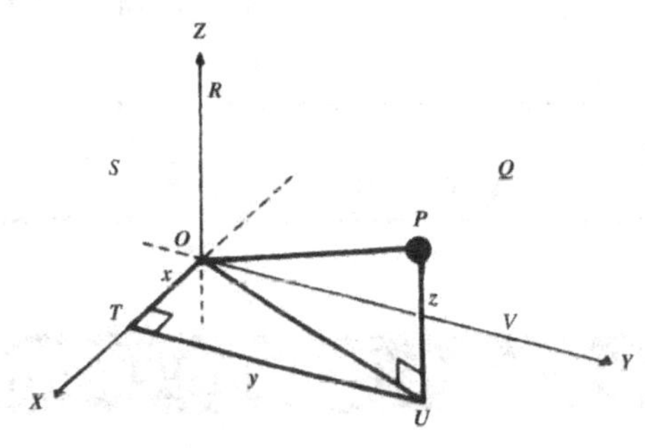

From the given diagram we see that point (x, y, z) is labeled P. Then OP is the distance of (x, y, z) from the origin 0. Thus, we wish to find OP. Referring to the figure, consider triangle OUP, in which the angle OUP is a right angle. From Pythagoras's theorem,

$$OP^2 = OU^2 + UP^2$$

Consider the triangle OTU in which the angle OTU is a right angle. Using Pythagoras' theorem again,

$$OU^2 = OT^2 + TU^2$$

Substituting this value OU^2 in the first equation

$$OP^2 = (OT^2 + TU^2) + UP^2$$

But OT = x, TU = y, UP = z, and so

$$OP^2 = x^2 + y^2 + z^2$$

The distance of the points (x, y, z) from the origin 0 is therefore.

$$OP = \sqrt{x^2 + y^2 + z^2}.$$

12.1.2 The Formula of a Midpoint

The midpoint of the line segment that connects two points is,

$$x_m = \frac{x_1 + x_2}{2}, \quad y_m = \frac{y_1 + y_2}{2}, \quad z_m = \frac{z_1 + z_2}{2}$$

Problem Solving Examples:

Show that the point $P_1(2, 2, 3)$ is equidistant from the points $P_2(1, 4, -2)$ and $P_3(3, 7, 5)$.

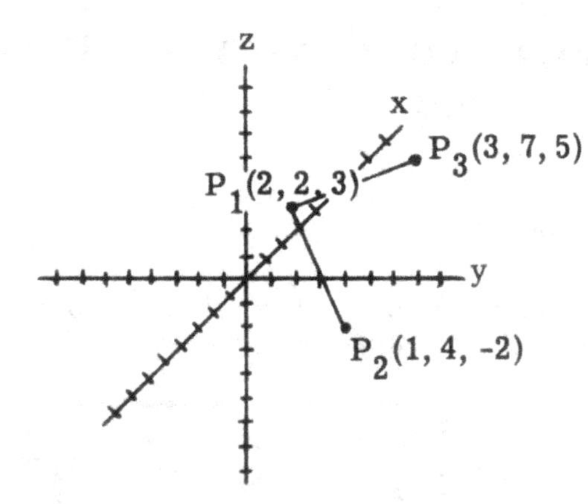

The distance, d, between any two points (x_1, y_1, z_1) and (x_2, y_2, z_2) is given by the formula

$$d = \sqrt{(x_2 - x_1)^2 + (y_2 - y_1)^2 + (z_2 - z_1)^2}$$

As a visual aid, let us plot the three points and draw the segments whose lengths we wish to show equal.

We are asked to show $P_1P_2 = P_1P_3$. By substituting into the formula given,

$$P_1P_2 = \sqrt{(2-1)^2 + (2-4)^2 + (3-(-2))^2}$$

$$= \sqrt{(1)^2 + (-2)^2 + (5)^2} = \sqrt{30}$$

$$P_1P_3 = \sqrt{(2-3)^2 + (2-7)^2 + (3-5)^2}$$

$$= \sqrt{(-1)^2 + (-5)^2 + (-2)^2} = \sqrt{30}$$

Hence, $P_1P_2 = P_1P_3$.

12.2 Equations of a Line and Plane in Space

12.2.1 Line

The equations of a line joining two points (x_0, y_0, z_0) and (x_1, y_1, z_1) are expressed parametrically as

$$x = x_0 + (x_1 - x_0)t$$
$$y = y_0 + (y_1 - y_0)t$$
$$z = z_0 + (z_1 - z_0)t$$

If a, b and c are the direction numbers of a line through the point $P_0(x_0, y_0, z_0)$, then the parametric equation is

$$x = x_0 + at, \quad y = y_0 + bt, \quad z = z_0 + ct$$

Utilizing direction cosines, the equation can be written as

$$x = x_0 + t\cos\alpha; \quad y = y_0 + t\cos\beta; \quad z = z_0 + t\cos\gamma$$

The equation of a line may also be written symmetrically as

$$\frac{x - x_0}{x_1 - x_0} = \frac{y - y_0}{y_1 - y_0} = \frac{z - z_0}{z_1 - z_0}$$

A line, therefore, is represented as the intersection of two planes.

Problem Solving Examples:

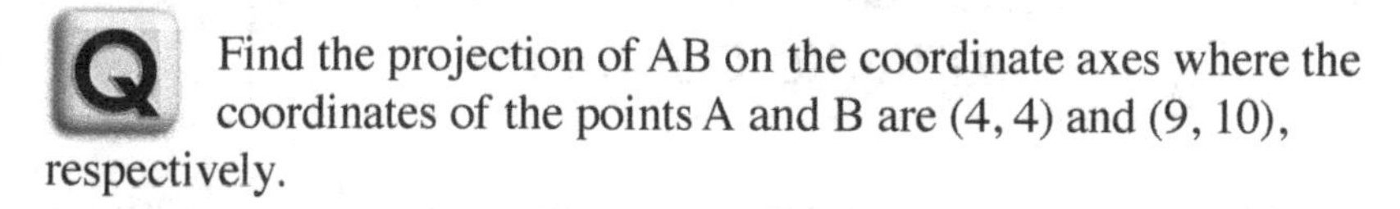

Q Find the projection of AB on the coordinate axes where the coordinates of the points A and B are (4, 4) and (9, 10), respectively.

A

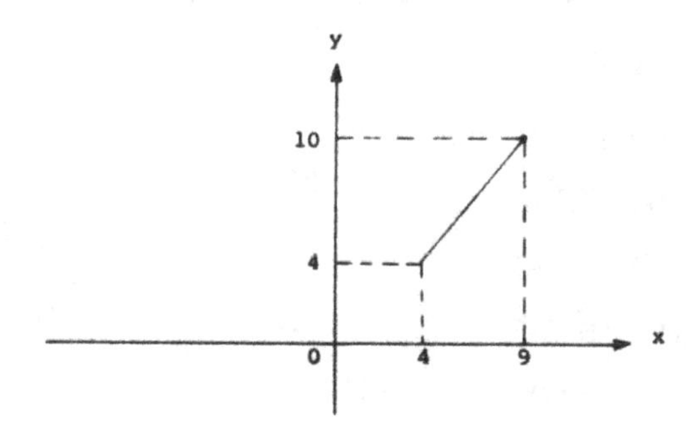

By definition, the projection of the line segment joining points $P_1(x_1, y_1)$ and $P_2(x_2, y_2)$ onto the x-axis or onto any line parallel to the x-axis is (x_2-x_1), and the projection of P_1P_2 onto the y-axis or any line parallel to the y-axis is (y_2-y_1).

Therefore, the projection of AB onto the x-axis is $(9-4) = 5$, and the projection of AB onto the y-axis is $(10-4) = 6$.

Quiz: The Parametric Equations, Polar Coordinates, and Analytic Geometry

1. The area of a region enclosed by the polar curve $r = \cos(2\theta)$ for $0 \le \theta \le \frac{\pi}{2}$ is

   (A) 1.571

   (B) 3.142

   (C) 0.393

   (D) 0.785

   (E) 1.00

2. If a particle moves on the x - axis according to $x(t) = 2t - t^2$, how far to the right will it get?

   (A) 1

   (B) 0

   (C) –1

   (D) 2

   (E) –2

3. The area of one leaf of the rose $r = \sin 3\theta$ is

   (A) $\frac{\pi}{12}$

   (B) $\frac{\pi}{6}$

   (C) $\frac{\pi}{4}$

(D) $\frac{\pi}{3}$

(E) $\frac{\pi}{2}$

4. If the movement of a particle in the plane is $x(t) = \sin t$, $y(t) = \cos^2 t$, if t is in the interval $[0, \pi]$, when is it stationary?

(A) 0

(B) $\frac{\pi}{2}$

(C) $\frac{\pi}{4}$

(D) $\frac{3\pi}{4}$

(E) π

5. Let $x(t) = \frac{1}{2}\left(e^t + e^{-t}\right)$, $y(t) = \frac{1}{2}\left(e^t + e^{-t}\right)$ generate a curve from $t = -1$ to $t = 1$. What is the length of this arc?

(A) $\sqrt{2}\int_0^1 \sqrt{e^{2t} + e^{-2t}}\,dt$

(B) $\int_{-1}^1 \sqrt{e^{2t} - e^{-2t}}\,dt$

(C) $\sqrt{2}\int_0^1 \sqrt{e^{2t} - 1}\,dt$

(D) $\sqrt{2}\int_0^1 \sqrt{e^{2t} - e^{-2t}}\,dt$

(E) $\sqrt{2}\int_{-2}^2 \sqrt{e^t - e^{-t}}\,dt$

6. If the motion of a particle on the x-axis has acceleration $\frac{d^2x}{dt^2}$ $= t^2 - 2t$, and it is stationary at 1 when $t = 1$, then $12x(t) =$

   (A) $t^4 + 4t^3$

   (B) $t^4 + 4t^3 + 8t + 7$

   (C) $4t^4 + 8t^3$

   (D) $t^4 - 4t^3 + 15t^2$

   (E) $t^4 - 2t^3 + 3t^2 - 4t + 14$

7. The length of arc joining (1, 0) to (0, 1) on the curve $x = \cos t$, $y = \sin^2 t$ is

   (A) $\int_0^1 (1 + x^2)\, dx$

   (B) $\int_0^1 \sqrt{1 + x^2}\, dx$

   (C) $\int_0^1 (1 + 4x^2)\, dx$

   (D) $\int_0^1 \sqrt{1 + 4x^2}\, dx$

   (E) None of the above

8. The equation $r = 2 \sin \theta - \cos \theta$ in rectangular coordinates is given by

   (A) $x^2 + y^2 + x - 2y = 0$

   (B) $x^2 - x + 2y = 0$

   (C) $x^2 + y^2 + 2x - y = 0$

   (D) $x^2 - y^2 - x + 2y = 0$

   (E) $y^2 - x^2 - x + 2y = 0$

9. A rectangular box resides in 3-space with one vertex at the origin, $(0,0,0)$ and three faces in the coordinate planes. If another vertex is $(x,2,3), x>0$, and the angle in radians between the diagonals from $(0,0,0)$ to $(x,2,3)$ and the other vertex in the xz plane is $\frac{\pi}{2}$, find x.

    (A) 1

    (B) $\sqrt{2}$

    (C) 2

    (D) 3

    (E) $\sqrt{3}$

10. Which of the following could be the equation for the curve in the figure below?

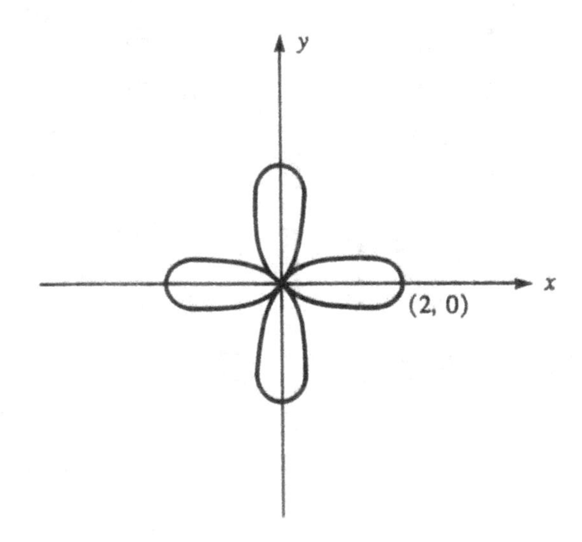

(A) $r = 1 - 2 \sin 2\theta$

(B) $r = 1 + 2 \sin 2\theta$

(C) $r = 4 \cos 2\theta$

(D) $r = 2 \sin 4\theta$

(E) $r = 2 \cos 4\theta$

ANSWER KEY

1. (C)
2. (A)
3. (A)
4. (B)
5. (A)
6. (B)
7. (D)
8. (A)
9. (E)
10. (E)

Vector Analysis

13.1 Two-Dimensional Vectors

Definition 1:

A scalar is a quantity that can be specified by a real number. It has only magnitude.

Definition 2:

A vector is a quantity that has both magnitude and direction. Velocity is an example of a vector quantity.

A vector (AB) is denoted by $\vec{AB}$, where B represents the head and A represents the tail. This is illustrated in Fig. 13-1.

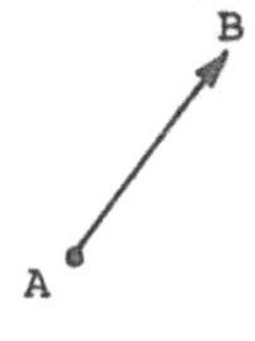

Fig. 13-1

The length of a line segment is the magnitude of a vector.

If the magnitude and direction of two vectors are the same, then they are equal.

Definition 3:

Vectors that can be translated from one position to another without any change in their magnitude or direction are called free vectors.

Definition 4:

The unit vector is a vector with a length (magnitude) of one.

Definition 5:

The zero vector has a magnitude of zero.

Definition 6:

The unit vector $\vec{i}$ is a vector with magnitude of one in the direction of the x-axis.

Definition 7:

The unit vector $\vec{j}$ is a vector with magnitude of one in the direction of the y-axis.

13.1.1 Vector Properties

1) When two vectors are added together, the resultant force of the two vectors produces the same effect as the two combined forces. This is illustrated in Fig. 13-2

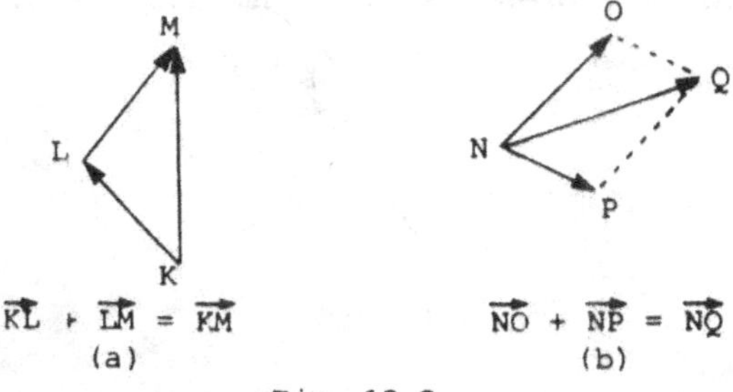

Fig. 13-2

In these diagrams, the vectors $\vec{KM}$ and $\vec{NQ}$ are the resultant forces.

13.1.2 Addition of Two Vectors

Let vector A be $<a_1,a_2>$ and vector B be $<b_1,b_2>$. Then

$$A + B = (a_1+b_1)\vec{i} + (a_2+b_2)\vec{j}$$

13.1.3 Multiplication of Vector by a Scalar

Let vector A be $a\vec{i} + b\vec{j}$ and let c be a constant. Then,

$$cA = c(a\vec{i}+b\vec{j}) = ca\vec{i} + cb\vec{j}$$

13.1.4 Additive and Multiplicative Properties of Vectors

Let s, t and u represent vectors and d and c represent real constants. All of the following are true:

1. $s + t = t + s$
2. $(s+t) + u = s + (t+u)$
3. $s + 0 = s$
4. $s + (-s) = 0$
5. $(c+d)s = cs + sd$
6. $c(s+u) = cs + cu$
7. $c(st) = (cs)t$
8. $1 \cdot s = s$
9. $0 \cdot s = 0$
10. $s \cdot s = |s|^2$
11. $c(ds) = (cd)s$

The magnitude $|s|$ of a vector $\vec{s} = a_1 i + a_2 j$ is

$$|s| = \sqrt{a_1^2 + a_2^2}$$

The difference between vectors $\vec{a}$ and $\vec{b}$ is given by the formula

$$\vec{a} - \vec{b} = \vec{a} + (-\vec{b})$$

13.1.5 Scalar (DOT) Product

Two vectors are parallel if(a) one is a scalar multiple of the other; and (b) neither is zero.

Definition:

If vector A = $\langle a_1, a_2 \rangle$ and vector B = $\langle b_1, b_2 \rangle$, then the scalar product of A and B is given by the formula

$$A \cdot B = a_1 b_1 + a_2 b_2$$

Theorem:

If θ is the angle between the vectors $A = a_1\vec{i} + a_2\vec{j}$ and $B = b_1\vec{i} + b_2\vec{j}$ then

$$\cos\theta = \frac{a_1 b_1 + a_2 b_2}{|A||B|}$$

Definition:

Let vector $A = a_1\vec{i} + a_2\vec{j}$ and vector $B = b_1\vec{i} + b_2\vec{j}$ The projection of vector A on B ($\text{Proj}_B A$) is given by the quantity $|A|\cos\theta$, where θ is the angle between the two vectors.

Therefore, $$\text{Proj}_B A = |A|\cos\theta = \frac{a_1 b_1 + a_2 b_2}{|B|} = \frac{A \cdot B}{|B|}$$

If the angle θ is acute, then $|A|\cos\theta$ is positive; if θ is obtuse, then $|A|\cos\theta$ is negative.

The scalar product of two non-zero vectors A and B is now redefined by the formula

$$\mathbf{A} \cdot \mathbf{B} = |\mathbf{A}|\,|\mathbf{B}| \cos\theta = a_1 b_1 + a_2 b_2$$

Problem Solving Examples:

Which of the following vectors are equal to $\overrightarrow{MN}$ if $M = (2, 1)$ and $N = (3, -4)$?

(a) $\overrightarrow{AB}$, where $A = (1, -1)$ and $B = (2, 3)$
(b) $\overrightarrow{CD}$, where $C = (-4, 5)$ and $D = (-3, 10)$
(c) $\overrightarrow{EF}$, where $E = (3, -2)$ and $F = (4, -7)$.

A

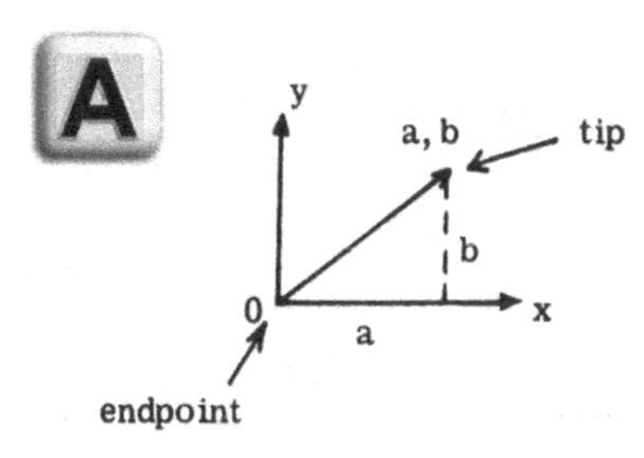

Figure A: (a-0, b-0) represents the vector.

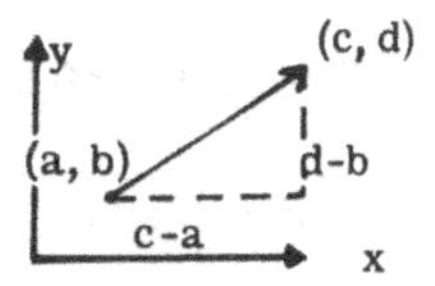

Figure B: (c-a, d-b) represents the vector.

With each ordered pair in the plane there can be associated a vector from the origin to that point.

The vector is determined by subtracting the coordinates of the endpoint from the corresponding coordinates of the tip. As for $\overrightarrow{MN}$, the tip is the point corresponding to the second letter of the alphabetical notation, N, while the endpoint is the point corresponding to the first, M. In this problem the vectors are of a general nature wherein their endpoints do not lie at the origin.

We first find the ordered pair which represents MN .

$$\overrightarrow{MN} = (3-2, -4-1) = (1, -5)$$

Now, we find the ordered pair representing each vector.

(a) $\overrightarrow{AB} = (2-1, 3-(-1)) = (1, 4)$

(b) $\overrightarrow{CD} = ((-3)-(-4), 10-5) = (1, 5)$

(c) $\overrightarrow{EF} = (4-3, -7-(-2)) = (1, -5)$

Only $\overrightarrow{EF}$ and $\overrightarrow{MN}$ are equal.

13.2 Three–Dimensional Vectors

A vector in three-dimensional space is denoted by

$$\vec{v} = ai + bj + ck$$

where i is the unit vector in the direction of the x-axis, j is the unit vector in the direction of the y-axis, and k is the unit vector in the direction of the z-axis.

The magnitude of vector $\vec{v}$ is given by the formula

$$|v| = \sqrt{a^2 + b^2 + c^2}$$

The unit vector $\vec{u}$ in the direction of vector $\vec{v}$ is represented by

$$\vec{u} = \frac{1}{|v|}\vec{v}$$

13.2.1 Vector Properties

1) Two vectors are proportional (parallel) if each vector is a scalar multiple of the other (there is a number c such that $\vec{u} = c\vec{v}$).

2) Two vectors A and B are orthogonal (perpendicular) if and only if $A \cdot B = 0$.

3) Two vectors $\vec{AB}$ and $\vec{CD}$ are equal (have the same magnitude and direction) if and only if one of the following conditions is satisfied:

   a) The two vectors are on the same directed line $\vec{L}$ and their directed lengths are equal; or

   b) The points A, B, C, D are the vertices of a parallelogram. This is illustrated in Fig. 13-3.

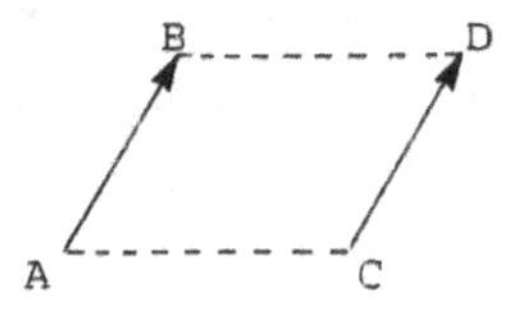

Fig. 13-3

13.2.2 Linear Dependence and Independence

Let $\vec{v}_1, \vec{v}_2, \ldots, \vec{v}_n$ represent a set of vectors and $c_1, c_2, \ldots, c_n$ represent numbers.

An expression of the form

$$\boxed{c_1\vec{v}_1 + c_2\vec{v}_2 + \ldots + c_n\vec{v}_n}$$

is the linear combination of the vectors.

Two proportional vectors have a linear combination equal to the zero vector.

A set of vectors is linearly dependent if and only if there exists a set of constants such that

$$c_1\vec{v}_1 + c_2\vec{v}_2 + \ldots + c_n\vec{v}_n = 0$$

If these constants are all zero, then the set of vectors is said to be linearly independent. Two proportional (parallel) vectors are linearly dependent; thus one member of the set can be expressed as a linear combination of the remaining members.

A set of vectors, r, s, t, are linearly independent if

$$r = a_{11}i + a_{12}j + a_{13}k,$$

$$s = a_{21}i + a_{22}j + a_{23}k,$$

$$t = a_{31}i + a_{32}j + a_{33}k,$$

and the determinant $D = \begin{vmatrix} a_{11} & a_{12} & a_{13} \\ a_{21} & a_{22} & a_{23} \\ a_{31} & a_{32} & a_{33} \end{vmatrix}$

is not equal to zero.

The determinant of three vectors is found by expanding the vectors as follows:

$$\det A = \sum_{j=1}^{n} (-1)^{1+j} a_{ij} \det M_{ij}.$$

or, for the three vectors listed above,

$$\det \begin{bmatrix} a_{11} & a_{12} & a_{13} \\ a_{21} & a_{22} & a_{23} \\ a_{31} & a_{32} & a_{33} \end{bmatrix} = a_{11}\det M_{11} - a_{12}\det M_{12} + a_{13}\det M_{13}.$$

for an illustrated example, take the vectors

$$u = 2i + j - k,$$

$$\vec{v} = -i + j + 2k, \text{ and}$$

$$\vec{w} = 2i + j + 3k.$$

$$D = \begin{vmatrix} 2 & 1 & -1 \\ -1 & 1 & 2 \\ 2 & 1 & 3 \end{vmatrix} = 2\begin{vmatrix} 1 & 2 \\ 1 & 3 \end{vmatrix} - 1\begin{vmatrix} -1 & 2 \\ 2 & 3 \end{vmatrix} - 1\begin{vmatrix} -1 & 1 \\ 2 & 1 \end{vmatrix}$$

$= 2(3-2) - (-3-4) - (-1-2)$

$= 2 + 7 + 3 = 12$

$D = 12 \neq 0$ thus the set of vectors is linearly independent.

Remember: The determinant of a 2×2 matrix $\begin{vmatrix} a & b \\ c & d \end{vmatrix}$ is found by $D = ad - bc$.

13.3 Vector Multiplication

13.3.1 Scalar (DOT) Product

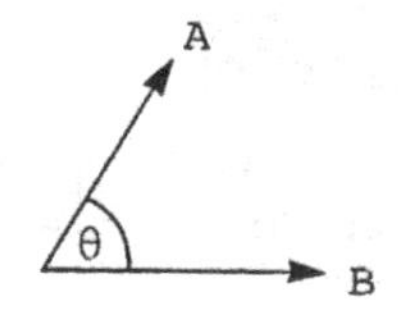

Let $A = a_1i + a_2j + a_3k$ and $B = b_1i + b_2j + b_3k$. If θ is the angle between these two vectors, then

$$\cos\theta = \frac{a_1b_1 + a_2b_2 + a_3b_3}{|A|\,|B|}$$

The scalar product of vectors A and B is

$$A \cdot B = |A|\ |B| \cos\theta = a_1b_1 + a_2b_2 + a_3b_3$$

Definition:

Let θ represent the angle between vectors A and B.

The component of A along B, or the projection of A on B ($|A|\cos\theta$), is given by the formula

$$\text{Proj}_B A = |A|\cos\theta = |A| \frac{A \cdot B}{|A|\,|B|} = \frac{A \cdot B}{|B|}$$

Scalar products are used to calculate the work done by a constant force when its point of application moves along a segment from C to D. The work done is the product of the distance from C to D and the projection of the constant force F on vector $\overrightarrow{CD}$.

$$W = \text{Proj}_{\overrightarrow{CD}} F = \frac{F \cdot \overrightarrow{CD}}{|CD|}$$

The work done by F is given by the formula

$$W = F \cdot s$$

13.3.2 Vector (CROSS) Product

The linearly independent vectors illustrated in Fig. 13-4 are said to form a right-handed triple. The vectors in Fig. 13-5 form a left-handed triple.

Fig. 13-4 Fig. 13-5

If two sets of ordered triples of vectors are both right-handed or left-handed, then they are said to be similarly oriented. If they are not, they are said to be oppositely oriented.

Theorem:

If the ordered triple $\langle A, B, C \rangle$ is right-handed, then the ordered triples $\langle A, B, C \rangle$ and $\langle c_1 A, c_2 B, c_3 C \rangle$ are also right-handed, provided that $c_1, c_2, c_3 > 0$.

Definition:

If A and B are vectors, then the vector product $A \times B$ is defined as follows:

1) If either A or B is 0, then

$$\boxed{A \times B = 0}$$

2) If A is parallel to B, then

$$\boxed{A \times B = 0}$$

3) Otherwise,

$$\boxed{A \times B = C}$$

where vector C has the following properties:

a) It is orthogonal to both A and B.

b) It has magnitude $|C| = |A||B| \sin\theta$, where θ is the angle between A and B.

c) It is directed so that $\langle A, B, C \rangle$ is a right-handed triple.

Theorem:

Let

1) A and B represent any vector;

2) $\langle i, j, k \rangle$ represent a right-handed triple; and

3) t represent any number.

Then:

1) $A \times B = -(B \times A)$

2) $(tA) \times B = t(A \times B) = A \times (tB)$

3) $i \times j = -j \times i = k$

4) $j \times k = -k \times j = i$

5) $k \times i = -i \times k = j$

6) $i \times i = j \times j = k \times k = 0$

Theorem:

If A, B and D are any vectors, then:

1) $A \times (B+D) = A \times B + A \times D$

2) $(A+D) \times B = A \times B + A \times D$

Theorem:

If

$$A = a_1 i + a_2 j + a_3 k \quad \text{and} \quad B = b_1 i + b_2 j + b_3 k$$

then the vector cross product $A \times B$ is given by

$$A \times B = (a_2 b_3 - a_3 b_2)i + (a_3 b_1 - a_1 b_3)j + (a_1 b_2 - a_2 b_1)k,$$

$$A \times B = \begin{vmatrix} i & j & k \\ a_1 & a_2 & a_3 \\ b_1 & b_2 & b_3 \end{vmatrix}$$

Example:

Find the cross product $A \times B$, if

$A = -2i + 4j + 5k$ and $B = 4i + 5k$.

$$\begin{array}{lcccccc} & i & j & k & i & j & k \\ A \times B = & -2 & 4 & 5 & -2 & 4 & 5 \\ & 4 & 0 & 5 & 4 & 0 & 5 \end{array}$$

$$\begin{array}{rcrcr} 20i & + & 20j & + & 0k \\ 0i & + & 10j & - & 16k \\ \hline \end{array}$$

$$A \times B = 20i + 30j - 16k$$

When moving upwards, multiply by -1.

13.3.3 Product of Three Vectors

Theorem:

If $\vec{v}_1, \vec{v}_2, \vec{v}_3$ are vectors and the points P,Q,R,S are chosen so that

$$\vec{u}(\vec{PQ}) = \vec{v}_1, \quad \vec{u}(\vec{PR}) = \vec{v}_2 \text{ and } \vec{u}(\vec{PS}) = \vec{v}_3,$$

then:

1) $|(\vec{v}_1 \times \vec{v}_2)\vec{v}_3|$ is the volume of the parallelepiped with a vertex at P and adjacent vertices at Q, R and S. The volume is zero if and only if the four points P, Q, R and S lie in the same plane.

2) If <i,j,k> is a right-handed coordinate triple and if

   $\vec{v}_1 = a_1 i + b_1 j + c_1 k,$

   $\vec{v}_2 = a_2 i + b_2 j + c_2 k,$

   $\vec{v}_3 = a_3 i + b_3 j + c_3 k,$

   then

$$(\vec{v}_1 \times \vec{v}_2) \cdot \vec{v}_3 = \begin{vmatrix} a_1 i & b_1 j & c_1 k \\ a_2 & b_2 & c_2 \\ a_3 & b_3 & c_3 \end{vmatrix}$$

3) $(\vec{v}_1 \times \vec{v}_2) \cdot \vec{v}_3 = \vec{v}_1 \cdot (\vec{v}_2 \times \vec{v}_3).$

Theorem:

Suppose A, B, and D are any vectors. Then:

1) $(A \times B) \times D = (A \cdot D) \cdot B - (B \cdot D) \cdot A$

2) $A \times (B \times D) = (A \cdot D) \cdot B - (A \cdot B) \cdot D$

Problem Solving Examples:

Two forces of 50 lbs. and 30 lbs. have an included angle of 60°. Find the magnitude and direction of their resultant.

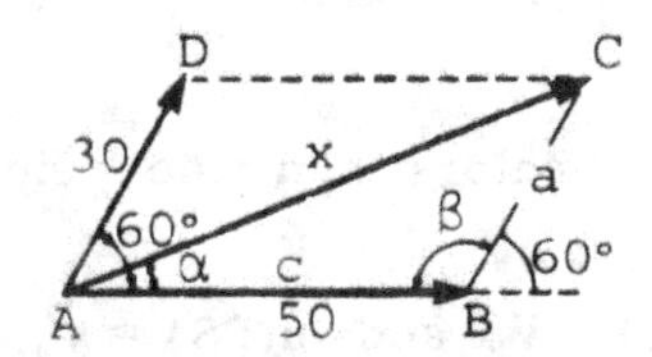

Construct the parallelogram and label it as in the figure. Since $\overline{AD}$ is parallel to $\overline{BC}$ we have
$\angle ABC = \beta = 180° - 60° = 120°$.
By the law of cosines:

$$x^2 = c^2 + a^2 - 2\,ac\cos\beta$$
$$= 2500 + 900 - 2(50)(30)(-\tfrac{1}{2})$$
$$= 2500 + 900 + 1500 = 4900.$$
$$x = 70 \text{ lbs.}$$

$$\cos\alpha = \frac{x^2 + c^2 - a^2}{2xc} = \frac{4900 + 2500 - 900}{2(70)(50)} = \frac{13}{14} = .9286.$$

$$\alpha = 21°47'.$$

Q Two forces act simultaneously on a body free to move. One force of 112 lbs. is acting due east, while the other of 88 lbs. is acting due north. Find the magnitude and direction of their resultant.

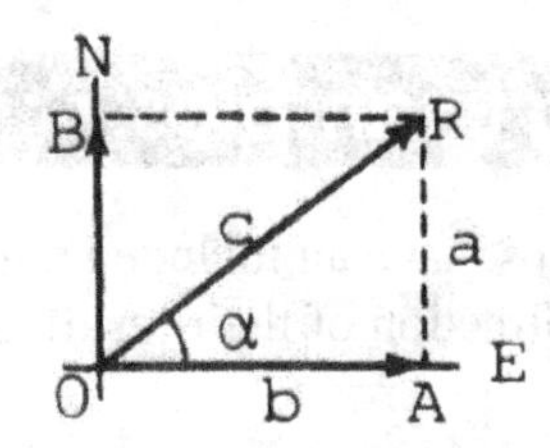

Construct the figure shown.

OA = b = 112 lbs.
OB = 88 lbs. = RA = a.

In ΔOAR: a = 88; b = 112.		
$\frac{a}{b} = \tan\alpha.$	$\log a = 1.94448$ $\log b = 2.04922$ $\log\tan\alpha = 9.89526 - 10$	$\alpha = 38°9'25''$
$\frac{a}{c}\sin\alpha$, or $c = \frac{a}{\sin\alpha}$	$\log a = 11.94448 - 10$ $\log\sin\alpha = 9.79086 - 10$ $\log c = 2.15362$	$c = 142.44$
Therefore the resultant is 142.44 lbs. and its direction is 38°9'25″ north of east.		

13.4 Limits and Continuity

A vector function is a function that associates a unique element of a set of vectors with each element of a set of real numbers. The vector function is represented by v or $\vec{v}$.

If v is a vector function, then for each real number n in the domain of v, there exists a unique vector v(t) $\leq$ <x,y,z>.

A curve in three-dimensional space is a set of ordered triples of the form <f(n),g(n),h(n)> where the functions f, g and h are continuous on the interval I.

13.4.1 Limit of a Vector Function

Definition:

If $V(n) = \langle f(n), g(n), h(n) \rangle$, then

$$\lim_{n \to a} v(n) = \langle \lim_{n \to a} f(n), \lim_{n \to a} g(n), \lim_{n \to a} h(n) \rangle$$

provided that f, g and h have limits as n approaches a.

Definition:

A vector function v is continuous at a if

$$\lim_{n \to a} v(n) = v(a).$$

13.5 Differentiation (Velocity, Acceleration and Arc Length)

Definition:

If v is a vector function then we define the derivative v' as

$$v'(t) = \lim_{h \to 0} \frac{v(t+h) - v(t)}{h}$$

whenever the limit exists. If v'(t) exists, then we say that the function v is differentiable at t.

We may also denote the derivative as

$$v'(t) = D_t v(t) = \frac{d}{dt} v(t)$$

Theorem:

Let v and w represent differentiable vector functions and let c represent a scalar.

1) $D_t[v(t)+w(t)] = v'(t) + w'(t)$

2) $D_t[cv(t)] = cv'(t)$

3) $D_t[v(t) \cdot w(t)] = v(t) \cdot w'(t) + w(t) \cdot v'(t)$

4) $D_t[v(t) \times w(t)] = v(t) \times w'(t) + v'(t) \times w(t)$

5) $D_t[\,|v(t)|\,] = \dfrac{1}{|v(t)|}[v(t) \cdot v'(t)]$

(In all cases, $v(t) \neq 0$.)

The velocity vector $v(t_0)$ is a tangent vector to a curve in a plane.

The curve is represented parametrically as $x = x(t)$ at time $t = t_0$.

At any instant of time the velocity vector points in the direction of motion.

Thus,

$$v(t_0) = x'(t_0) = \lim_{t \to t_0} \frac{x(t) - x(t_0)}{t - t_0}$$

The speed at $t = t_0$ (the magnitude of the velocity) is represented by the formula

$$|v(t_0)| = \sqrt{x'_1(t_0)^2 + x'_2(t_0)^2 + \ldots + x^1_n(t_0)^2}$$

Definition:

The acceleration of a particle at time t_0 is

$$a(t_0) = \lim_{t \to t_0} \frac{v(t) - v(t_0)}{t - t_0} = v'(t_0) = x''(t_0)$$

Definition:

The arc length of a curve between t = a and t = b is

$$L = \int_a^b |x'(t)|\, dt$$

Speed measures the rate of change in arc length with respect to time.

13.6 Curvatures, Tangential and Normal Components

The unit tangent vector to a curve x(t) is defined as

$$T(t) = \frac{1}{|x'(t)|} x'(t) \qquad \text{if } x'(t) \neq 0$$

Definition:

The curvature k of a curve f(t) is defined as the magnitude of the rate of change of the direction of the curve with respect to arc length.

$$k = \left| \frac{dT(t)}{ds} \right|_{t = t_0} = \left| \frac{T'(t)}{f'(t)} \right|$$

where T(t) is the limit tangent vector.

f(t) is the equation of the curve given parametrically.

If $T' \neq 0$, then the unit normal vector to the curve is given by the formula

$$N = \frac{1}{|T'|} T'$$

The radius of a curvature is $R(t) = \frac{1}{k}$. The osculating plane is the plane k which contains the tangent line to the path and the center of curvature at t. This quantity is defined by only those values of t for which $k \neq 0$.

Thus,

$$N_1 = \frac{1}{k} \frac{dT}{ds}$$

The acceleration vector can be resolved into two components, the tangent to the curve and the normal to the curve.

$$a_T = \frac{d^2s}{dt^2} T$$

This formula expresses the acceleration in terms of its tangential component.

$$a_N = \left(\frac{ds}{dt}\right)^2 kN$$

This is the expression for the normal component of the acceleration.

Thus, the formula that expresses the acceleration in terms of its tangential and normal component is

$$a = \frac{d^2s}{dt^2} T + \left(\frac{ds}{dt}\right)^2 kN$$

13.7 Kepler's Laws

Kepler's laws were formulated to describe the motion of planets about the sun.

13.7.1 First Law

The orbit of each planet is an ellipse with the sun at one focus.

13.7.2 Second Law

The vector from the sun to a moving planet sweeps over an area at a constant rate.

13.7.3 Third Law

If the time required for a planet to travel once around its elliptical orbit is T, and if the major axis of ellipse is 2a, then $T^2 = ka^3$ for some constant k.

Kepler's laws can be proven by the use of vector techniques since the force of gravity which the sun exerts on a planet exceeds that exerted by other celestial bodies. The orbit of a planet is a plane curve.

CHAPTER 14

Real Valued Functions

14.1 Open and Closed Sets

Boundary points: A point (a,b) in two-dimensional space is a boundary point of region R if every possible circle with center (a,b) contains both points which are in R and points which are not in R.

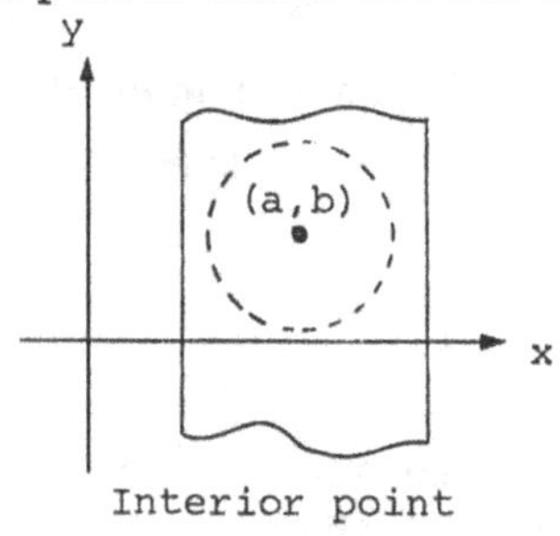

Interior point

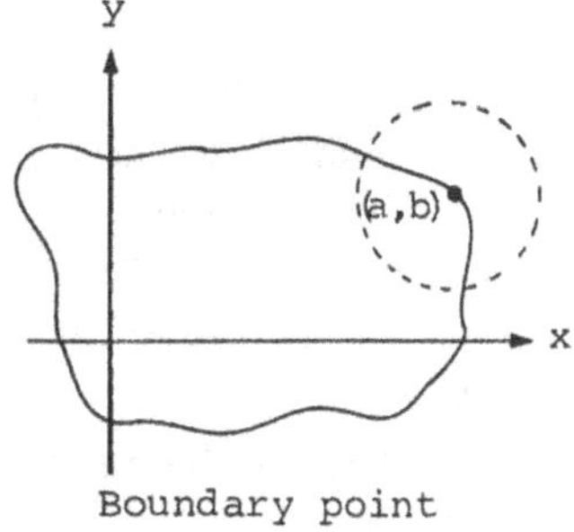

Boundary point

Interior Points: A point is an interior point of region R in two-dimensional space. If there exists a circle which has the point (a,b) as its center and which contains only points in the region R.

Open set: An open set is a set in which none of its points are boundary points.

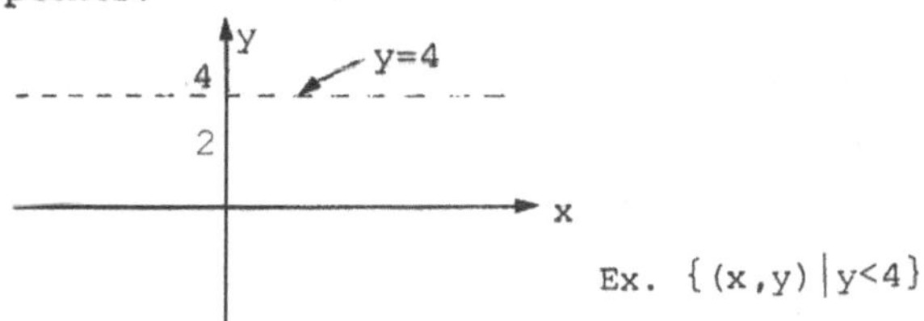

Ex. $\{(x,y) | y<4\}$

Closed set: A closed set includes all of its boundary points.

Ex. $\{(x,y) \mid x^2+y^2 \leq 4\}$

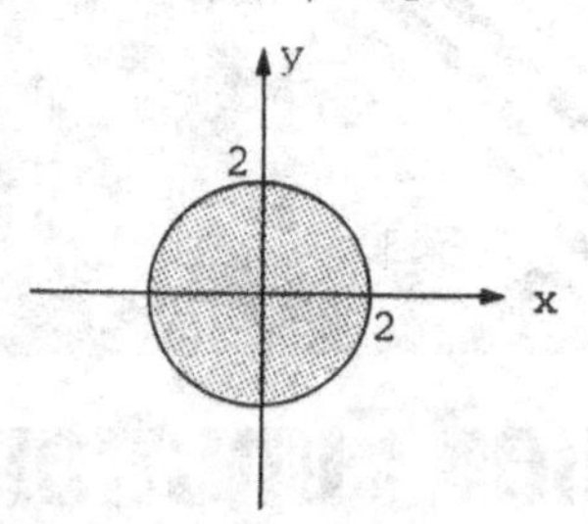

Problem Solving Examples:

Q Give an example of a subset C of R which contains no open interval and is not dense in any interval containing itself, but yet is uncountable.

A It is known that a subset of R containing an open interval is uncountable, so such an example is somewhat unusual. To construct such an example, let

$$C_0 = [0, 1] .$$

Obtain C_1 by deleting the middle third of this interval. Obtain each successive C_n by deleting the middle thirds of all intervals in C_{n-1}. Thus,

$$C_1 = [0, \frac{1}{3}] \cup [\frac{2}{3}, 1]$$

$$C_2 = [0, \frac{1}{9}] \cup [\frac{2}{9}, \frac{1}{3}] \cup [\frac{2}{3}, \frac{7}{9}] \cup [\frac{8}{9}, 1] \ldots .$$

Let the Cantor set

$$C = \bigcap_{n=1}^{\infty} C_n .$$

The Cantor set can be described by using ternary (base 3) notation. Let

$$x = \sum_{k=1}^{\infty} \frac{x_k}{3^k} = .x_1 x_2 \ldots \qquad \text{(base 3)},$$

where

$$x_k = 0, 1, \text{or } 2.$$

Then for all x in C_1, $x_1 = 0$ or 2. [Note, for example, that

$$\frac{1}{3} = .1000 \ldots = .0222 \ldots \text{(base 3)}.]$$

For all x in C_2, $x_1, x_2 = 0$ or 2. In general, for all x in C_n, $x_m = 0$ or 2 for m = 1,2, ... n. Therefore, C consists of all $x \in [0, 1]$ whose ternary expansion contains only zeroes and twos. Define

$$f : C \rightarrow C_0 = [0, 1] \text{ by}$$

$$f(x) \equiv f(.x_1\, x_2\, x_3 \ldots) = .y_1\, y_2\, y_3 \ldots$$

$$= \sum_{k=1}^{\infty} \frac{y_k}{2^k}, \quad \text{(base 2)}$$

where

$$y_k = \frac{1}{2} x_k \; (k = 1,2,3,\ldots).$$

This function is subjective (i.e., this function is onto) and therefore, the Cantor set is uncountable.

Since C is the complement of an open set (the union of the deleted intervals), it is closed. From the ternary representation of C it is clear that if

$$x = \sum_{1}^{\infty} \frac{x_k}{3^k} \quad (x_k = 0 \text{ or } 2)$$

is in C and if $\varepsilon > 0$ is given, then the point

$$y = \sum_{k=1}^{n} \frac{x_k}{3^k} + \frac{2}{3^m}$$

(where m is such that

$$\frac{2}{3^m} < \frac{\varepsilon}{2}$$

and n is such that

$$\sum_{k=n+1}^{\infty} \frac{x_k}{3^k} \le 2 \sum_{k=n+1}^{\infty} \frac{1}{3^k} < \frac{\varepsilon}{2})$$

is in C and

$$|y - x| < \varepsilon .$$

This means that every point of C is an accumulation point. Such a set is called a perfect set: namely a closed set for which every point is an accumulation point. Note that C cannot contain any open interval because any open interval in [0, 1] will contain a deleted interval, which is not in C. The same reasoning shows that if I is any interval contained in [0, 1] and which contains C, then C cannot be dense in I. For density would imply that a deleted point is a limit point of C and must be therefore in C which is false, since by definition a set E is dense in X, (X is a metric space) if every point of X is a limit point of E, or a point of E (or both). Hence, since a neighborhood around $\frac{1}{2}$ which is contained in [0, 1] contains no point of C, it is not a limit point of C.

Q Let X be a topographical space and let C and U be subsets of X. Define C to be closed if C contains all its limit points and define U to be open if every point $p \in U$ has a neighborhood which is contained in U. Assuming these definitions show that the following statements are equivalent for a subset S of X,

i) S is closed in X;
ii) X – S is open in X;
iii) $S = \overline{S}$.

(i) ⇒ (ii) : From i) follows that S contains all its limit points. Hence, any neighborhood of such a limit point contains a point of S other than itself. Now suppose X – S is not open. Then there exists some

$$x \in X - S$$

such that every neighborhood of x contains a $y \neq x$ such that

$$y \in X - (x - S) = S.$$

But then x is a limit point of S and therefore $x \in S$. So, X – S is open since x cannot belong to both S and X – S.

(ii) ⇒ (iii): By definition $\overline{S} = S \cup S'$ where S′ is the set of all limit points of S. Evidently,

$$S \subseteq \overline{S}$$

for any set S. Let x be any limit point of S (i.e., $x \in S'$). Since any neighborhood of x contains a different point of S and X – S is open, $x \in S$. Hence, $S' \leq S$ and therefore $\overline{S} \subseteq S$. Thus $S = \overline{S}$.

(iii) ⇒ (i): This is obvious, since by (iii) S contains all of its limit points so S is closed.

Note that $S' \subseteq S$ or even $\partial S \subseteq S$, where ∂S is the boundary of S, is equivalent to the conditions above. [$S' \subseteq \partial S$ since ∂S is the set of points whose neighborhoods have non-empty intersections with both S and X – S.] Therefore, $\overline{S}$ can be thought of as the smallest closed set containing S.

14.2 Limits and Continuity

The limit of f(x,y) as (x,y) approaches (a,b) is L. This is written:

$$\lim_{(x,y) \to (a,b)} f(x,y) = L$$

It means that for every $\varepsilon > 0$ there corresponds a $\delta > 0$ such that if $0 < \sqrt{(x-a)^2+(y-b)^2} < \delta$, then $|f(x,y)-L| < \varepsilon$.

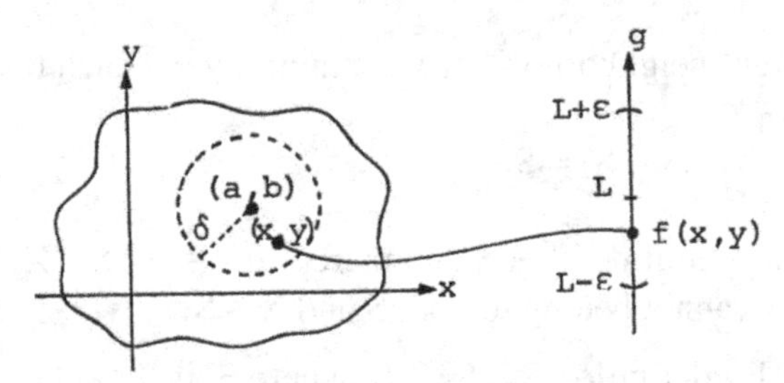

The preceding figures illustrate that, given that the definition of a limit is true, there is a circle of radius $\delta > 0$, such that for every point (x,y) inside the circle with radius δ and center (a,b), the number corresponding to f(x,y) is in the interval $(L-\varepsilon, L+\varepsilon)$.

If the limiting values obtained by taking two different paths to a point M(a,b) are different, then the limit of the function as (x,y) approaches (a,b) does not exist.

Problem Solving Examples:

Show that $\lim_{(x,y) \to (0,0)} \frac{2x^3 - y^3}{x^2 + y^2} = 0$.

A Let f: $R^2 \to R$ be a real-valued function from the plane to the real line. To show that f is continuous at a given point $(a_1\ a_2)$ in the plane it suffices to show:

1) $\lim_{(x,y) \to (a_1,a_2)} f(x,y)$ exists

2) $$\lim_{(x,y) \to (a_1,a_2)} f(x,y) = f(a_1,a_2)$$

If only the first condition is satisfied, f is said to have a removable discontinuity or discontinuity of the first kind. If neither condition is satisfied f has an essential discontinuity or discontinuity of the second kind.

In the given problem it is required to show that f approaches 0 as $(x,y) \to (0,0)$. This means that if $\varepsilon > 0$ is given, there exists a $\delta > 0$ (depending on ε) such that

$$\left| \frac{2x^3 - y^3}{x^2 + y^2} \right| < \varepsilon \text{ if } 0 < (x^2 + y^2)^{\frac{1}{2}} < \delta \qquad (1)$$

(Thus x,y are in an open disk of radius δ, with center at the origin; this point is deleted since what happens at (0,0) is not of concern here.)

Now

$$|2x^3 - y^3| \leq 2|x|^3 + |y|^3, (|a - b| \leq |a| + |b|)$$

$$2|x|^3 + |y|^3 = 2|x|x^2 + |y|y^2.$$

Now

$$|x| \leq (x^2 + y^2)^{\frac{1}{2}}, |y| \leq (x^2 + y^2)^{\frac{1}{2}}$$

and thus

$$|2x^3 - y^3| \leq (x^2 + y^2)^{\frac{1}{2}} x^2 + (x^2 + y^2)^{\frac{1}{2}} y^2$$

or,

$$|2x^3 - y^3| \leq (x^2 + y^2)^{\frac{1}{2}} [2x^2 + y^2] \leq 2(x^2 + y^2)^{\frac{3}{2}}$$

But

$$|2x^3 - y^3| \leq 2(x^2 + y^2)^{\frac{3}{2}}$$

implies $$\left|\frac{2x^3 - y^3}{x^2 + y^2}\right| \le 2(x^2 + y^2)^{\frac{1}{2}} \text{ if } 0 < x^2 + y^2 . \qquad (2)$$

Let $\varepsilon > 0$ given Choose $\delta = \frac{\varepsilon}{2}$. Then comparing (1) and (2), observe that (1) will hold for all positive ε. Thus

$$\lim_{(x,y) \to (0,0)} \frac{2x^3 - y^3}{x^2 + y^2} = 0.$$

14.2.1 Definition of Continuity

$$\lim_{(x,y)\to(a,b)} f(x,y) = f(a,b)$$

for f continuous at (a,b)

If a function f of two variables is continuous at (a,b) and a function g of one variable is continuous at f(a,b), then the function h, defined by h(x,y) = g(f(x,y)) is continuous at (a,b).

Problem Solving Examples:

Investigate the continuity of the expression:

$$y = \frac{x^2 - 9}{x - 3} \text{ at } x = 2.$$

For the function to be continuous at a point, in this case, 2, it must satisfy three conditions: (1) f(2) is defined, (2) $\lim_{x \to 2} f(x)$ exists, (3) $\lim_{x \to 2} f(x) = f(2)$.

For $x = 2$, $y = f(x) = f(2) = \frac{(2)^2 - 9}{2 - 3} = 5$.

Also,

$$\lim_{x \to 2} \frac{x^2 - 9}{x - 3} = 5 .$$

Therefore, the function is continuous at x = 2.

Let $f:R^2 \to R$ be given by $f(x,y) = x^2 + y^2$. Show that f is continuous at (0,0)

A real-valued function

$$f:R^2 \to R$$

is continuous at point (a_1, a_2) in its domain if

1) $\lim_{(x,y) \to (a_1,a_2)} f(x,y)$ exists

2) $\lim_{(x,y) \to (a_1,a_2)} f(x,y) = f(a_1,a_2)$

The definitions of the limit of a function of two variables is as follows: A function f(x,y) approaches a limit A as x approaches a and y approaches b,

$$\lim_{\substack{x \to a \\ y \to b}} f(x,y) = A$$

if, and only if, for each $\varepsilon > 0$, there is another $\delta > 0$ such that whenever $|x-a| < \delta$, $|y-b| < \delta$ and $0 < (x-a)^2 + (y-b)^2$, then $|f(x,y) - A| < \varepsilon$.

This means that when (x,y) is at any point inside a certain square with center at (a,b) and width 2δ (except at the center), f(x,y) differs from A by less than ε.

The function whose limit is to be evaluated as (x,y) approaches the origin is $f(x,y) = x^2 + y^2$. Let $\varepsilon > 0$ be given and choose $\delta = \sqrt{\frac{\varepsilon}{2}}$.

Then the inequalities $|x| < \sqrt{\frac{\varepsilon}{2}}$, $|y| < \sqrt{\frac{\varepsilon}{2}}$ imply $(x^2 + y^2) < \varepsilon$.
Hence,

$$\lim_{\substack{x \to 0 \\ y \to 0}} (x^2 + y^2) = 0$$

and condition 1) is satisfied.

For condition 2) note that

$$f(0,0) = 0^2 + 0^2 = 0$$

and thus f is continuous at (0,0).

14.3 Graphing

Definition:

A cylinder is the set of all points on all lines which intersect a curve C in a plane and are parallel to a line L that is not in the plane.

Definition:

A surface of revolution is the surface which results from the revolution of a plane curve about a line in the plane.

Example: When a circle is revolved about a line along a diameter of the circle a sphere results.

If a parabola is revolved about its principal axis, the resulting surface is a paraboloid.

If a hyperbola is revolved about its transverse axis, the resulting surface is a hyperboloid of one sheet. If it is revolved about its conjugate axis, a hyperboloid of two sheets results.

When an ellipse is revolved about its major or minor axis the resulting surface is an ellipsoid.

Level Curves: Graphical representation of curves of the form $f(x,y) = k$, where k is a constant.

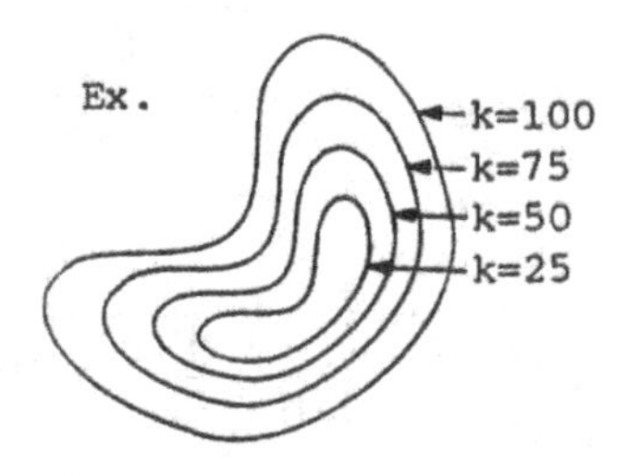

Level Surfaces: Graphical representation of curves of the form $f(x,y,z) = c$, where c is a constant.

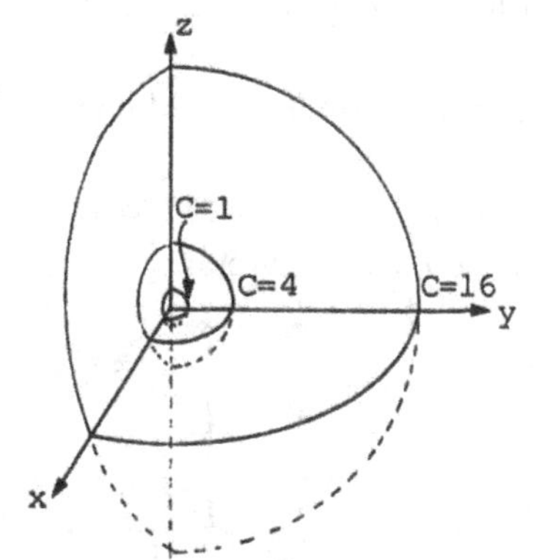

For the function $f(x,y,z,)=x^2+y^2+z^2=$ a constant C. As C becomes larger, we can picture an enlarging sphere.

Problem Solving Examples:

Determine the continuity of the following function and sketch its graph:

$$M(x) = \frac{|x|}{x}(x^2 - 1).$$

For $x > 0$, $|x| = x$ and for $x < 0$, $|x| = -x$. Therefore for $x > 0$

$$M(x) = \frac{x}{x}(x^2 - 1) = x^2 - 1 \qquad (1)$$

for $x < 0$

$$M(x) = \frac{-x}{x}(x^2 - 1) = -(x^2 - 1) = 1 - x^2 \qquad (2)$$

for $x = 0$

$$M(x)\Big|_{x=0} = \frac{x}{x}(x^2 - 1)\Big|_{x=0} = \frac{0}{0}, \text{ undefined} \qquad (3)$$

To test the continuity of the function at $x = 0$, we might test whether

$$\lim_{x \to 0^+} M(x) = \lim_{x \to 0^-} M(x) = M(0).$$

It turns out that the former $= -1$, the latter $= +1$.

However, from (3), we know that $M(0)$ is not even defined, therefore there is no need to calculate the limits, $M(x)$ is not continuous at $x = 0$.

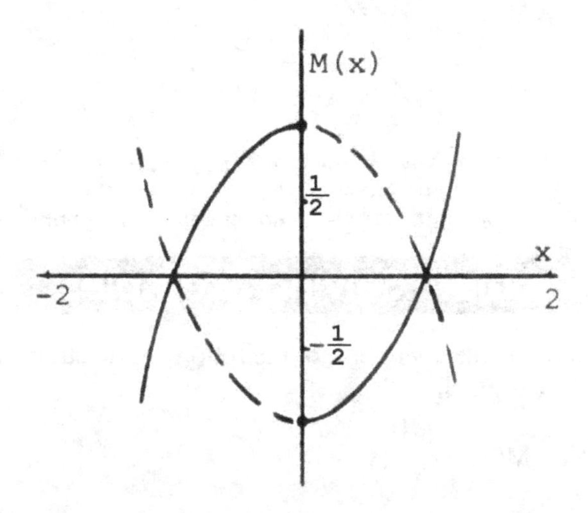

14.4 Quadric Surfaces

Definition:

In three-dimensional analytic geometry, a quadric surface is the graph of a second-degree equation in x, y, and z.

The graph of

$$\frac{x^2}{a^2} + \frac{y^2}{b^2} + \frac{z^2}{c^2} = 1$$

is an ellipsoid where a, b, and c are positive real numbers.

The graph of

$$\frac{x^2}{a^2} + \frac{y^2}{b^2} - \frac{z^2}{c^2} = 1$$

is a hyperboloid of one sheet.

The graph of

$$\frac{x^2}{a^2} - \frac{y^2}{b^2} - \frac{z^2}{c^2} = 1$$

is a hyperboloid of two sheets.

The graph of

$$\frac{x^2}{a^2} + \frac{y^2}{b^2} - \frac{z^2}{c^2} = 0$$

is a cone with z-axis as its axis.

The graph of

$$\frac{x^2}{a^2} + \frac{y^2}{b^2} = cz$$

is a paraboloid with the z-axis as its axis.

If $c > 0$, the paraboloid opens upward.

If $c < 0$, the paraboloid opens downward.

Graphs of equations of the form

$$\frac{x^2}{a^2} + \frac{z^2}{b^2} = cy \quad \text{and} \quad \frac{y^2}{a^2} + \frac{z^2}{b^2} = cx$$

are paraboloids whose axes are the y- and x-axis, respectively.

The graph of

$$\frac{y^2}{a^2} - \frac{x^2}{b^2} = cz$$

is a hyperbolic paraboloid.

Diagrams for the graphs of the above given equations can be found in any standard text covering the relevant material.

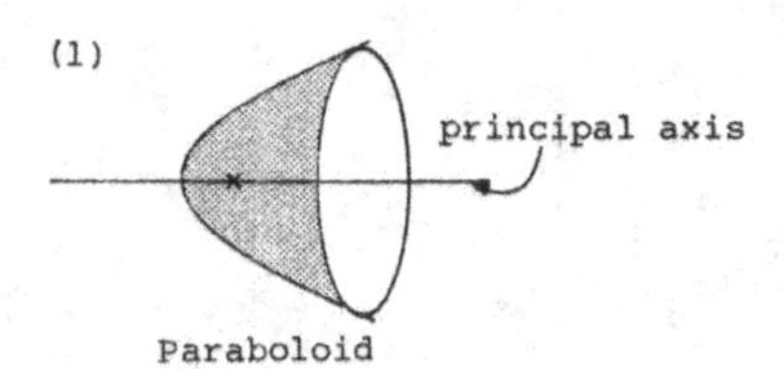

Paraboloid

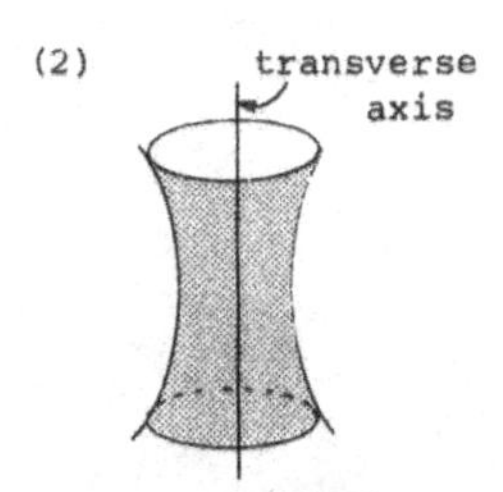

Hyperboloid of one sheet

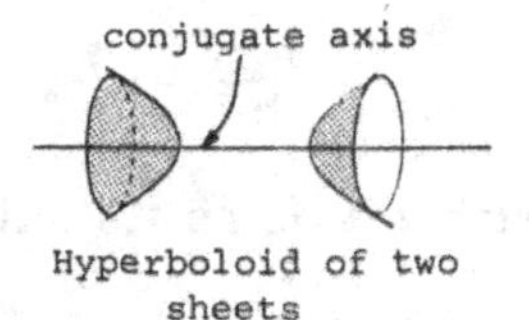

Hyperboloid of two sheets

(3) major axis

Ellipsoid

Quiz: Vector Analysis and Real Valued Functions

1. A particle moves in the xy - plane so that at any time, t, its coordinates are $x = t^3 - t^2$ and $y = t^4 - 5t^2$. At $t = 1$ its acceleration vector is:

   (A) $(0, 4)$

   (B) $(0, 2)$

   (C) $(2, 4)$

   (D) $(-4, -2)$

   (E) $(4, 2)$

2. A particle moves in the plane according to $x = t \cos t$, $y = t \sin t$. Which of the following vectors is orthogonal to the acceleration vector at $t = \pi$?

   (A) $(1, \pi)$

   (B) $(2, -\pi)$

   (C) $(2, \pi)$

   (D) $(\pi, 2)$

   (E) $(\pi, 1)$

3. A particle moves along the x -axis so that its position at time t is given by $x(t) = \left(1+t^2\right) e^{-t}$. For what values of t will the particle be at rest.

   (A) No values

   (B) 0 only

   (C) $\frac{1}{2}$ only

(D) 1 only

(E) 2 only

4. A particle moves in the xy-plane so that, at any time t its coordinates are $x = \dfrac{t^3 - 2t^2}{4}$ and $y = t^2 - t$. At $t = 2$, its acceleration vector is

(A) $(0, 0)$

(B) $(1, 2)$

(C) $(2, 0)$

(D) $(2, 2)$

(E) $(-2, -2)$

5. The maximum value of the directional derivative on the surface $z = f(x, y) = xe^{xy} + y \cos x$ at $P(0, 1)$ is

(A) 1

(B) $\sqrt{2}$

(C) $\sqrt{3}$

(D) $\sqrt{4}$

(E) $\sqrt{5}$

6. The cross product $\vec{u} \times \vec{v}$ of the vectors

$$\vec{u} = 2\vec{i} - \vec{j} + 3\vec{k}$$
$$\vec{v} = \vec{i} - 2\vec{j} + \vec{k}$$

is given by

(A) $7\vec{i} + \vec{j} + 3\vec{k}$

(B) $5\vec{i} + 5\vec{j} + 3\vec{k}$

(C) -3

(D) $5\vec{i} + \vec{j} - 3\vec{k}$

(E) $7\vec{i} + \vec{j}$

7. Given two vectors $\vec{U} = 2\vec{i} - 3\vec{j} + 5\vec{k}$, $\vec{V} = -\vec{i} + 4\vec{j} + 2\vec{k}$ then their vector product $\vec{U} \times \vec{V} =$

(A) $\vec{i} + \vec{j} + 7\vec{k}$

(B) $-2\vec{i} - 12\vec{j} + 10\vec{k}$

(C) $-26\vec{i} - 9\vec{j} + 5\vec{k}$

(D) $3\vec{i} + 4\vec{j} + 5\vec{k}$

(E) $3\vec{i} - 4\vec{j} - 5\vec{k}$

8. In an $x - y$ plane, if $\vec{a} = (x, y), \vec{b} = (x', y')$, then their scalar product $\vec{a} \bullet \vec{b} =$

(A) $xx' + yy'$

(B) $xx' - yy'$

(C) $xy - x'y'$

(D) $xy' + x'y$

(E) $xy' - x'y$

9. The equation of the tangent line to the graph of the equation

$$\begin{cases} x = t^3 - 4 \\ y = 2t^2 + 1 \end{cases} \text{ at } t = 2 \text{ is}$$

(A) $2x - 3y - 19 = 0$

(B) $2x - 3y + 19 = 0$

(C) $3x - 2y - 6 = 0$

(D) $3x - 2y + 6 = 0$

(E) $3x + 2y - 6 = 0$

10. Suppose $\vec{a}$ and $\vec{b}$ are vectors in R^2 and further, that $\vec{a}$ and $\vec{b}$ are linearly independent.

Let $\vec{c} = \dfrac{\vec{a} \bullet \vec{b}}{\|\vec{a}\|} \vec{a} - \vec{b}$.

(A) $\vec{a}$ and $\vec{c}$ are linearly dependent.

(B) $\vec{a}$ and $\vec{c}$ are orthogonal.

(C) $\vec{b}$ and $\vec{c}$ are linearly dependent.

(D) $\vec{b}$ and $\vec{c}$ are orthogonal.

(E) $\{\vec{a}, \vec{b}, \vec{c}\}$ form a linearly independent set.

ANSWER KEY

1. (D)
2. (C)
3. (D)
4. (D)
5. (B)
6. (D)
7. (C)
8. (A)
9. (B)
10. (B)

CHAPTER 15

Partial Differentiation

15.1 Limits and Continuity

A function is a mapping which takes each element of the domain into one and only one element of the range.

A function is a function of two variables if the domain consists of ordered pairs of numbers. A function of two variables is a polynomial function if $f(x,y)$ can be expressed as a sum of terms $cx^a y^b$, where c is a real number and a and b are non-negative integers.

Definition:

Let $f: x \to R$ represent a function of n variables and let a represent a point in the domain of f. Then f is continuous at point a if

$$\lim_{x \to a} f(x) = f(a).$$

15.2 Partial Derivatives

Definition:

Let f represent a function of two variables. The first partial derivative of f with respect to x and y are the functions f_x and f_y, respectively. These functions are defined as

$$f_x(x,y) = \lim_{h \to 0} \frac{f(x+h,y) - f(x,y)}{h}$$

and

$$f_y(x,y) = \lim_{h \to 0} \frac{f(x,y+h) - f(x,y)}{h}$$

This holds true provided that the limits exist.

The two first partial derivatives are found in the manner prescribed below:

a) The derivative $f_x(x,y)$ is found by differentiating $f(x,y)$ with respect to x in the usual manner, while y is considered to be constant.

b) The derivative $f_y(x,y)$ is found by differentiating $f(x,y)$ with respect to y, keeping x constant.

15.2.1 Notation of Partial Derivatives

1. $f_x = \frac{\partial f}{\partial x} = D_x f$ 2. $f_y = \frac{\partial f}{\partial y} = D_y f$

The derivative $f_x(x,y)$ is the measure of the rate of change of the function $f(x,y)$ as (x,y) moves in the horizontal direction.

The derivative $f_y(x,y)$ is the measure of the rate of change of the function $f(x,y)$ as (x,y) moves in the vertical direction.

The first partial derivative of functions of three or more variables is defined in the same manner except that all variables, except one, are held constant and differentiation takes place with respect to the remaining variable.

15.2.2 Higher-Order Partial Derivatives

The notations for the second partial derivatives are as follows:

1) $\frac{\partial}{\partial x} f_x = (f_x)_x = f_{xx} = \frac{\partial}{\partial x}\left(\frac{\partial f}{\partial x}\right) = \frac{\partial^2 f}{\partial x^2}$

2) $\frac{\partial}{\partial y} f_x = (f_x)_y = f_{xy} = \frac{\partial}{\partial y}\left(\frac{\partial f}{\partial x}\right) = \frac{\partial^2 f}{\partial y \partial x}$

3) $\frac{\partial}{\partial x} f_y = (f_y)_x = f_{yx} = \frac{\partial}{\partial x}\left(\frac{\partial f}{\partial y}\right) = \frac{\partial^2 f}{\partial x \partial y}$

4) $\frac{\partial}{\partial y}(f_y) = (f_y)_y = f_{yy} = \frac{\partial}{\partial y}\left(\frac{\partial f}{\partial y}\right) = \frac{\partial^2 f}{\partial y^2}$

The notations for the third partial derivatives are as follows:

1. f_{xxx} 2. f_{xxy} 3. f_{xyy} 4. f_{yyy}

5. f_{yyx} 6. f_{yxx} 7. f_{xyx} 8. f_{yxy}

The symbol f_{xyy} means that the order of the partial derivative is taken from left to right. The derivative with respect to x is taken first, then the derivative with respect to y is taken twice.

Problem Solving Examples:

Let u = u(x,y) be implicitly defined as a function of x and y by the equation $u + \ell n\, u = xy$. Find

$$\frac{\partial u}{\partial x}, \frac{\partial u}{\partial y}, \frac{\partial^2 u}{\partial x \partial y} \text{ and } \frac{\partial^2 u}{\partial y \partial x}.$$

A

The partial derivative of a function is defined as follows: Let z = f(x,y) be defined in a domain D of the xy-plane and let (x_1,y_1) be a point of D. Then $f(x,y_1)$ is a function depending only on x and if its derivative at the point x_1 exists, it is called the partial derivative of f with respect to x at (x_1,y_1) and is denoted by

$$\frac{\partial f}{\partial x}(x_1,y_1) \quad \text{or} \quad \left.\frac{\partial z}{\partial x}\right|_{(x_1,y_1)}.$$

If the point (x_1,y_1) is now allowed to vary, one obtains a new function of x and y (wherever the derivative exists) denoted by

$$\frac{\partial f}{\partial x}(x,y) = f_x(x,y) = \frac{\partial z(x,y)}{\partial x}.$$

Apparently, then, $\frac{\partial f}{\partial x}(x,y)$ may be obtained by simply treating y as a constant and differentiating f with respect to its only remaining variable, x. In the case at hand, u is not defined explicitly as a function of x and y but the partial derivatives may be obtained by differentiating both sides of the defining equation with respect to x or y recalling that u depends on both x and y. Thus

$$\frac{\partial u(x,y)}{\partial x} + \frac{\partial \ell n[u(x,y)]}{\partial x} = \frac{\partial(xy)}{\partial x}. \qquad (1)$$

Using the chain rule of single variable calculus, (1) becomes

$$\frac{\partial u}{\partial x} + \frac{d\,[\ell n\, u]}{dx}\,\frac{\partial u}{\partial x} = y \qquad (2)$$

or $u_x + \frac{1}{u} u_x = y$. Therefore, $u_x\left(\frac{u+1}{u}\right) = y$, so that,

$$u_x = \frac{\partial u}{\partial x}\;\frac{uy}{u+1} = \frac{u(x,y)\cdot y}{u(x,y)+1}. \qquad (3)$$

Also from the defining equation,

$$\frac{\partial u(x,y)}{\partial y} + \frac{\partial \ell n[u(x,y)]}{\partial y} = \frac{\partial(x,y)}{\partial y},$$

and the single variable chain rule can be used again to obtain

$$u_y + \frac{1}{u} u_y = x,$$

so that

$$u_y = \frac{\partial u}{\partial y} \frac{ux}{u+1} = \frac{u(x,y)\cdot x}{u(x,y)+1} . \tag{4}$$

The second partial derivative $\frac{\partial^2 u}{\partial x \partial y}$ is defined as $\frac{\partial(u_y)}{\partial x}$; that is, the partial derivative of the new function $\frac{\partial u}{\partial y}$ with respect to x. Thus, from (4), one obtains

$$\frac{\partial^2 u}{\partial x \partial y} = \frac{\partial}{\partial x}\left[\frac{ux}{u+1}\right] = \frac{u}{u+1} + x\frac{\partial}{\partial x}\left[\frac{u(x,y)}{u(x,y)+1}\right]$$

$$= \frac{u}{u+1} + x\frac{d}{du}\left[\frac{u}{u+1}\right] \cdot u_x$$

$$= \frac{u}{u+1} + \frac{x}{(u+1)^2} \cdot u_x = \frac{u}{u+1} + \frac{uxy}{(u+1)^3} . \tag{5}$$

Similarly,

$$\frac{\partial^2 u}{\partial y \partial x} = \frac{\partial}{\partial y}\left[\frac{uy}{u+1}\right] = \frac{u}{u+1} + y\frac{\partial}{\partial y}\left[\frac{u(x,y)}{u(x,y)+1}\right]$$

$$= \frac{u}{u+1} + y\frac{d}{du}\left[\frac{u}{u+1}\right] \cdot u_y$$

$$= \frac{u}{u+1} + \frac{y}{(u+1)^2} \cdot u_y = \frac{u}{u+1} + \frac{uxy}{(u+1)^3} . \tag{6}$$

It can be seen from (5) and (6) that $\frac{\partial^2 u}{\partial x \partial y} = \frac{\partial^2 u}{\partial y \partial x}$. This is a relation that can be proved to be true for all continuous functions u(x,y).

Find z_{xy} and z_{yx} from the expression:

$$z = x^2y + 2xe^{\frac{1}{y}} \text{ and show that } z_{xy} = z_{yx}$$

The first step of finding second partial derivatives is finding the first partial derivatives.

$$z_x = 2xy + 2e^{\frac{1}{y}}$$

and

$$z_y = x^2 - \frac{2xe^{\frac{1}{y}}}{y^2}.$$

To find z_{xy}, we differentiate z_y with respect to x.

$$z_{xy} = 2x - \frac{2e^{\frac{1}{y}}}{y^2}.$$

To find z_{yx} we differentiate z_x with respect to y, and obtain:

$$z_{yx} = 2x - \frac{2e^{\frac{1}{y}}}{y^2}.$$

Therefore $z_{yx} = z_{xy}$.

15.3 Increments and Differentials

The increment of x and y is denoted by Δx and Δy, respectively, if f is a function of the two variables x and y.

Definition:

Notations of the increments:

$$f_x(x,y) = \lim_{\Delta x \to 0} \frac{f(x+\Delta x,y) - f(x,y)}{\Delta x}$$

$$f_y(x,y) = \lim_{\Delta y \to 0} \frac{f(x,y+\Delta y) - f(x,y)}{\Delta y}$$

If $z = f(x,y)$, then the differentials dx and dy of the independent variables x and y are defined as

$$dx = \Delta x \text{ and } dy = \Delta y.$$

The differential dz of the dependent variable z is defined as

$$dz = f_x(x,y)dx + f_y(x,y)dy$$

$$= \frac{\partial z}{\partial x} dx + \frac{\partial z}{\partial y} dy$$

If the first partial derivatives f_x and f_y are continuous on a rectangular region, then the function f is differentiable and therefore continuous.

Let us consider w, x and y independent variables and z, the dependent variable.

The differential of z is defined as

$$dz = \frac{\partial z}{\partial x} dx + \frac{\partial z}{\partial y} dy + \frac{\partial z}{\partial w} dw$$

Problem Solving Examples:

Q What is the increment in y, (Δy) and the differential of y,(dy) for an increment of $\Delta x = 1$ at $x = 2$, if $y = x^2$? For an increment of $\Delta x = 0.01$?

If $y = x^2$, $x = 2$, and $\Delta x = 1$, then

$$y + \Delta y = (x + \Delta x)^2,$$

so that $\Delta y = (x + \Delta x)^2 - y$, or,

$$\Delta y = (x + \Delta x)^2 - x^2 = (2 + 1)^2 - (2)^2 = 9 - 4 = 5.$$

Therefore the increment in y is 5. To obtain the differential of y, we differentiate and find: $dy = 2x\,dx = (2)(2)(1) = 4$.

If $y = x^2$, $x = 2$, and $\Delta x = 0.01$, then

$$\Delta y = (x + \Delta x)^2 - x^2 = (2.01)^2 - 2^2 = 0.0401 \text{, the increment,}$$

and

$$dy = 2x\,dx = (2)(2)(0.01) = 0.0400 \text{, the differential.}$$

In the second case, Δy is very close to dy, but in the first case the difference is very substantial, so that the relationship: $\Delta y \approx dy$, holds only when $\frac{\Delta x}{x}$ is small.

Q The side of a square, x, is increased by an amount $dx = \Delta x$. See accompanying figure. Analyze and interpret the increases in the area in terms of the increment of the area and the differential.

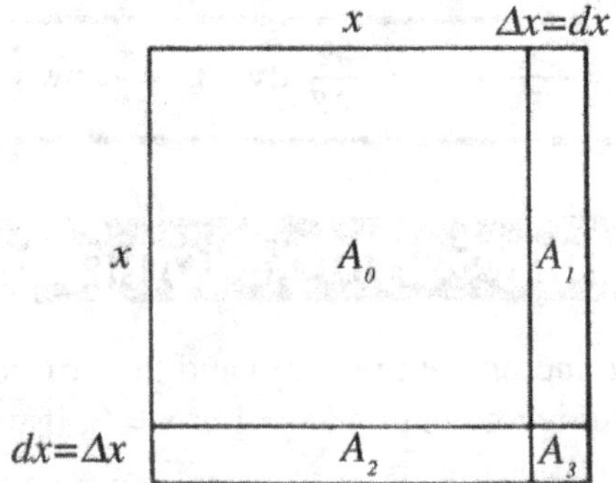

For the original area, we have $A_0 = x^2$. The enlarged area is:

$$A = (x + \Delta x)^2 = x^2 + 2x\Delta x + (\Delta x)^2 .$$

The term x^2 is, of course, A_0. The term $2x\Delta x$ corresponds to the two areas A_1 and A_2. The term $(\Delta x)^2$ corresponds to A_3. The increment, $A_1 + A_2 + A_3$, is equivalent to $2x\Delta x + (\Delta x)^2$; and the differential of the area, $dA = 2x\,dx$, corresponds to $A_1 + A_2$.

15.4 Application of the Chain Rule

Partial derivatives are computed by using the chain rule.

Let $z = f(x,y)$ represent a continuous function and let $\frac{\partial f}{\partial x}$ and $\frac{\partial f}{\partial y}$ also be continuous.

If $x = x(a,b)$ and $y = y(a,b)$ are functions of a and b so that $\frac{\partial x}{\partial a}, \frac{\partial x}{\partial b}, \frac{\partial y}{\partial a}, \frac{\partial y}{\partial b}$ all exist, then z is a function of a and b and the following are true:

$$\frac{\partial z}{\partial a} = \left(\frac{\partial f}{\partial x}\right)\left(\frac{\partial x}{\partial a}\right) + \left(\frac{\partial f}{\partial y}\right)\left(\frac{\partial y}{\partial a}\right)$$

$$\frac{\partial z}{\partial b} = \left(\frac{\partial f}{\partial y}\right)\left(\frac{\partial y}{\partial b}\right) + \left(\frac{\partial f}{\partial x}\right)\left(\frac{\partial x}{\partial b}\right)$$

In the chain rule, a and b are considered independent variables; x and y are the intermediate variables.

The chain rule may be remembered as the rule by which the derivative appears in a fractional form.

This is illustrated by

$$\frac{dy}{dx} = \frac{dy}{dr} \cdot \frac{dr}{dx}$$

Problem Solving Examples:

Find the derivative of: $y = (2x^3 - 5x^2 + 4)^5$.

$D_x = \frac{d}{dx}$. This problem can be solved by simply applying the theorem for $d(u^n)$. However, to illustrate the use of the chain rule, make the following substitutions:

$$y = u^5 \quad \text{where } u = 2x^3 - 5x^2 + 4$$

Therefore, from the chain rule,

$$D_x y = D_u y \cdot D_x u = 5u^4(6x^2 - 10x)$$

$$= 5(2x^3 - 5x^2 + 4)^4 (6x^2 - 10x).$$

Find the derivative of: $y = (x^2 + 2)^3$.

Method 1. We may expand the cube and write:

$$\frac{dy}{dx} = \frac{d}{dx}[(x^2+2)^3] = \frac{d}{dx}(x^6 + 6x^4 + 12x^2 + 8)$$

$$= 6x^5 + 24x^3 + 24x.$$

Method 2. Let $u = x^2 + 2$, then $y = (x^2 + 2)^3 = u^3$; Using the chain rule we have:

$$\frac{dy}{dx} = \frac{dy}{du} \cdot \frac{du}{dx} = \frac{d(u^3)}{du} \cdot \frac{d(x^2+2)}{dx} = 3u^2(2x)$$

$$= 3(x^2 + 2)^2 \cdot (2x) = 3(x^4 + 4x^2 + 4) \cdot (2x)$$

$$= 6x^5 + 24x^3 + 24x.$$

15.5 Directional Derivative and Gradients

15.5.1 Directional Derivative

Let $f(x,y)$ represent a function and $P(x,y)$ represent a point in the xy plane. A particular direction is found by specifying the angle θ, that the line through the point makes with the x-axis.

Definition:

If f is a function of x and y, and $\vec{a} = \cos\theta i + \sin\theta j$ is the unit vector, then the directional derivative of f in the direction of a, denoted D_{af}, is given by the formula

$$D_{af} = \lim_{h \to 0} \frac{f(x+h\cos\theta, y+h\sin\theta) - f(x,y)}{h}$$

whenever the limit exists.

When $\theta = 0$, then $\cos\theta = 1$, $\sin\theta = 0$ and the direction is the positive x-direction. The directional derivative is $\partial f/\partial x$.

If θ is equal to $\pi/2$, we have $\cos\theta = 0$, $\sin\theta = 1$, and the directional derivative is $\frac{\partial f}{\partial y}$.

Theorem:

If f(x,y) and its partial derivatives are continuous, and the unit vector a is equal to $\cos\theta i + \sin\theta j$, then $D_{af}(x,y) = f_x(x,y)\cos\theta + f_y(x,y)\sin\theta$.

An alternate notation for directional derivatives of functions of two variables is given by the formula

$$d\theta f(x,y)$$

where θ is the angle the direction makes with the positive x-axis.

The directional derivative is the function θ if the variables x and y have fixed values.

Problem Solving Examples:

Compute the directional derivative of $f(x,y) = x^2 + y^3$ at the point $(-1,3)$ in the direction of maximal increase of f. Explain your reasoning.

A As was proven in a previous problem, the directional derivative of a real-valued function f of n variables at the point P in the direction $\vec{A}$ is given by

$$D_{\vec{A}} f(P) = \text{grad } f(P) \cdot \vec{A} \tag{1}$$

where $\vec{A}$ must be a unit vector. This dot product may be written as

$$\text{grad } f(P) \cdot \vec{A} = \|\text{grad } f(P)\| \|\vec{A}\| \cos\theta \tag{2}$$

where θ is the angle between the vectors grad f(P) and $\vec{A}$. Since $\vec{A}$ is a unit vector, $\|\vec{A}\|=1$ and (1) and (2) give

$$D_{\vec{A}} f(P) = \|\text{grad } f(P)\| \cos\theta. \tag{3}$$

The directional derivative is thus seen to be a maximum when $\theta = 0$; i.e., the direction of maximal increase of f is the direction of the gradient and the directional derivative in that direction is simply

$$D_{\vec{A}} f(P)\Big|_{max} = \|\text{grad } f(P)\|. \tag{4}$$

Now in the case at hand, $f = x^2 + y^3$ so $\frac{\partial f}{\partial x} = 2x$, $\frac{\partial f}{\partial y} = 3y^2$ and at the point (–1,3),

$$\|\text{grad } f\| = \sqrt{\left(\frac{\partial f}{\partial x}\right)^2 + \left(\frac{\partial f}{\partial y}\right)^2} = \sqrt{(2x)^2 + (3y^2)^2}$$

$$= \sqrt{4x^2 + 9y^4} = \sqrt{4(-1)^2 + 9(3)^4}$$

$$= \sqrt{733}.$$

15.5.2 Gradient

The gradient of a function is a vector containing the partial derivatives of the function.

Definition:

If f(x,y) has partial derivatives, then the gradient is defined as

$$\text{grad } f(x,y) = f_x(x,y)i + f_y(x,y)j.$$

The symbol "del" (∇) is used to denote a gradient.

The gradient f(x,y) may also be denoted by $\nabla f(x,y)$, where

$$\nabla = i\frac{\partial}{\partial x} + j\frac{\partial}{\partial y}$$

The dot product of a gradient and a unit vector is illustrated below:

$$\vec{a} \cdot \nabla f = |a|\,|\nabla f| \cos\phi = D_{af}$$

where a is the unit vector,

∇f is the gradient

ϕ is the angle between the vector $\vec{a}$ and the gradient ∇f.

We may now conclude that the directional derivative is maximum when ϕ is zero, that is when the unit vector is in the direction of the gradient.

15.6 Tangent Planes

Let P represent a point on the graph of the function f(x,y,z), where the first partial derivatives of this function are continuous and are not all zero at point p.

The derivative at point P is the slope of the line tangent to the curve at this point. The plane through point P, with normal vector ∇f, is the tangent plane at that point.

Definition:

The equation of the tangent plane to the graph of $f(x,y,z) = 0$ at the point $P(x_0,y_0,z_0)$ is

$$f_x(x_0,y_0,z_0)(x-x_0)+f_y(x_0,y_0,z_0)(y-y_0)+f_z(x_0,y_0,z_0)(z-z_0) = 0$$

The normal line is the line perpendicular to the tangent plane at point P. The equation of the normal line is

$$\frac{x - x_0}{f_x(x_0,y_0,z_0)} = \frac{y - y_0}{f_y(x_0,y_0,z_0)} = \frac{z - z_0}{f_z(x_0,y_0,z_0)}$$

We conclude that the normal line is perpendicular to the tangent plane.

Problem Solving Examples:

Find the equation of the tangent plane to the graph of

$$z = f(x,y) = x^2 + 2y^2 - 1$$

at the points

(a) $(x_0,y_0) = (0,0)$

(b) $(x_0,y_0) = (1,1)$.

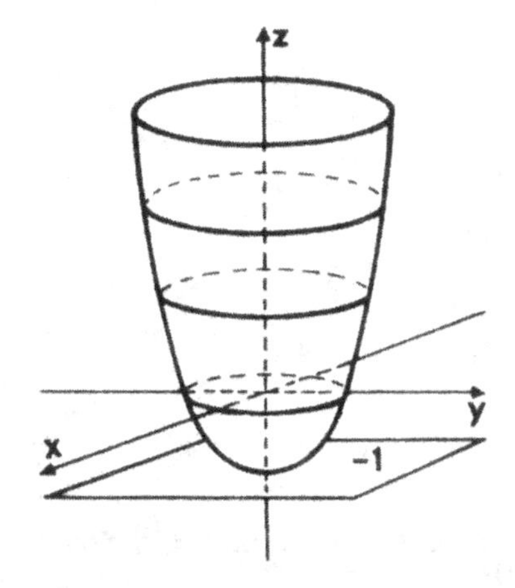

Since points (x,y,z) on the tangent plane of the graph of $F(x,y,z) = 0$ at (a_0,y_0,z_0) must be such that the line segment or vector connecting (x,y,z) with (x_0,y_0,z_0) is perpendicular to the normal to this surface, the equation of this plane is

$$\vec{\nabla} F(x_0,y_0,z_0) \cdot \begin{pmatrix} x - x_0 \\ y - y_0 \\ z - z_0 \end{pmatrix} = 0. \qquad (1)$$

This is so because the gradient of F at (x_0,y_0,z_0), i.e.,$\vec{\nabla} F(x_0,y_0,z_0)$, is indeed normal to the surface described by $F(x,y,z) = 0$. In the particular case where the surface is described by $z = f(x,y)$, we have

$$F(x,y,z) = f(x,y) - z = 0$$

so that writing out (1) gives

$$\vec{\nabla}F(x_0,y_0,z_0) \cdot \begin{pmatrix} x-x_0 \\ y-y_0 \\ z-z_0 \end{pmatrix} = \frac{\partial F}{\partial x}\bigg|_0 (x-x_0) + \frac{\partial F}{\partial y}\bigg|_0 (y-y_0) + \frac{\partial F}{\partial z}\bigg|_0 (z-z_0)$$

$$= \frac{\partial f}{\partial x}\bigg|_0 (x-x_0) + \frac{\partial f}{\partial y}\bigg|_0 (y-y_0) - (z-z_0)$$

$$= 0 \tag{2}$$

where the symbol $\Big|_0$ means that the partial derivative is to be evaluated at the point (x_0,y_0,z_0). In the case at hand, (2) yields

$$2x\Big|_0 (x-x_0) + 4y\Big|_0 (y-y_0) - (z-z_0) = 0 \tag{3}$$

for the equation of the tangent plane at the point (x_0,y_0,z_0).

(a) Here $(x_0,y_0z_0) = (0,0,-1)$ so that (3) gives

$$z = -1 \tag{4}$$

as the equation of the tangent plane at $(0,0,-1)$. This is a horizontal plane as shown in the figure.

(b) Here $(x_0,y_0,z_0) = (1,1,2)$ so that (3) gives

$$2(x-1) + 4(y-1) - (z-2) = 0$$

or

$$z = 2x + 4y - 4. \tag{5}$$

Equation (5) then gives the equation of the tangent plane to the graph of

$$z = f(x,y) = x^2 + 2y^2 - 1 \text{ at the point } (1,1,2).$$

Q Find the equation of the tangent plane to the surface described by the equation $F(x,y,z) = 0$ at the point (x_0,y_0,z_0). What is this equation if $z = f(x,y)$ is a solution to $F(x,y,z,) = 0$?

A The key point in the argument to be given is that the gradient vector of a function $F(x,y,z)$ at a point is normal to the surface given by $F(x,y,z) = 0$ at that point. This result was proved in a previous problem but the proof may be sketched as follows: let (x_1,y_1,z_1) be a point on the surface S given by $F(x,y,z) = 0$ and let

$$\vec{r}(t) = \begin{pmatrix} x(t) \\ y(t) \\ z(t) \end{pmatrix}$$

be a position vector which traces out a curve C on S with

$$\vec{r}(t_0) = \vec{r}_0 = \begin{pmatrix} x_0 \\ y_0 \\ z_0 \end{pmatrix}$$

Then $F(x(t),y(t),z(t)) = F(t) = 0$ for all t in the domain of C and by the chain rule

$$\left.\frac{\partial F}{\partial t}\right|_{\vec{r}_0} = \left.\frac{\partial F}{\partial x}\right|_{\vec{r}_0} \left. x'(t)\right|_{\vec{t}_0} + \left.\frac{\partial F}{\partial y}\right|_{\vec{r}_0} \left. y'(t)\right|_{\vec{t}_0} + \left.\frac{\partial F}{\partial z}\right|_{\vec{r}_0} \left. z'(t)\right|_{\vec{t}_0} . \quad (1)$$

This may be written in vector notation as

$$\vec{\nabla} F(\vec{r}_0) \cdot \vec{r}'(t_0) = 0 , \quad (2)$$

where $\vec{\nabla} F = (F_x,F_y,F_z)$ is the gradient of F and $\vec{r}'(t) = (x'(t),y'(t),z'(t))$ is the velocity vector of C. Since $\vec{r}'(t)$ is always tangent to C, (2) implies that $\vec{\nabla} F$ is normal or perpendicular to C at r_0 since the dot product of two nonzero vectors is zero only if they are perpendicular. But C can be any curve on S through $\vec{r}_0$ so (2) implies that $\vec{\nabla} F$ is normal to S at $\vec{r}_0$. Now, a point (x,y,z) if the vector directed from (x_0,y_0,z_0) to (x,y,z) is perpendicular to the normal $\vec{\nabla}F(x_0,y_0,z_0)$. That is, (x,y,z) is on the tangent plane if

$$\vec{\nabla} F(\vec{r}_0) \cdot (\vec{r} - \vec{r}_0) = 0\,, \tag{3}$$

or, writing this out, the equation of the tangent plane to S at r_0 is given by

$$\left.\frac{\partial F}{\partial x}\right|_{\vec{r}_0}(x-x_0) + \left.\frac{\partial F}{\partial y}\right|_{\vec{r}_0}(y-y_0) + \left.\frac{\partial F}{\partial z}\right|_{\vec{r}_0}(z-z_0) = 0\,. \tag{4}$$

If the surface is described by $z = f(x,y)$, then $F(x,y,z) = f(x,y) - z = 0$ so that

$$\frac{\partial F}{\partial x} = \frac{\partial f}{\partial x}\,,\ \frac{\partial F}{\partial y} = \frac{\partial f}{\partial y}\,,\ \frac{\partial F}{\partial z} = -1 \tag{5}$$

and (4) becomes

$$\left.\frac{\partial f}{\partial x}\right|_{(x_0,y_0)}(x-x_0) + \left.\frac{\partial f}{\partial y}\right|_{(x_0,y_0)}(y-y_0) - (z-z_0) = 0\,. \tag{6}$$

Find the line orthogonal to the graph of $f(x,y) = xy$ at the point $(x_0,y_0,z_0) = (-2,3,-6)$.

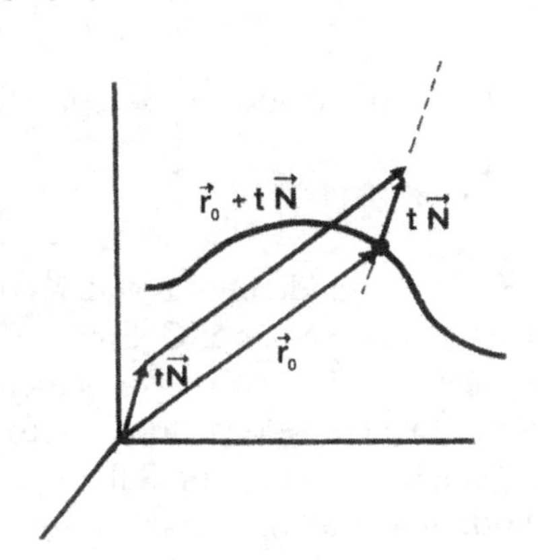

A It is possible to solve this problem if one of two facts is known. The first fact, which has been derived in a previous problem, is that the equation of the tangent plane to the graph of $f(x,y) = xy$ at a point $(x_0,y_0,z_0) = r_0$ is given by

$$\left.\frac{\partial f}{\partial x}\right|_{(x_0,y_0)} (x-x_0) + \left.\frac{\partial f}{\partial y}\right|_{(x_0,y_0)} (y-y_0) - (z-z_0) = 0 . \quad (1)$$

In the case at hand,

$$\left.\frac{\partial f}{\partial x}\right|_{(x_0,y_0)} = \left.\frac{\partial f}{\partial x}\right|_{(-2,3)} = y\Big|_{(-2,3)} = 3 \quad (1a)$$

$$\left.\frac{\partial f}{\partial y}\right|_{(x_0,y_0)} = \left.\frac{\partial f}{\partial y}\right|_{(-2,3)} = x\Big|_{(-2,3)} = -2, \quad (1b)$$

so that from (1)

$$3(x + 2) - 2(y - 3) - (z + 6) = 0$$

or

$$z = 3x + 6 - 2y \quad (2)$$

is the equation of the tangent plane to the graph of $f(x,y) = xy$ at the point $(-2,3, -6)$. Equation 2 can also be written as

$$3x - 2y - z + 6 = 0 . \quad (3)$$

Now recall that for a plane given by the equation $Ax + By + Cz + D = 0$ a vector normal to this plane is given by $\vec{n} = (A,B,C)$. Hence it can be seen from (3) that the vector $\vec{N} + (3,-2,-1)$ is normal to this tangent plane. Thus, the line orthogonal to the graph of $f(x,y) = xy$ at $\vec{r_0}$, that is to say the line orthogonal to the tangent plane to this graph at $\vec{r_0}$, is given by

$$(x,y,z) = r_0 + t\ N = (-2,3,-6) + t(3,-2,-1) \quad (4)$$

or

$$\begin{matrix} x \\ y \\ z \end{matrix} = \begin{pmatrix} -2+3t \\ 3-2t \\ -6-t \end{pmatrix} \tag{5}$$

Justification of (4) may be made if one remembers the parallelogram rule for adding vectors, as shown in the figure.
The second fact which gives an immediate solution to the problem and which was also derived in a previous problem is that the vector which is normal to a surface given by f = f(x,y) at a point (x_0,y_0) has direction numbers $(f_x(x_0,y_0), f_y(x_0,y_0), -1)$. Thus, from (1a) and (1b), the normal to the surface f(x,y) = xy at (–2,3,–6) has direction numbers

$$\vec{N} = (3,-2,-1).$$

This is the same result as was found before, so that using equation (4) will give the same result (i.e., equation (5)), for the line orthogonal to f(x,y) = xy at (–2,3,–6) as we previously found.

15.7 Total Differential

Given the function y = f(x), the quantity df is the differential of f defined as

$$df = f'(x)h,$$

Definition:

If F is a function of three variables x,y,z, then the total differential is defined as

$$dF(x,y,z) = F_x(x,y,z)h + F_y(x,y,z)k + F_z(x,y,z)\ell$$

Theorem:

Suppose that z = f(x,y) and x and y are functions of the same variable. Then,

$$dz = \frac{\partial z}{\partial x} dx + \frac{\partial z}{\partial y} dy.$$

This is similar to the results obtained for the function $w = f(x,y,z)$.

$$dw = \frac{\partial w}{\partial x} dx + \frac{\partial w}{\partial y} dy + \frac{\partial w}{\partial z} dz$$

This equation holds when x, y and z are either independent or intermediate variables.

Example: Find the derivative dy/dx of the following equation by methods of partial differentiation:

$$e^{xy} + \sin xy + 1 = 0$$

$$f_x = ye^{xy} + y \cos xy$$

$$f_y = xe^{xy} + x \cos xy$$

$$\frac{dy}{dx} = -\frac{ye^{xy} + y \cos xy}{xe^{xy} + x \cos xy} = -\frac{y(e^{xy} + \cos xy)}{x(e^{xy} + \cos xy)}$$

$$\frac{dy}{dx} = \frac{-y}{x}$$

Problem Solving Examples:

Find the total differential of the function:

$$z = x^3y + x^2y^2 - 3xy^3.$$

By definition,

$$dz = \frac{\partial z}{\partial x} dx + \frac{\partial z}{\partial y} dy.$$

$$\frac{\partial z}{\partial x} = 3x^2y + 2xy^2 - 3y^3,$$

and

$$\frac{\partial z}{\partial y} = x^3 + 2x^2y - 9xy^2 .$$

Then,

$$dz = (3x^2y + 2xy^2 - 3y^3)dx + (x^3 + 2x^2y - 9xy^2)dy.$$

If $u = x^2 - xy + y^2$, and if $x = 1 + t^2$, $y = 1 - t^2$, what is the total derivative of u?

u is a function of x and y which are each, in turn, functions of t. The total derivative of u with respect to t is given by,

$$\frac{du}{dt} = \frac{\partial u}{\partial x}\frac{dx}{dt} + \frac{\partial u}{\partial y}\frac{dy}{dt} .$$

Note that if u were a function of three variables, $u = F(x,y,z)$, which are all, in turn, functions of t, the total derivative of u with respect to t would be written as:

$$\frac{du}{dt} = \frac{\partial u}{\partial x}\frac{dx}{dt} + \frac{\partial u}{\partial y}\frac{dy}{dt} + \frac{\partial u}{\partial z}\frac{dz}{dt} .$$

To solve this problem, we need to find $\frac{\partial u}{\partial x}$, $\frac{\partial u}{\partial y}$, $\frac{dx}{dt}$ and $\frac{dy}{dt}$, and then to substitute in the equation for the total derivative, $\frac{du}{dt}$. Hence

$$\frac{\partial u}{\partial x} = 2x - y \qquad \frac{dx}{dt} = 2t$$

$$\frac{\partial u}{\partial y} = 2y - x \qquad \frac{dy}{dt} = -2t$$

and

$\frac{du}{dt} = (2x - y)\,(2t) + (2y - x)\,(-2t) = 2t(3x - 3y)$ which, to obtain a relation in terms of t alone, is:

$$2t[3(1 + t^2) - 3(1 - t^2)]$$

$$= 2t(3 + 3t^2 - 3 + 3t^2)$$

$$= 12t^3.$$

15.8 Taylor's Theorem with Remainder

The expansion of the function F(x) with n+1 derivatives on an interval containing the value x_0 is

$$F(x) = F(x_0) + F'(x_0)(x-x_0) + \ldots + \frac{F^{(n)}(x_0)(x-x_0)^n}{n!} + Rn$$

Rn is the remainder represented by the formula

$$Rn = \frac{F^{(n+1)}(\varepsilon_0)(x-x_0)^{n+1}}{(n+1)!}$$

where ε_0 is some number between x and x_0.

15.8.1 Taylor's Theorem

Let f be a continuous function of two variables and also let all of its partial derivatives be continuous about the point P(a,b).

The expansion for f is

$$F(x,y) = F(a,b) + \sum_{1 \leq r+s \leq p} \frac{\partial^{r+s} f(a,b)}{\partial x^r \, \partial y^s} \frac{(x-a)^r}{r!} \frac{(y-b)^s}{s!} + Rp$$

The remainder R_p is represented by the formula

$$R_p = \sum_{r+s = p+1} \frac{\partial^{r+s} F(\varepsilon, \eta)}{\partial_x^r \, \partial_y^s} \frac{(x-a)^r}{r!} \frac{(y-b)^s}{s!}$$

The value of (ε, η) is located on the line segment between points (a,b) and (x,y).

The remainder term demonstrates that if $0 < r < d$, then

$$\varepsilon = a + \frac{(x-a)}{d}\tau, \quad \eta = b + \frac{(y-b)}{d}\tau.$$

Another mode of expression of Taylor's rule is

$$F(x,y) = f(a,b) + \sum_{q=1}^{p} \frac{1}{q!}\left[\sum_{r=0}^{q} \frac{q!}{(q-r)!r!} \frac{\partial^q f(a,b)}{\partial x^{q-r}\partial y^r} (x-a)^{q-r}(y-b)^r\right] + Rp$$

If P tends to infinity for some function f, the remainder approaches 0 ($R_p \to 0$) and we obtain a representation of f as an infinite series in x and y. This series is called a double series and f is said to have expanded about the point (a,b).

Problem Solving Examples:

Expand the function: cos x, in powers of x – a, where $a = -\frac{\pi}{4}$, and determine the interval of convergence.

Expanding the given function in powers of x-a is equivalent to finding the Taylor Series for the function. To find the Taylor Series we determine f(x), f(a), f′(x), f′(a), f″(x), f″(a), etc. We find:

$$f(x) = \cos x; \qquad f(a) = f\left(-\frac{\pi}{4}\right) = \frac{\sqrt{2}}{2}.$$

$$f'(x) = -\sin x; \qquad f'(a) = f'\left(-\frac{\pi}{4}\right) = \frac{\sqrt{2}}{2}.$$

This is a positive value because the value of sin x in the second quadrant is negative, therefore –sin x is positive.

$$f''(x) = \cos x; \qquad f''(a) = f''\left(-\frac{\pi}{4}\right) = \frac{\sqrt{2}}{2}.$$

$$f'''(x) = -\sin x; \qquad f'''(a) = f'''\left(-\frac{\pi}{4}\right) = \frac{\sqrt{2}}{2}.$$

We develop the series as follows:

$$f(x) = f(a) + f'(a)\,[x-a] + \frac{f''(a)}{2!}[x-a]^2 + \frac{f'''(a)}{3!}[x-a]^3 + \dots .$$

By substitution:

$$\cos x = \frac{\sqrt{2}}{2} + \frac{\sqrt{2}}{2}\left(x + \frac{\pi}{4}\right) + \frac{\frac{\sqrt{2}}{2}}{2!}\left(x + \frac{\pi}{4}\right) + \frac{\frac{\sqrt{2}}{2}}{3!}\left(x = \frac{\pi}{4}\right)^3 + \cdots.$$

We examine the terms of this series to determine the law of formation. We find the nth term of the series to be:

$$\frac{\frac{\sqrt{2}}{2}}{(n-1)!}\left(x + \frac{\pi}{4}\right)^{n-1};$$

Then, the (n+1)th term is:

$$\frac{\frac{\sqrt{2}}{2}}{n!}\left(x + \frac{\pi}{4}\right)^{n}.$$

Therefore the Taylor Series is:

$$\cos x = \frac{\sqrt{2}}{2} + \frac{\sqrt{2}}{2}\left(x + \frac{\pi}{4}\right) + \frac{\frac{\sqrt{2}}{2}}{2!}\left(x + \frac{\pi}{4}\right)^2 + \frac{\frac{\sqrt{2}}{2}}{3!}\left(x + \frac{\pi}{4}\right)^3 + \cdots$$
$$+ \frac{\frac{\sqrt{2}}{2}}{(n-1)!}\left(x + \frac{\pi}{4}\right)^{n-1} + \frac{\frac{\sqrt{2}}{2}}{n!}\left(x + \frac{\pi}{4}\right)^{n} + \cdots$$

To find the interval of convergence we use the Ratio Test. We set up the ratio $\frac{U_{n+1}}{U_n}$, obtaining:

$$\frac{\sqrt{2}\left(x+\frac{\pi}{4}\right)^n}{2(n!)} \cdot \frac{2(n-1)!}{\sqrt{2}\left(x+\frac{\pi}{4}\right)^{n-1}} =$$

$$\frac{\left(x+\frac{\pi}{4}\right)^n}{n(n-1)!} \cdot \frac{(n-1)}{\left(x+\frac{\pi}{4}\right)^{n-1}} = \frac{x+\frac{\pi}{4}}{n}$$

Now we find $\lim_{n\to\infty}\left|\frac{x+\frac{\pi}{4}}{n}\right| = |0| = 0$. By the ratio test we know that if $\lim_{n\to\infty}\left|\frac{U_{n+1}}{U_n}\right| < 1$ the series converges. Since $0<1$, the series converges for all values of x.

Prove Taylor's Theorem for $f \in C^r(E)$ where $E \subseteq R^n$ is an open convex set.

A Let $f{:}E \to R$ where $E \subseteq R^n$ is open and convex have continuous partial derivatives through order r (i.e., $f \in C^r(E)$). Let a be any point in E, then there exists an $\bar{x} \in E$ such that

$$0 < \|\bar{x} - a\| < \|x - a\|$$

and

$$f(x) = f(a) + J_f(a)(x-a) + \frac{1}{2}\langle H_f(a)(x-a), x-a\rangle + \ldots$$

$$= f(a) + \sum_{k=1}^{r-1} \frac{1}{k!} \sum_{\substack{i_j=1 \\ 1\le j\le k}}^{n} \frac{\partial^k f}{dx_{i_1} \cdots dx_{i_k}}(a)(x_{i_1} - a_{i_1}) \ldots (x_{i_k} - a_{i_k}) +$$

$$\frac{1}{r!} \sum_{\substack{i_j=1 \\ 1\le j\le r}}^{n} \frac{\partial^r f}{dx_{i_1} \cdots dx_{i_r}}(\bar{x})(x_{i_1} - a_{i_1}) \ldots (x_{i_r} - a_{i_r}),$$

where $J_f(a)$ and $H_f(a)$ are, respectively, the Jacobian and Hessian matrices evaluated at a, and $<, >$ denotes the Euclidean inner product. This is the version of Taylor's Theorem which is to be proved. In order to prove this version, the one-dimensional version of the theorem is needed. Let

$$f \in C^r[a,b]$$

and let $f^{(r+1)}$ exist in (a,b). If $\alpha, \beta \in [a,b]$, then there exists $\Upsilon \in (\alpha,\beta)$ such that

$$f(\beta) = f(\alpha) + f'(\alpha)(\beta-\alpha) + \frac{1}{2} f''(\alpha)(\beta-\alpha)^2 + \ldots + \frac{1}{r!} f^{(r)}(\alpha)(\beta-\alpha)^r + \frac{1}{(r+1)!} f^{(r+1)}(\Upsilon)(\beta-\alpha)^{r+1}.$$

To prove the latter version, let $\zeta \in R$ be defined by (1)

$$\frac{(\beta-\alpha)^{r+1}}{(r+1)!} = \zeta\, f(\beta) - \left[f(\alpha) + f'(\alpha)(\beta-\alpha) + \ldots + \frac{1}{r!} f^{(r)}(\alpha)(\beta-\alpha)^r\right]$$

and define ϕ on [a,b] by

$$\phi(x) = f(\beta) - \left(f(x) + f'(x)(\beta-x) + \ldots + \frac{1}{r!} f^{(r)}(x)(\beta-x)^r + \frac{\zeta}{(r+1)!}(\beta-x)^{r+1}\right)$$

which is continuous on [a,b] and differentiable on (a,b). Evidently $\phi(\beta) = 0$ and, by the definition of ζ, $\phi(\alpha) = 0$. Applying Rolle's Theorem, there exists $\Upsilon \in (\alpha,\beta)$ such that $\phi'(\Upsilon) = 0$. Differentiating ϕ gives

$$\phi'(x) = -\left(f'(x) - f'(x) + f''(x)(\beta-x) - \ldots - \frac{(\beta-x)^{r-1}}{(r-1)!} f^{(r)}(x) + \frac{1}{r!} f^{(r+1)}(x)\right.$$

$$\left.(\beta-x)^r - \frac{\zeta}{r!}(\beta-x)^r\right)$$

$$= \frac{\zeta - f^{(r+1)}(x)}{r!}(\beta-x)^r \qquad (2)$$

Since $\phi'(\Upsilon) = 0$, (2) implies that $\zeta = f^{(r+1)}(\Upsilon)$ and hence (1) is the desired result. Now to apply the n-dimensional version of the theorem,

Let $F \in C^r[0,1]$ be defined by

$$F(t) = f(a + t(x - a))$$

By the version which was just proved, there exists a

$$t_0 \in (0,1)$$

such that

$$F(1) = F(0) + F'(0) + \frac{1}{2}F''(0) + \ldots + \frac{1}{(r-1)!}F^{(r-1)}(0) + \frac{1}{r!}F^{(r)}(t_0).$$

Note from the above definition of F(t) that $F(1) = f(x)$, $F(0) = f(a)$,...,

$$F^{(k)}(0) = \sum_{\substack{i_j=1 \\ 1\le j\le k}}^{n} \frac{\partial^k f}{\partial x_{i_1} \ldots \partial x_{i_k}}(a)(x_{i_1} - a_{i_1}) \ldots (x_{i_k} - a_{i_k}),$$

and

$$F^{(r)}(t_0) = \sum_{\substack{i_j=1 \\ 1 \le j \le k}}^{n} \frac{\partial^r f}{\partial x_{i_1} \dots \partial x_{i_r}} (a + t_0(x-a))\,(x_{i_1} - a_{i_1})\dots(x_{i_r} - a_{i_r}).$$

Since E is convex,

$$\bar{x} = a + t_0\,(x-a) \in E \text{ and } 0 < \|\bar{x}-a\| < \|x-a\|.$$

15.9 Maxima and Minima

A function f (x,y) of two variables is said to have a relative (local) maximum at (x_0, y_0) if there is an open region (R) containing (x,y) such that $(x,y) \le f(x_0, y_0)$ for all pairs (x,y) in the open region.

The relative maximum can be attained geometrically by drawing the graph of the function. The high points on the graph represent the relative maxima.

Theorem:

Let f(x,y) represent a function in the rectangular region R. Let $f_x(x_0, y_0)$ and $f_y(x_0, y_0)$ be defined, and let

$f(x,y) \le f(x_0, y_0)$ for all (x,y) in R.

If

$$f_x(x_0, y_0) = f_y(x_0, y_0) = 0,$$

then $f(x_0, y_0)$ is the relative maximum.

The function f(x,y) has a relative minimum at (a,b) if there is a region containing (a,b) such that $f(x,y) \ge f(a,b)$ for all (x,y) in the region. The relative minimum corresponds to the low point on the graph of the function.

Let f(x,y) represent a continuous function on the closed rectangular region. The function has an absolute maximum f(a,b) and an absolute minimum f(c,d) for some (a,b) and (c,d) in the region. Thus,

$$f(c,d) \leq f(x,y) \leq f(a,b)$$

for all (x,y) in the rectangular region.

Definition:

The point on a graph at which both f_x and f_y vanish is called a critical point.

A critical point at which the function f is neither a maximum nor a minimum is a saddle point. This is illustrated in the figure below

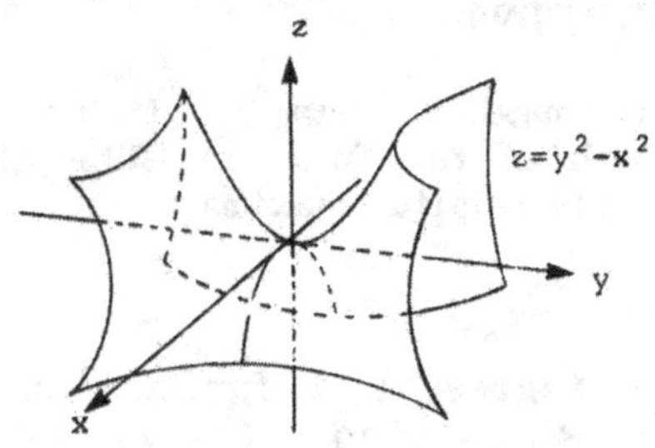

In this graph, the saddle point is (0,0,0)

Theorem: Test for Extrema

Let f(x,y) and its partial derivatives to the third order be continuous near the point (a,b).

The point (a,b) is a critical point if $f_x(a,b) = f_y(a,b) = 0$. Thus,

1) f(a,b) is a local minimum if $f_{xx}(a,b)f_{yy}(a,b) - [f_{xy}(a,b)]^2 > 0$ and $f_{xx}(a,b) > 0$.
2) f(a,b) is a local maximum if $f_{xx}(a,b)f_{yy}(a,b) - [f_{xy}(a,b)]^2 > 0$ and $f_{xx}(a,b) < 0$.

3) f(a,b) is a saddle point if $f_{xx}(a,b)f_{yy}(a,b) - [f_{xy}(a,b)]^2 < 0$.

4) No information is attained if $f_{xx}(a,b)f_{yy}(a,b) - [f_{xy}(a,b)]^2 = 0$.

If the function f(x,y,z) has first derivatives, then a critical point exists at the point where $f_x = f_y = f_z = 0$.

Problem Solving Examples:

Find the point (x,y,z) obeying $g(x,y,z) = 2x + 3y + z - 12 = 0$ for which $f(x,y,z) = 4x^2 + y^2 + z^2$ is a minimum.

This problem is an extremal problem subject to a constraint. For the function f(x,y,z), the variables x,y,z are not independent of each other, but are restricted by some relation existing between them, this relation being expressed by an equation $g(x,y,z) = 0$. There are several methods to deal with these extremal problems with constraints. For this problem we use the method of direct elimination. This method uses the equation or equations of constraint to express certain of the variables in terms of the remaining variables. Then these latter are regarded as the independent variables and the function f(x,y,z) is then expessed in terms of these independent variables only. The solution for the extreme values of the function is then carried out by standard methods. For this problem the constraint is $g(x,y,z) = 2x + 3y + z - 12 = 0$. Solving for z, then expresses z as a function of x,y, the latter being regarded as the independent variables. This gives

$$z = 12 - 2x - 3y, \qquad (1)$$

so that $f(x,y,z) = 4x^2 + y^2 + z^2$ is rewritten as

$$F_1(x,y) = 4x^2 + y^2 + (12 - 2x - 3y)^2 \qquad (2)$$

Hence, the problem now is to find the values of x,y for which F(x,y) is a minimum; then substitution of these values into (1) finds the value of z for the minimum of f(x,y,z). Therefore, from equation (2), the critical points are solutions of

$$F_1(x,y) = 8x + 2(12 - 2x - 3y)(-2) = 0 \tag{3}$$

and

$$F_2(x,y) = 2y + 2(12 - 2x - 3y)(-3) = 0. \tag{4}$$

Simplifying gives the two equations $16x + 12y = 48$; $12x = 20y = 72$, which when solved simultaneously give one solution, namely

$$P_0 = \frac{6}{11}, \frac{36}{11}. \text{ Next let } \Delta = [F_{12}(p_0)]^2 - F_{11}(p_0)F_{22}(p_0). \tag{5}$$

From (3), (4); $F_{11}(x,y) = 16$ so that $F_{11}(p_0) = 16$,

$F_{22}(x,y) = 20$ so that $F_{22}(p_0) = 20$,

and

$F_{12}(x,y) = 12$ so that $F_{12}(p_0) = 12$.

Then by equation (5); $\Delta = (12)^2 - (16)(20) = 144 - 320 < 0$. Hence, since $\Delta < 0$ and $F_{11}(p_0) = 16 > 0$, the point $P_0 = \left(\frac{6}{11}, \frac{36}{11}\right)$ is a relative minimum for F(x,y). To find z, substitution in (1) yields $z = 12 - 2\left(\frac{6}{11}\right) - 3\left(\frac{36}{11}\right)$; giving $z = \frac{12}{11}$. Thus, the point (x,y,z) for which f(x,y,z) is minimum subject to the constraint $g(x,y,z) = 0$ is $P_1 = \left(\frac{6}{11}, \frac{36}{11}, \frac{12}{11}\right)$.

Find the values of (x,y,z) that minimize F(x,y,z) = xy + 2yz + 2xz given the condition G(x,y,z) = xyz = 32.

This is an extremal problem given the constraint G(x,y,z) = k (k is a given constant). The method of direct elimination can be used to find the minimum of f(x,y,z) given the constraint. However, an alternate procedure can be applied, namely the method of implicit functions. Assume that F(x,y,z) and G(x,y,z) are given

functions with continuous first partial derivatives and that $\frac{\partial G}{\partial z} \neq 0$. If the equation

$$G(x,y,z) - k = 0 \tag{1}$$

has a solution z = f(x,y,z), then it is desired to make the quantity u =

F(x,y,f(x,y)) a maximum or minimum. So, it is needed that the equations

$$\frac{\partial u}{\partial x} = 0 \text{ and } \frac{\partial u}{\partial y} = 0 \tag{2}$$

be solved. Now,

$$\frac{\partial u}{\partial x} = \frac{\partial F}{\partial x} + \frac{\partial F}{\partial z}\frac{\partial f}{\partial x}\,;\; \frac{\partial u}{\partial y} = \frac{\partial F}{\partial y} + \frac{\partial F}{\partial z}\frac{\partial f}{\partial y} \tag{3}$$

where z is replaced by f(x,y) after the differentiations are performed. But, since G(x,y,f(x,y)) – k = 0, it follows that

$$\frac{\partial G}{\partial x} + \frac{\partial G}{\partial z}\frac{\partial f}{\partial x} = 0\,;\; \frac{\partial G}{\partial y} + \frac{\partial G}{\partial z}\frac{\partial f}{\partial y} = 0. \tag{4}$$

Then solving (4) for $\frac{\partial f}{\partial x}$ and $\frac{\partial f}{\partial y}$ and substituting into (3) yields

$$\frac{\partial u}{\partial x} = \frac{\frac{\partial F}{\partial x}\frac{\partial G}{\partial z} - \frac{\partial F}{\partial z}\frac{\partial G}{\partial x}}{\frac{\partial G}{\partial z}}$$

(and a similar equation for $\frac{\partial u}{\partial y}$). Equations (2) now become

$$\frac{\partial F}{\partial x}\frac{\partial G}{\partial z} - \frac{\partial F}{\partial z}\frac{\partial G}{\partial x} = 0\ ;\ \frac{\partial F}{\partial y}\frac{\partial G}{\partial z} - \frac{\partial F}{\partial z}\frac{\partial G}{\partial y} = 0 \qquad (5)$$

in which after the differentiations are performed, f(x,y) replaces z. Now, for the implicit function method, z is not solved for at the outset as in the method of direct elimination. Instead, equations (5) are arrived at as equations in all three variables. These two equations, along with the constraint give three equations which when solved simultaneously as equations in x,y,z give points among which the required points of extreme are present.

For this problem, think of y as a function of x and z (note: the preceding discussion utilized z as a function of x and y). The given function is $f(x,y,z) = xy + 2yz + 2xz$, $G(x,y,z) = xyz = 32$. (6)
Differentiating both equations with respect to x gives

$$0 = \frac{\partial f}{\partial x} = y + x\frac{\partial y}{\partial x} + 2z\frac{\partial y}{\partial x} + 2z\ ;\ 0 = \frac{\partial G}{\partial x} = yz + xz\frac{\partial y}{\partial x}\ .$$

Now eliminating $\frac{\partial y}{\partial x}$ between the two equations yields

$\frac{\partial y}{\partial x} = \frac{-y}{x}$ (from the second equation), and

$$0 = 2z + \left(\frac{-y}{x}\right) + y + 2z\left(\frac{-y}{x}\right)$$

so that

$$0 = 2z - \frac{2zy}{x} \text{ from which } x = y. \qquad (7)$$

Then differentiating both equations in (6) with respect to z gives

$$0 = \frac{\partial f}{\partial z} = x\frac{\partial y}{\partial z} + 2y + 2z\frac{\partial y}{\partial z} + 2x;\ 0 = \frac{\partial G}{\partial z} = xy + xz\frac{\partial y}{\partial z}\ .$$

Now eliminating $\frac{\partial y}{\partial z}$ yields $\frac{\partial y}{\partial z} = \frac{-y}{z}$ so that

$$0 = x\left(\frac{-y}{z}\right) + 2y + 2z\left(\frac{-y}{z}\right) + 2x \text{ which simplifies to } 0 = \frac{-xy}{z} + 2x$$

from which $z = \frac{y}{2}$. (8)

Now to get values of x,y,z substitute into $G(x,y,z) = xyz = 32$ using (7), (8) and obtain $\frac{x \cdot x \cdot x}{2} = 32 \Rightarrow x^3 = 64$. Therefore $x = 4$, $y = 4$, and $z = 2$. Hence, the value of (x,y,z) that minimizes f(x,y,z) under the constraint $xyz = 32$ is (4,4,2).

15.10 Lagrange Multipliers

Problems of free maxima and minima are problems in which the maxima and minima are functions of several variables without any added conditions.

A problem involving an added or side condition is called a problem in constrained maxima and minima.

Let f and g be functions of x, y and z. To find the local extrema of f(x,y,z) subject to the constraint $\phi(x,y,z) = 0$:

1) **We solve the equation $\phi(x,y,z)$ to obtain $z = g(x,y)$, where the functions f, ϕ and g have continuous first partial derivatives throughout a suitable domain.**
2) **We define a function of four variables by $w = f(x,y,z) + h\phi(x,y,z)$, where h is the Lagrange multiplier.**
3) **The values of x,y,z which give the extrema of f are obtained by solving four equations in four unknowns, simultaneously.**

For example, if $w = u + h\nu$, then the extremas are given by the solution of the following four equations in four unknowns:

$$w_x = u_x + h\nu_x = 0,$$

$$w_y = u_y + h\nu_y = 0,$$

$$w_z = u_z + h\nu_z = 0,$$

and $wh = \nu = 0$.

We assume that the partial derivatives of ν are not zero, and solve the first three equations for h to obtain

$$h = \frac{-u_x}{\nu_x} = \frac{-u_y}{\nu_y} = \frac{-u_z}{\nu_z}$$

Example: Find the minimum of $f(x,y,z) = x^2 + y^2 + z^2$, subject to the condition $x + 3y - 2z - 4 = 0$.

$$w(x,y,z,h) = x^2 + y^2 + z^2 + h(x+3y-2z-4) = 0 \quad (1)$$

$$w_x = 2x + h = 0 \quad (2)$$

$$w_y = 2y + 3h = 0 \quad (3)$$

$$w_z = 2z - 2h = 0 \quad (4)$$

From equations (2), (3) and (4) we obtain

$$h = -2x = -\frac{2}{3}y = z,$$

which gives

$$x = \frac{1}{3}y \quad \text{and} \quad z = -\frac{2}{3}y.$$

Substitute these into

$$x + 3y - 2z - 4 = 0,$$

and we have

$$\frac{1}{3}y + 3y - 2\left(-\frac{2}{3}y\right) - 4 = 0,$$

$$y = \frac{6}{7}.$$

Thus, $\quad x = \frac{1}{3}y = \frac{1}{3} \times \frac{6}{7} = \frac{2}{7} \quad$ and

$$z = -\frac{2}{3}y = -\frac{2}{3} \times \frac{6}{7} = -\frac{4}{7}.$$

The minimum of f(x,y,z) is then found by:

$$f(x,y,z) = x^2 + y^2 + z^2$$

$$= \left(\frac{2}{7}\right)^2 + \left(\frac{6}{7}\right)^2 + \left(-\frac{4}{7}\right)^2$$

$$= \frac{4}{49} + \frac{36}{49} + \frac{16}{49} = \frac{56}{49} = \frac{8}{7}$$

Problem Solving Examples:

Find the maximum of $f(x,y) = xy$ on the curve $G(x,y) = (x+1)^2 + y^2 = 1$, assuming that such a maximum exists.

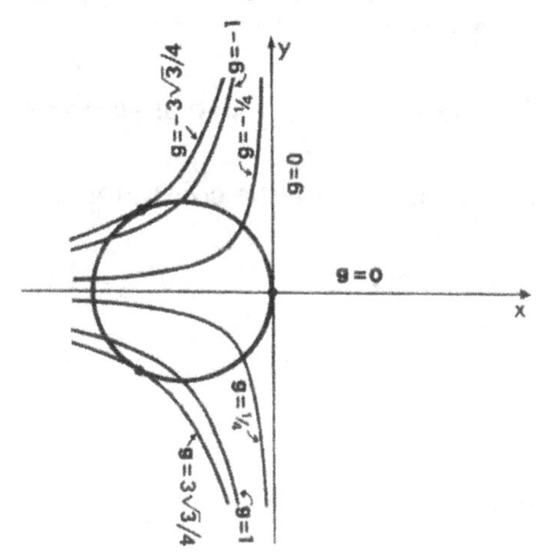

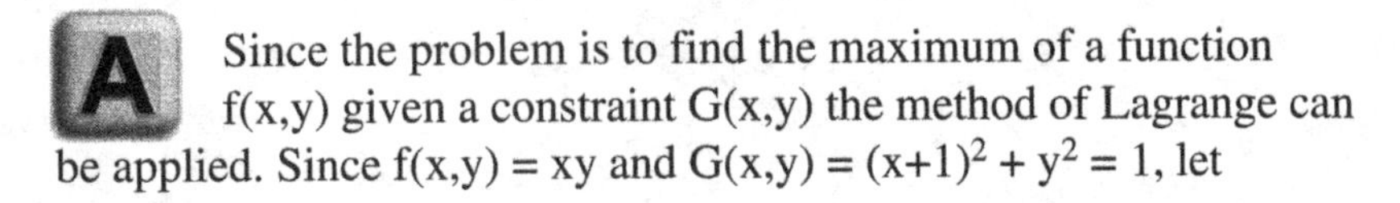

Since the problem is to find the maximum of a function f(x,y) given a constraint G(x,y) the method of Lagrange can be applied. Since $f(x,y) = xy$ and $G(x,y) = (x+1)^2 + y^2 = 1$, let

$$u = xy + \lambda((x+1)^2) + y^2). \qquad (1)$$

Then to use this method compute

$$\frac{\partial u}{\partial x};\ \frac{\partial u}{\partial y}$$

and set each equal to zero. This gives

$$\frac{\partial u}{\partial x} = y + 2\lambda(x+1) = 0 \qquad (2)$$

and

$$\frac{\partial u}{\partial y} = x + 2\lambda y = 0. \qquad (3)$$

Now the equations (2), (3) above are to be solved simultaneously with the constraint curve to give solutions of x,y such that extremal values are among them. One solution is $x = 0$; $y = 0$ so that $\lambda = 0$; hence $P_1 = (0,0)$ is one point of possible extrema. For any other solution of (2), (3) $y \neq 0$; thus from (3), $\lambda = -\frac{x}{2y}$; then by (2)

$$y = (2)\left(\frac{-x}{2y}\right)(x+1) = \frac{x^2 + x}{y} \text{ so that } y^2 = x^2 + x.$$

Solving this simultaneously with the constraint

$$(x+1)^2 + y^2 = 1 \text{ gives } (x+1)^2 + x^2 + x = 1 \text{ or } 2x^2 + 3x = 0,$$

which equals $x(2x+3) = 0$ and has solutions $x = 0$; $x = \frac{-3}{2}$. The first value, $x = 0$, gives $y = 0$.

The other solution $x = \frac{-3}{2}$ gives the two points $\left(\frac{-3}{2}, \pm\frac{\sqrt{3}}{2}\right)$ (from substitution into G(x,y,z)). Therefore, the extreme values of the function are among the values given by the points $P_1 = (0,0)$;

$P_2 = \left(\frac{-3}{2}, \pm\frac{\sqrt{3}}{2}\right)$ and $P_3 = \left(\frac{-3}{2}, \pm\frac{\sqrt{3}}{2}\right)$. But direct substitution into

$f(x,y) = xy$ shows the point $\left(\frac{-3}{2}, \pm\frac{-\sqrt{3}}{2}\right)$ to give a maximum, the point $\left(\frac{-3}{2}, \pm\frac{\sqrt{3}}{2}\right)$ to give a minimum, and (0,0) to give neither. This is

because $f(0,0) = 0$, $f\left(\frac{-3}{2}, \frac{-\sqrt{3}}{2}\right) = 3\frac{\sqrt{3}}{4}$ and $f\left(\frac{-3}{2}, \frac{\sqrt{3}}{2}\right) =$

$-3\frac{\sqrt{3}}{4}$. Level curves of f(x,y) are given in the figure.

Find the extrema for the function $x^2 + y^2 + z^2$ subject to the constraint $x^2 + 2y^2 - z^2 - 1 = 0$.

A The problem is to find the extreme values of a function f(x,y,z) given a constraint G(x,y,z) = k (k a constant). The methods of direct elimination or implicit functions can be tried to solve this problem. However, another method is applied, namely Lagrange's method. This method consists of the formation of the function $u = f(x,y,z) + \lambda G(x,y,z)$ subject to the conditions

$$\frac{\partial u}{\partial x} = 0, \ \frac{\partial u}{\partial y} = 0, \ \frac{\partial u}{\partial z} = 0. \qquad (1)$$

Here x,y,z are treated as independent variables and λ is a constant, independent of x,y,z, called Lagrange's multiplier.

For this method solve the three equations in (1) along with the equation of constraint G(x,y,z) = k to find the values of x, y, z,λ.

More than one point (x,y,z) may be found in this way, but among the points so found will be the points of extremal values of F. For this problem

$$f(x,y,z) = x^2 + y^2 + z^2 \text{ and } G(x,y,z) = x^2 + 2y^2 - z^2 = 1.$$

To apply the method of Lagrange let

$$u = x^2 + y^2 + z^2 + \lambda(x^2 + 2y^2 - z^2).$$

Then

$$\frac{\partial u}{\partial x} = 2x + \lambda 2x = 0 \quad (2)$$

$$\frac{\partial u}{\partial y} = 2y + \lambda 4y = 0 \quad (3)$$

$$\frac{\partial u}{\partial z} = 2z - \lambda 2z = 0 \quad (4)$$

Next, let (x_0,y_0,z_0) be a solution. If $z_0 \neq 0$, then from equation (4), $\lambda =$ 1. For this to be true in (2) and (3), it is needed $x = y = 0$. In that case from (1), this gives $z_0^2 = -1$ which is impossible. Hence, any solution must have $z_0 = 0$.

If $x \neq 0$, then from (2), $\lambda = -1$ and from (3) and (4) for this to be true $y = z = 0$. Then, from (1), this yields $x_0^2 = 1$; $x_0 \pm 1$. Therefore, two solutions satisfying the conditions have been obtained, namely

(1,0,0) and (-1,0,0). similarly, if $y \neq 0$, then from (3), $\lambda = -\frac{1}{2}$ and

from (2) and (4), this gives $x = z = 0$. Hence, from (1), $y_0^2 = \frac{1}{2}$; $y_0 =$ $\pm\sqrt{\frac{1}{2}}$ and two more solutions, namely $(0,\sqrt{\frac{1}{2}},0)$ and $(0,-\sqrt{\frac{1}{2}},0)$,

are found. These four points are therefore the extrema of the function f(x,y,z) subject to the constraint g. If a minimum of f is desired, then direct computation shows that the two points $(0, \pm\sqrt{\frac{1}{2}}, 0)$ are

the only possible solutions. This is because the function $f(x,y,z) = x^2 + y^2 + z^2$ is the square of the distance from the origin. Since the constraint defines a surface, a minimum for f(x,y,z) given the constraint is a point on the surface which is at a minimum distance

from the origin. Hence because $1 > \frac{1}{2}$ the points $(0, \pm\sqrt{\frac{1}{2}}, 0)$ are a

shorter distance from the origin than the points $(\pm 1,0,0)$.

15.11 Exact Differentials

The differential of the function f(x,y) is

$$df = \frac{\partial f}{\partial x} dx + \frac{\partial f}{\partial y} dy$$

The quantity df is a function of four variables, x, dx, y and dy.

Definition:

Suppose there is a function f(x,y,z) such that

$$df = P(x,y,z)dx + Q(x,y,z)\,dy + R(x,y,z)dz$$

for all (x,y,z) in some region, and for all values of dx, dy, dz, we say that

$$Pdx + Qdy + Rdz$$

is the exact differential.

Example: Show that $(2xe^{x^2} \sin y)dx + (e^{x^2} \cos y)dy$ is an exact differential and find the function f of which it is the total differential.

$$(2xe^{x^2} \sin y)dx + (e^{x^2} \cos y)dy$$

Solution:

Let $P = 2xe^{x^2} \sin y$, and $Q = e^{x^2} \cos y$. Differentiate to find P_y and Q_x:

$$P_y = Q_x = 2xe^{x^2} \cos y.$$

Thus, the function is an exact differential.

Next, set $f_x = 2xe^{x^2} \sin y$. Thus,

$$f = e^{x^2} \sin y + c(y).$$

Differentiate with respect to y to obtain

$$f_y = e^{x^2} \cos y + c'(y)$$

Let f_y equal to Q:

$$e^{x^2} \cos y + c'(y) = e^{x^2} \cos y$$

Solve for $c'(y)$:

$$c'(y) = 0$$

Integrate to obtain $c(y)$:

$$c(y) = c = \text{constant}$$

Thus,

$$f(x,y) = e^{x^2} \sin y + c.$$

CHAPTER 16

Multiple Integration

16.1 Double Integrals: Iterated Integrals

Definition:

Suppose f is a function defined on a region R. The double integral of f over R, denoted by $\iint_R f(x,y)dA$, is given by the formula

$$\iint_R f(x,y)dA = \lim_{\|\Delta A\| \to 0} \Sigma f(u_i, v_i) \Delta A_i,$$

provided that the limit exists.

The double integral measures the volume under a surface region.

Theorem:

If the function f(x,y) is continuous for (x,y) in a closed region R, then f is said to be integrable over R. Furthermore, if $f(x,y) > 0$ for (x,y) in R, then

$$v = \iint_R f(x,y)dA,$$

where v is the volume of the solid defined by

$B = \{(x,y,z):(x,y) \text{ in } R, \text{ and}$
$0 \leq z \leq f(x,y)\}.$

16.1.1 Properties of the Double Integral

1) Suppose c is a number and f is integrable over a closed region R, then cf is integrable and

$$\iint_R cf(x,y)dA = c\iint_R f(x,y)dA$$

2) If f and g are two integrable functions over a closed region R, then

$$\iint_R [f(x,y)+g(x,y)]dA = \iint_R f(x,y)dA + \iint_R g(x,y)dA$$

3) Consider f is an integrable function on a closed region R, and $m \leq f(x,y) \leq M$ for all (x,y) in R. If A denotes the area of R, we have

$$mA \leq \iint_R f(x,y)dA \leq MA$$

4) If the function f and g are both integrable over the region R, and $f(x,y) \leq g(x,y)$ for all (x,y) in R, then

$$\iint_R f(x,y)dA \leq \iint_R g(x,y)dA$$

5) Suppose that the closed region R is decomposed into non-overlapping regions R_1 and R_2, and suppose that f

is continuous over R, then:

$$\iint_R f(x,y)dA = \iint_{R_1} f(x,y)dA + \iint_{R_2} f(x,y)dA$$

16.1.2 Iterated Integrals

The symbol $\int_c^d f(x,y)dy$ denotes the partial integration with respect to y.

For each x in the interval [a,b], there corresponds a unique value of this integral.

A function A is thus determined, where the value A(x) is given by the formula

$$A(x) = \int_c^d f(x,y)dy.$$

The function A is continuous on the interval [a,b], and its definite integral may be written

$$\int_a^b A(x)dx = \int_a^b \left[\int_c^d f(x,y)dy \right] dx$$

The expression on the right side of the equation is called an iterated (double) integral.

The terms "successive integrals" and "repeated integrals" may also be used.

$$\int_c^d \int_a^b f(x,y)dxdy = \int_c^d \left[\int_a^b f(x,y)dx \right] dy$$

$$\int_a^b \int_c^d f(x,y)dydx = \int_a^b \left[\int_c^d f(x,y)dy \right] dx$$

Iterated integrals may also be defined over regions that have curved boundaries.

Theorem:

Suppose R is a closed region given by

$$R = \{(x,y) : a \le x \le b, p(x) \le y \le g(x)\},$$

where p and g are continuous and $p(x) \le g(x)$ for $a \le x \le b$. Also suppose that f(x,y) is continuous on R. Then,

$$\iint_R f(x,y)dA = \int_a^b \int_{p(x)}^{q(x)} f(x,y)dydx$$

The corresponding results hold if the closed region R has the representation

$$R = \{(x,y) : c \le y \le d, \ r(y) \le x \le s(y)\},$$

where $r(y) \le s(y)$ for $c \le y \le d$. In such a case,

$$\iint_R f(x,y)dA = \int_c^d \int_{r(y)}^{s(y)} f(x,y)dxdy$$

Both iterated integrals are equal to the double integral and are therefore equal to each other.

Problem Solving Examples:

Find $u = \iint (x^3 + y^3)\, dy \cdot dx$.

We integrate first with respect to y, treating x as a constant; and then with respect to x, treating y as a constant. We write:

$$u = \int dx \int (x^3 + y^3)\, dy$$

$$= \int dx \left[x^3 y + \frac{y^4}{4} + \psi(x) \right] \text{ where } \psi(x) \text{ is}$$

an arbitrary function of x.

$$u = y \int x^3 dx + \frac{y^4}{4} \int dx + \int \psi(x) dx$$

$$= \frac{x^4 y}{4} + \frac{y^4 x}{4} + \varphi_1(x) + \varphi_2(y),$$

where $\varphi_1(x)$ and $\varphi_2(y)$ are arbitrary functions.

Calculate: $\int_0^1 \int_y^1 y e^{-x^3}\, dx\, dy$.

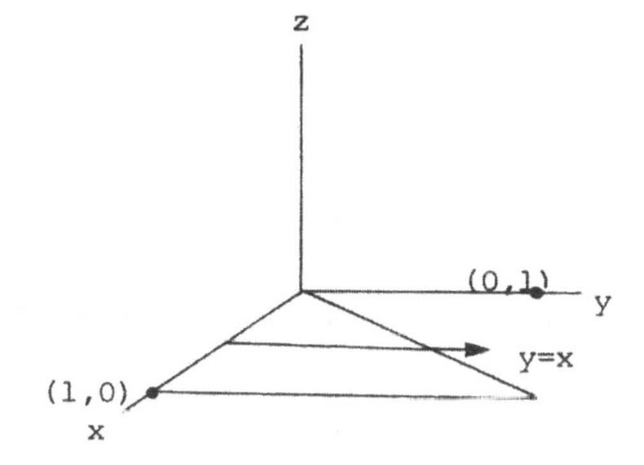

Here it is difficult to carry out the first or inner integration. However, we can tell from the limits of integration that this iterated integral represents the volume of the solid in the first octant bounded by the planes, $y = 0$, $y = 1$, $x = 1$ and $x = y$ and the surface $z = ye^{-x^3}$ (see the accompanying figure). We can therefore reverse the order of integration and integrate first with respect to y and then with respect to x. For the integral with respect to y, the limits of integration can be found by drawing a line parallel to the y-axis. The line first meets $y = 0$, the x-axis and lower limit, and then meets the line $y = x$, the upper limit. We can now write the new integral as:

$$\int_0^1 \int_0^x ye^{-x^3}\, dy\, dx$$

This is more easily evaluated.

$$\int_0^1 \int_0^x ye^{-x^3}\, dy\, dx = \int_0^1 dx \int_0^x ye^{-x^3} dy$$

$$= \int_0^1 dx \left[\frac{y^2}{2} e^{-x^3}\right]_0^x$$

$$= \int_0^1 \frac{x^2}{2} e^{-x^3} dx$$

$$= -\frac{1}{6} \int_0^1 e^{-x^3} (-3x^2)\, dx$$

$$= -\frac{1}{6} \left[e^{-x^3}\right]_0^1 = -\frac{1}{6}\left(\frac{1}{e} - 1\right) = \frac{e-1}{6e}.$$

Use double integration to find the area enclosed by $y = x^2$ and $x + y - 2 = 0$.

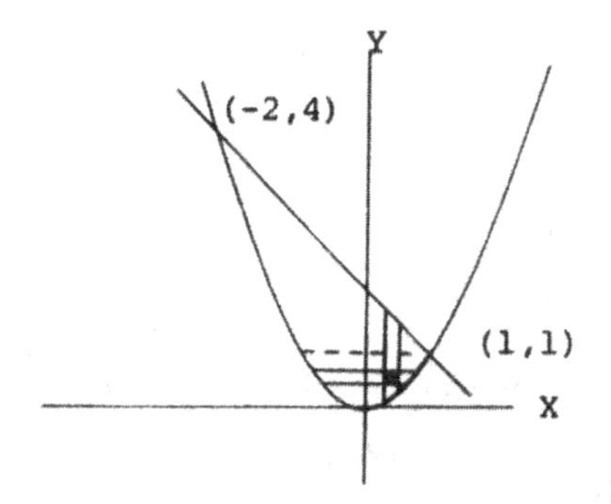

The formula for area in Cartesian coordinates, using double integrals is:

$$A = \iint dy\,dx = \int_a^b dx \int_{f(x)}^{F(x)} dy.$$

The limits a and b of the integral with respect to x are the x-coordinates of the points of intersection of the two curves. To find the points of intersection, we set $y = x^2$ equal to $y = 2-x$ and solve for x.

$$x^2 = 2 - x \quad (x+2)(x-1) = 0$$
$$x = -2 \quad x = 1$$

The limits of the integral with respect to y are the two functions. The lower limit is the lower function, the parabola $y = x^2$. The upper limit is the upper function, the line $y = 2 - x$. Therefore,

$$A = \int_{-2}^{1} dx \int_{x^2}^{2-x} dy$$

$$= \int_{-2}^{1} dx \Big[y \Big]_{x^2}^{2-x}$$

$$= \int_{-2}^{1} \left[2 - x - x^2\right] dx = \left[2x - \frac{x^2}{2} - \frac{x^3}{3}\right]_{-2}^{1}$$

$$= \frac{7}{6} + \frac{10}{3} = \frac{27}{6}$$

$$= \frac{9}{2} \text{ sq. units}$$

We can also find the area by reversing the order of integration. Instead of integrating first with respect to y and then x we employ:

$$A = \iint dx\,dy = \int_a^b dy \int_{f(y)}^{F(y)} dx.$$

However, from the diagram we can see that the total area must be considered as the sum of two areas, one above the dotted line in the diagram, and one below. The reason for this is that the limits for the integrals are different for each area. for the area above the dotted line, the limits of the integral with respect to x are $x = -\sqrt{y}$, the lower limit, and $x = 2 - y$, the upper limit. For the integral with respect to y for the area above the dotted line, y goes from $y = 1$, the lower limit, to $y = 4$, the upper limit. For the area below the dotted line, the curve $x = -\sqrt{y}$ is the lower limit and the curve $x = \sqrt{y}$ is the upper limit of the integral with respect to x. For the integral with respect to y, for the area below the dotted line, x goes from $y = 0$, the lower limit, to $y = 1$, the upper limit. We can now write,

$$A_{\text{Total}} = \int_1^4 dy \int_{-\sqrt{y}}^{2-y} dx + \int_0^1 dy \int_{-\sqrt{y}}^{\sqrt{y}} dx$$

$$= \int_1^4 dy \left[x\right]_{-\sqrt{2}}^{2-y} + \int_0^1 dy \left[x\right]_{\sqrt{2}}^{\sqrt{y}}$$

$$= \int_1^4 2 - y + \sqrt{y}\,dy + \int_0^1 \sqrt{y} + \sqrt{y}\,dy$$

$$= \left[2y - \frac{y^2}{2} + \frac{2y^{\frac{3}{2}}}{3} \right]_1^4 + \left[\frac{4y^{\frac{3}{2}}}{3} \right]_0^1$$

$$= \frac{9}{2} \text{ sq, units}$$

Clearly, the first method is preferable here.

16.2 Area and Volume

16.2.1 Volume

The volume between a surface and a region R is represented by the double integral of a non-negative function. This volume is expressed in terms of the volume of the cylinders which have generators parallel to the z-axis and located between the surface and the region R in the xy plane. Thus,

$$\text{Volume (v)} = \iint_R dA$$

Problem Solving Examples:

Sketch the volume represented by the iterated integral:

$\int_0^1 \int_0^{\sqrt{1-y^2}} 4y \, dx \, dy$, and compute its volume.

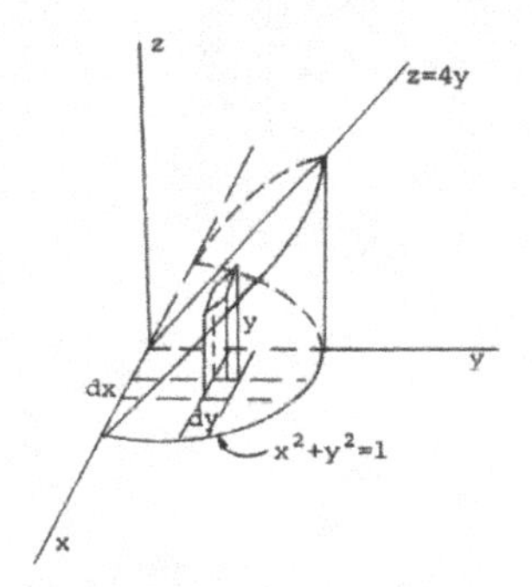

The volume of any solid can be represented as the sum of the volumes of all the approximate parallelepipeds, the dimensions of which are: the height z and the elementary area on the xy-plane (dx by dy). This volume of the parallelepiped is:

$$dV = z\,dx\,dy$$

Integration over the region R (the area of the surface projected on the xy-plane) gives the required volume:

$$V = \int_R z\,dA.$$

Comparing to the given expression, we see that $z = 4y$ represents the equation of the surface. It is a plane. Since, from the limits of the integration, x varies from 0 to $\sqrt{1-y^2}$, the variation of x in terms of y is:

$$x^2 + y^2 = 1,$$

which is the circle of unity radius on the xy-plane, namely the region R. $0 \le y \le 1$ indicates the positive y-axis, and the fact that x starts from zero indicates the positive x-axis. Thus, the volume to be calculated is the volume in the first octant.

According to the process of evaluating multiple integrals:

$$V = \int_0^1 4y \left[\int_0^{\sqrt{1-y^2}} dx \right] dy$$

$$= \int_0^1 4y \left(x \Big|_0^{\sqrt{1-y^2}} \right) dy$$

$$= 4\int_0^1 y\,\sqrt{1-y^2}\,dy$$

Using the substitution method,

$$V = 4\int_0^1 y\,\sqrt{1-y^2}\,dy$$

$$= 4\left(-\frac{1}{3}(1-y^2)^3\right|_0^1 = \frac{4}{3}.$$

Find the volume bounded by the cylinder $x^2 + y^2 = 4$ and the hyperboloid $x^2 + y^2 - z^2 = 1$.

A On the xy-plane, the required volume is bounded by the circle $r = 1$ (obtained by setting $z = 0$ in the equation of the hyperboloid), and the base radius of the cylinder is 2. The volume is then obtained by integrating

$$z = (x^2 + y^2 - 1)^{\frac{1}{2}} = \sqrt{r^2 - 1}$$

over the region bounded by the two circles. Since z can assume a positive or negative sign, and due to symmetry over the plane $z = 0$, the height for the required volume is

$$2z = 2\sqrt{r^2 - 1}$$

Hence,

$$V = \int_0^{2\pi}\int_1^2 2r\sqrt{r^2 - 1}\,dr\,d\theta.$$

The integral:

$$\int_1^2 2r\sqrt{r^2-1}\,dr = \int_1^2 2r(r^2-1)^{\frac{1}{2}}\,dr = \frac{2}{3}(r^2-1)^{\frac{3}{2}}\Big|_1^2 = 2\sqrt{3},$$

and

$$V = \int_0^{2\pi} 2\sqrt{3}\,d\theta = 4\pi\sqrt{3}.$$

16.2.2 Area

The area of a surface is found by using the iterated integral. The formula used to find the area of a region is

$$A = \iint_R dA = \int_a^b \int_{p(x)}^{q(x)} dydx = \lim_{\|\Delta\|} \sum_i \sum_j \Delta y_j \, \Delta x_i$$

Problem Solving Examples:

Find the area inside the cardioid: $r = a(1 + \sin \theta)$, and outside the circle: $r = a$.

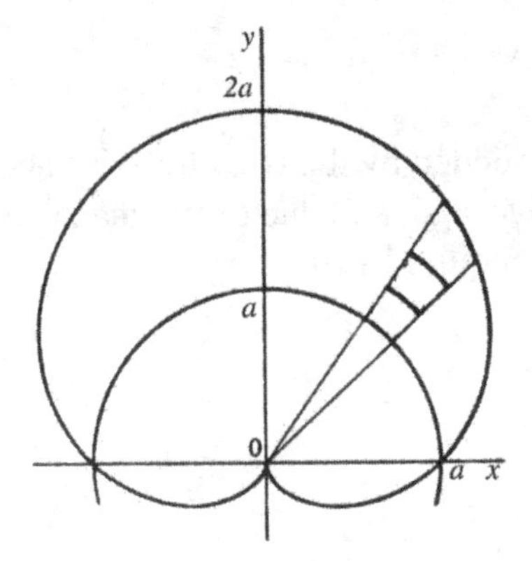

The area is the difference between the two areas, one bounded by the cardioid, and the other bounded by the circle. We can use the formula for area in polar coordinates, which is:

$$A = \int_\alpha^\beta \frac{1}{2} r^2 d\theta$$

The limits are the points of intersection of the two curves, which are at $\theta = 0$ and at $\theta = \pi$. Hence, we can write:

$$A = \int_0^{\pi} \frac{1}{2}(a(1 + \sin\theta))^2 \, d\theta - \int_0^{\pi} \frac{1}{2} a^2 \, d\theta$$

$$= \frac{a^2}{2}\int_0^{\pi} (1 + 2\sin\theta + \sin^2\theta)\, d\theta - \frac{a^2}{2}\int_0^{\pi} d\theta.$$

Now, substituting: $\sin^2\theta = \frac{1 - \cos 2\theta}{2}$, and combining the two integrals, we obtain:

$$A = \frac{a^2}{2}\int_0^{\pi}\left(2\sin\theta + \frac{1-\cos 2\theta}{2}\right) d\theta$$

$$= \frac{a^2}{2}\left[\frac{\theta}{2} - \sin 2\theta - 2\cos\theta\right]_0^{\pi}$$

$$= \frac{a^2}{4}(8 + \pi).$$

We can also obtain this value by using double integrals. The formula for area using double integrals in polar coordinates is:

$$A = \int\int r dr\, d\theta.$$

It is convenient in this problem to integrate first with respect to r. The lower limit on r is the circle: $r = a$ and the upper limit is the cardioid: $r = a(1 + \sin\theta)$. The limits on θ are the θ-coordinates of the points of intersection of the two curves, which in this example are on the rays $\theta = 0$ and $\theta = \pi$. Therefore, the desired area is given by

$$A = \int_0^{\pi} d\theta \int_a^{a(1+\sin\theta)} r dr$$

$$= \int_0^{\pi} d\theta\left[\frac{r^2}{2}\right]_a^{a(1+\sin\theta)}$$

$$= \int_0^{\pi} \left[\frac{(a(1 + \sin\theta))^2}{2} - \frac{a^2}{2} \right] d\theta$$

$$= \int_0^{\pi} \frac{a^2}{2}(1 + 2\sin\theta + \sin^2\theta - 1)\, d\theta$$

$$= \int_0^{\pi} \frac{a^2}{2}(2\sin\theta + \sin^2\theta)\, d\theta$$

$$= \frac{a^2}{2}(-2\cos\theta)\Big|_0^{\pi} + \frac{a^2}{2}\int_0^{\pi} \left(\frac{1 - \cos 2\theta}{2}\right) d\theta$$

$$= -a^2\cos\theta + \frac{a^2}{2}\left(\frac{\theta}{2} + \frac{\sin 2\theta}{4}\right)\Big|_0^{\pi}$$

$$= \frac{a^2}{4}\left[(1 + 1)\,4 + \pi + 0\right]$$

$$= \frac{a^2}{4}(8 + \pi).$$

16.2.3 Volume of Surface of Revolution

Theorem:

If a plane figure B lies on one side of a line L in its plane, the volume of the solid, generated by revolving the figure around the line, is equal to the product of the area (A) and the length of the path described by the center of mass. Thus, if the figure is in the xy plane and L is the x-axis, then,

$$v = 2\pi\bar{y}A = \iint 2\pi y\, dA$$

16.3 Moment of Inertia and Center of Mass

Let L denote a lamina having the shape of the region R as shown in the figure below.

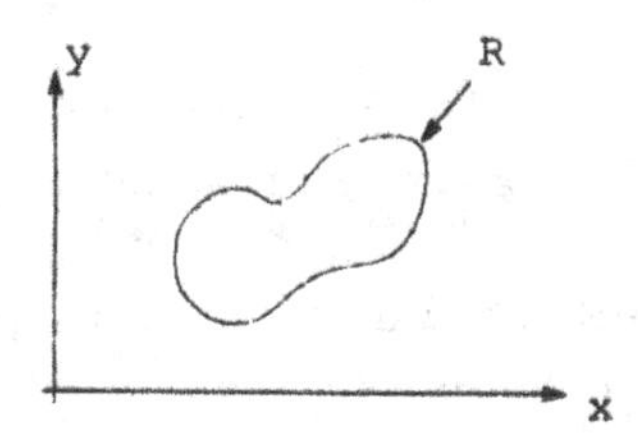

Let it be given that the density per unit area at a point (x,y) is $\rho(x,y)$ where ρ is a continuous function on R.

The mass of L is given by

$$M = \iint_R \rho(x,y)dA.$$

The moment of L with respect to the x-axis is given by

$$M_x = \iint_R y \, \rho(x,y)dA.$$

The moment of L with respect to the y-axis is given by

$$M_y = \iint_R x \, \rho(x,y)dA.$$

The center of mass of the Lamina, L, is the point $(\bar{x},\bar{y})$, where

$$\bar{x} = \frac{M_y}{M} \quad \text{and} \quad \bar{y} = \frac{M_x}{M}.$$

16.3.1 Moments of Inertia

The moment of inertia with respect to the x-axis is

$$I_x = \sum_{i=1}^{n} y_i^2 m_i,$$

for n particles of masses $m_1, m_2, \ldots, m_n$ located at points $(x_1, y_1), (x_2, y_2), \ldots, (x_n, y_n)$, respectively.

Also the moment of inertia with respect to the y-axis is

$$I_y = \sum_{i=1}^{n} x_i^2 m_i.$$

For lamina:

$$I_x = \iint_R y^2 \rho(x,y)\,dA$$

and

$$I_y = \iint_R x^2 \rho(x,y)\,dA$$

Problem Solving Examples:

Find the moment of inertia about the x-axis of the curve: $4x = 2y^2 - \ln y$, from $y = 2$ to $y = 4$.

If an arc length Δs, is between x and $x + \Delta x$, its moment of inertia with respect to the x-axis is

$$\Delta I_x = y^2 \Delta s,$$

where Δs is a one-dimensional arc. Therefore it is not necessary to find its area or volume. The sum of all such infinitesimal curves, each multiplied by the square of its ordinate, gives the required moment of inertia. From differential calculus, we have

$$ds = \sqrt{dx^2 + dy^2},$$

where $ds = \Delta s$ when Δs decreases to an infinitesimal dimension. The expression for ds can also be expressed as:

$$ds = \left[\left(\frac{dx}{dy}\right)^2 + 1\right]^{\frac{1}{2}} dy .$$

From the given equation,

$$x = \frac{2y^2}{4} - \frac{1}{4}\ln y, \text{ and}$$

$$\frac{dx}{dy} = y - \frac{1}{4y}$$

$$ds = \left[\left(y - \frac{1}{4y}\right)^2 + 1\right]^{\frac{1}{2}} dy = \left(y^2 - \frac{1}{2} + \frac{1}{16y^2} + 1\right)^{\frac{1}{2}} dy$$

$$= \left(y + \frac{1}{4y}\right) dy.$$

$$I_x = \int \Delta I_x = \int_2^4 \left(y + \frac{1}{4y}\right) y^2\, dy = \int_2^4 \left(y^3 + \frac{y}{4}\right) dy = \left.\frac{y^4}{4} + \frac{y^2}{8}\right]_2^4 = \left[64 + 2 - 4 - \frac{1}{2}\right]$$

$$= \frac{123}{2} .$$

Determine the center of gravity of the area bounded by $y^2 = 2x$, $x = 2$, and $y = 0$.

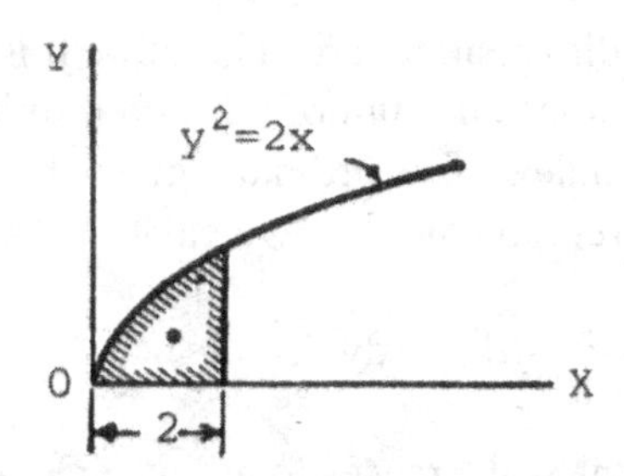

$\Delta M_y = x\Delta A$ and $\Delta M_x = y\Delta A$. Using the double integral over the given region,

$$M_y = \int_0^2 \int_0^{\sqrt{2x}} x \cdot dy \cdot dx,$$

the moment about the y-axis.

$$M_y = \int_0^2 \left[x \cdot y \right]_0^{\sqrt{2x}} \cdot dx = \int_0^2 x\sqrt{2x} \cdot dx$$

$$= \sqrt{2} \int_0^2 x^{\frac{3}{2}} \cdot dx .$$

Finally,

$$M_y = \left[\sqrt{2} \cdot x^{\frac{5}{2}} \cdot \frac{2}{5} \right]_0^2 = \frac{2\sqrt{2}}{5} \left[x^{\frac{5}{2}} \right]_0^2$$

$$= \frac{2\sqrt{2}}{5} \cdot \sqrt{2^5} = \frac{16}{5} .$$

Also,

$$M_x = \int_0^2 \int_0^{\sqrt{2x}} y \cdot dy \cdot dx,$$

the moment about the x-axis.

$$M_x = \int_0^2 \left[\frac{y^2}{2} \right]_0^{\sqrt{2x}} \cdot dx = \int_0^2 \frac{2x}{2} \cdot dx = \int_0^2 x \cdot dx = \left[\frac{x^2}{2} \right]_0^2 .$$

Finally

$$M_x = \frac{4}{2} = 2.$$

Now

$$A = \text{area} = \int_0^2 \int_0^{\sqrt{2x}} dy \cdot dx = \int_0^2 \Big[y \Big]_0^{\sqrt{2x}} \cdot dx .$$

$$A = \int_0^2 \sqrt{2x} \cdot dx = \sqrt{2} \int_0^2 x^{\frac{1}{2}} \cdot dx .$$

Thus,

$$A = \sqrt{2} \left[x^{\frac{3}{2}} \cdot \frac{2}{3} \right]_0^2 \; \frac{2\sqrt{2}}{3} \left[x^{\frac{3}{2}} \right]_0^2 = \frac{2\sqrt{2}}{3} \cdot \sqrt{2^3} = \frac{2}{3} \cdot 4 = \frac{8}{3},$$

and hence

$$\bar{x} + \frac{M_y}{A} = \frac{16}{5} \cdot \frac{3}{8} = \frac{6}{5}$$

$$\bar{y} = \frac{M_x}{A} = 2 \cdot \frac{3}{8} = \frac{3}{4} .$$

Therefore, the c.g. is at $\bar{x} = \frac{6}{5}, \bar{y} = \frac{3}{4}$.

Find the center of gravity of a homogeneous hemisphere of radius r.

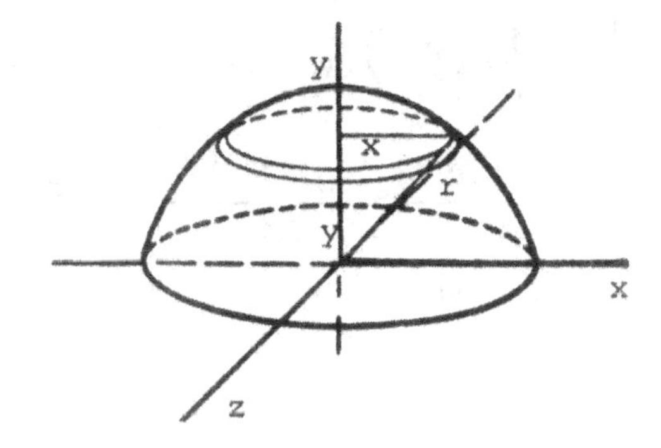

Since the hemisphere is symmetric on the y-z and x–y planes, the net moment with respect to these planes is zero, making $\overline{x} = \overline{z} = 0$. Then, the center of gravity will lie on the diameter perpendicular to the base at a distance $\overline{y}$ above the base.

To find the volume and moment in integral forms, the hemisphere can be thought of as a volume generated by a semi-circle in the first quadrant about the y-axis. Hence, the small strip made by a line drawn from the y-axis to an arbitrary point p(x,y) parallel to the x-axis and another parallel line Δy units above it will look like a disc with radius x units, thickness Δy, and volume

$$\Delta v = \pi x^2 \Delta y.$$

Passing to the limits, we obtain the volume expression in an integral form:

$$v = \int_0^r \pi x^2 \, dy.$$

The turning effect of the mass of the hemisphere with respect to the z–x plane can be expressed as:

$$M_{zx} = \rho \int_V y du, \text{ where } \rho \text{ is the density.}$$

We wish to obtain an analogous expression, where the mass would be thought of as concentrated at a point some distance $\overline{y}$ above the origin, such that $\overline{y}m = M_{zx}$, or, under constant density,

$$\overline{y}V = \frac{M_{zx}}{\rho} = \int_V y \, dv.$$

$$\overline{y} = \frac{\int_V y \, du}{V} = \frac{\int_0^r y\pi x^2 \, dy}{\int dv} \cdot x^2 = r^2 - y^2$$

Hence,

$$\overline{y} = \frac{\int_0^r \pi y(r^2 - y^2) \, dy}{\frac{2}{3}\pi r^3}$$

$$= \frac{3}{2r^3}\left[\frac{r^2y^2}{2} - \frac{y^4}{4}\right]_0^r$$

$$= \frac{3}{8} r .$$

Therefore the center of gravity of the given hemisphere with radius r is at $(0, \frac{3}{8}r, 0)$.

16.4 Polar Coordinates

Points in the Cartesian coordinate system can be transformed to the polar coordinate system by use of the formulas

$$x = r\cos\theta \quad \text{and} \quad y = r\sin\theta$$

The area of a region in polar coordinates is given by the formula

$$A = \iint_R dA_{r,\theta} = \int_{\theta_1}^{\theta_2}\left[\int_{r_1}^{r_2} r\,dr\right] d\theta$$

where $dA_{r,\theta} = r\,dr\,d\theta$

Theorem:

If a function $f(x,y)$ is continuous on a region R and if this region is related to another region R* by mapping $x = r\cos\theta$ and $y = r\sin\theta$, then the function $s(r,\theta) = f(r\cos\theta, r\sin\theta)$ is defined and continuous on R* and

$$\iint_R f(x,y)\,dA_{x,y} = \iint_{R^*} s(r,\theta)\,dA_{r,\theta}$$

This equation is transformed to

$$\iint_R f(x,y)\,dxdy = \iint_{R^*} s(r,\theta)r\,drd\theta.$$

Problem Solving Examples:

Let F(x,y) be differentiable in x and y and introduce polar coordinates r, θ by writing $x = r\cos\theta$, $y = r\sin\theta$.

Find $\frac{\partial F}{\partial r}$ and $\frac{\partial F}{\partial \theta}$ in terms $\frac{\partial F}{\partial x}$ and $\frac{\partial F}{\partial y}$

The chain rule for composite functions states that if F = F(x,y) and $x = g(u,v)$, $y = h(u,v)$, then at the points where all of the derivatives exist,

$$\frac{\partial F}{\partial u} = \frac{\partial F}{\partial x}\frac{\partial x}{\partial u} + \frac{\partial F}{\partial y}\frac{\partial y}{\partial u}, \tag{1}$$

$$\frac{\partial F}{\partial v} = \frac{\partial F}{\partial x}\frac{\partial x}{\partial v} + \frac{\partial F}{\partial y}\frac{\partial y}{\partial v}. \tag{2}$$

In the case at hand, $x = x(r,\theta) = r\cos\theta$ and $y = y(r,\theta) = r\sin\theta$, so that

$$\frac{\partial x}{\partial r} = \cos\theta,\ \frac{\partial x}{\partial \theta} = -r\sin\theta,\ \frac{\partial y}{\partial r} = \sin\theta,\ \frac{\partial y}{\partial \theta} = r\cos\theta. \tag{3}$$

Thus, replacing u and v with r and θ in equations (1) and (2) and using the results of (3) gives

$$\frac{\partial F}{\partial r} = \frac{\partial F}{\partial x}\cos\theta + \frac{\partial F}{\partial y}\sin\theta \tag{4}$$

$$\frac{\partial F}{\partial \theta} = -\frac{\partial F}{\partial x}r\sin\theta + \frac{\partial F}{\partial y}r\cos\theta. \tag{5}$$

Let F = F(x,y) have second partial derivatives in a region of the xy-plane and introduce polar coordinates in this region by writing $x = r\cos\theta$, $y = r\sin\theta$. Find $\frac{\partial^2 F}{\partial r \partial \theta}$ in terms of derivatives of F with respect to x and y.

Here we have a composite function $F = F(x(r,\theta),y(r,\theta))$ so that the chain rule must be used. First, the chain rule gives

$$\frac{\partial F}{\partial \theta} = \frac{\partial F}{\partial x}\frac{\partial x}{\partial \theta} + \frac{\partial F}{\partial y}\frac{\partial y}{\partial \theta}$$

$$= -\frac{\partial F}{\partial x} r\sin\theta + \frac{\partial F}{\partial y} r\cos\theta \,. \tag{1}$$

Now differentiate both sides of (1) with respect to r, keeping in mind that $\frac{\partial F}{\partial x}$ and $\frac{\partial F}{\partial y}$ depend on x and y and are thus composite functions of r and θ. Therefore, the chain rule must be used when differentiating these functions with respect to r. Hence

$$\frac{\partial^2 F}{\partial r \partial \theta} = \frac{\partial}{\partial r}\left(\frac{\partial F}{\partial \theta}\right) = \frac{\partial}{\partial r}\left(\frac{-\partial F}{\partial x} r\sin\theta\right) + \frac{\partial}{\partial r}\left(\frac{\partial F}{\partial y} r\cos\theta\right)$$

$$= -\frac{\partial F}{\partial x}\frac{\partial}{\partial r}(r\sin\theta) - \frac{\partial}{\partial r}\left(\frac{\partial F}{\partial x}\right) r\sin\theta$$

$$+ \frac{\partial F}{\partial y}\frac{\partial}{\partial r}(r\cos\theta) + \frac{\partial}{\partial r}\left(\frac{\partial F}{\partial y}\right) r\cos\theta \,. \tag{2}$$

Now, the chain rule must be used on the composite functions $\frac{\partial F}{\partial x}$ $[x(r,\theta),y(r,\theta)]$ and $\frac{\partial F}{\partial y}$ $[x(r,\theta)\ y(r,\theta)]$ Hence, it is found that

$$\frac{\partial}{\partial r}\left(\frac{\partial F}{\partial x}\right) = \frac{\partial}{\partial x}\left(\frac{\partial F}{\partial x}\right)\cdot\frac{\partial x}{\partial r} + \frac{\partial}{\partial y}\left(\frac{\partial F}{\partial x}\right)\cdot\frac{\partial y}{\partial r} \tag{3}$$

and
$$\frac{\partial}{\partial r}\left(\frac{\partial F}{\partial y}\right)=\frac{\partial}{\partial x}\left(\frac{\partial F}{\partial y}\right)\cdot\frac{\partial x}{\partial r}+\frac{\partial}{\partial y}\left(\frac{\partial F}{\partial y}\right)\cdot\frac{\partial y}{\partial r}\,. \tag{4}$$

Thus, (3) and (4) become

$$\frac{\partial}{\partial r}\left(\frac{\partial F}{\partial x}\right)=\frac{\partial^2 F}{\partial x^2}\cos\theta+\frac{\partial^2 F}{\partial y\partial x}\sin\theta\,, \tag{5}$$

$$\frac{\partial}{\partial r}\left(\frac{\partial F}{\partial y}\right)=\frac{\partial^2 F}{\partial x\partial y}\cos\theta+\frac{\partial^2 F}{\partial y^2}\sin\theta\,. \tag{6}$$

Using (5) and (6) in (2) yields

$$\frac{\partial^2 F}{\partial r\partial\theta}=-\frac{\partial F}{\partial x}\sin\theta-\left(\frac{\partial^2 F}{\partial x^2}\cos\theta+\frac{\partial^2 F}{\partial y\partial x}\sin\theta\right)r\sin\theta$$

$$+\frac{\partial F}{\partial y}\cos\theta+\left(\frac{\partial^2 F}{\partial x\partial y}\cos\theta+\frac{\partial^2 F}{\partial y^2}\sin\theta\right)r\cos\theta. \tag{7}$$

Now, if the second partial derivatives of a function F are continuous, as shall be assumed in this problem, then a theorem which states that

$\frac{\partial^2 F}{\partial x\partial y}=\frac{\partial^2 F}{\partial y\partial x}$ is valid. Using this relation, equation (7) may be

simplified to yield

$$\frac{\partial^2 F}{\partial r\partial\theta}=r\sin\theta\cos\theta\left(\frac{\partial^2 F}{\partial y^2}-\frac{\partial^2 F}{\partial x^2}\right)-\sin\theta\frac{\partial F}{\partial x}$$

$$+\cos\theta\frac{\partial F}{\partial y}+r(\cos^2\theta-\sin^2\theta)\frac{\partial^2 F}{\partial x\partial y}\,.$$

16.5 The Triple Integrals

The triple integral of a function over the region R is represented by the formula

$$\iiint f(x,y,z)dv = \int_k^e \int_c^d \int_a^b f(x,y,z)dxdydz.$$

The integral on the right is the iterated integral.

Triple integrals may be defined on regions bounded by a parallelepiped and regions in the xy plane.

Theorem:

Suppose that function f is a continuous function on the region R, which is defined by the inequalities

$$R = \{(x,y,z) : a \le x \le b, p(x) \le y \le q(x), \\ r(x,y) \le z \le s(x,y)\},$$

where the functions p, q, r and s are continuous. Then,

$$\iiint_R f(x,y,z)dv = \int_a^b \left\{ \int_{p(x)}^{q(x)} \left[\int_{r(x,y)}^{s(x,y)} f(x,y,z)dz \right] dy \right\} dx$$

Problem Solving Examples:

Find the volume bounded by the surfaces:
$x^2 + y^2 = z$, and $x^2 + y^2 = 4$ and the xy-plane.

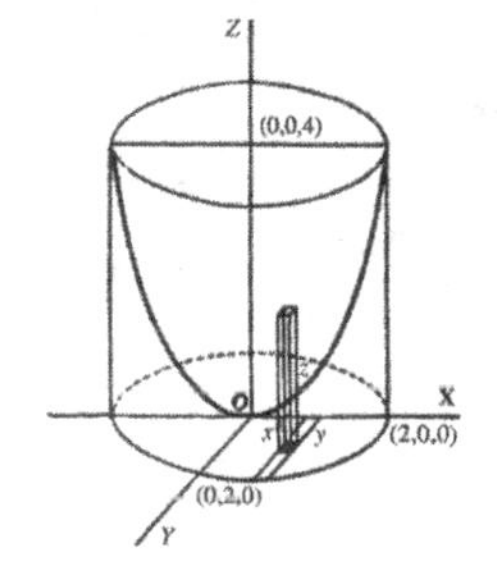

The surface: $x^2 + y^2 = z$, is a paraboloid passing through the origin and symmetric about the z-axis, while the surface: $x^2 + y^2 = 4$, is a cylinder about the z-axis, as shown. Using the general definition of triple integral for volumes, we have:

$$V = \int_S dV = \int_a^b \int_{g_1(x)}^{g_2(x)} \int_{f_1(x,y)}^{f_2(x,y)} dz\, dy\, dx,$$

where the symbol $\int_S dV$ means integration over the given surface S.

We can also express this in the following form

$$V = \int_R f(x,y)\, dA,$$

which means integration over the region R projected on the x-y plane. This is an alternative form of the triple integral. Now, $f(x,y) = z = x^2 + y^2$, y varies in the interior of a circle $x^2 + y^2 = 4$, or $-\sqrt{4-x^2} \le y \le \sqrt{4-x^2}$. $x = -2$ is the lower limit, and $x = 2$ is the upper limit.

By symmetry, the required volume is 4 times the volume in the first octant. Hence,

$$v = 4\int_0^2 \left(\int_0^{\sqrt{4-x^2}} (x^2 + y^2)\, dy \right) dx$$

$$= 4\int_0^2 \left[x^2\sqrt{4-x^2} + \frac{\sqrt{4-x^2}^{\frac{3}{2}}}{3} \right] dx$$

$$= 4\left[4 \text{ arc sin}\frac{x}{2} + \sqrt{4-x^2}\,(x)\left(\frac{x^2}{6} + \frac{1}{3}\right) \right|_0^2$$

$$= 4\left(\frac{4\pi}{2} + 0 - 0 - 0\right)$$

$$= 8\pi.$$

16.5.1 Application of Triple Integrals

The mass of a solid is represented by the formula

$$m = \iiint_R \rho(x,y,z)dv \quad \text{if:}$$

1) The solid has the shape of a three-dimensional region R.
2) The density at (x,y,z) is $\rho(x,y,z)$ where ρ is continuous throughout the region.

The moment of a particle of mass at the point (x,y,z) is

$$M_{xy} = \iiint_R z\,\rho(x,y,z)dv,$$

$$M_{xz} = \iiint_R y\,\rho(x,y,z)dv,$$

and $$M_{yz} = \iiint_R x\,\rho(x,y,z)dv,$$

where the moment was taken with respect to the xy-, xz- and yz-planes.

The center of mass is represented by the point $(\bar{x},\bar{y},\bar{z},)$, where

$$\bar{x} = \frac{M_{yz}}{m}, \quad \bar{y} = \frac{M_{xz}}{m} \quad \text{and} \quad \bar{z} = \frac{M_{xy}}{m}.$$

The center of mass of a homogeneous solid is dependent only on the shape of the region, since the function ρ is constant and therefore is cancelled.

A centroid is defined as the graph of corresponding points for geometric solids in two dimensions.

The moments of inertia (I) of a mass with respect to the x-, y-, and z-axes are given by the following formulas:

$$I_x = \iiint_R (y^2+z^2)\,\rho(x,y,z)dv$$

$$I_y = \iiint_R (x^2+z^2)\,\rho(x,y,z)dv$$

$$I_z = \iiint_R (x^2+y^2)\,\rho(x,y,z)dv$$

The radius of gyration of a solid of mass m which has a moment of inertia (I) with respect to a line, is defined as a number r, such that

$$I = mr^2$$

Thus we conclude that the radius of gyration is the distance from the line at which all the mass could be concentrated without changing the moment of inertia of the solid.

16.6 Cylindrical and Spherical Coordinates of Triple Integrals

16.6.1 Cylindrical Coordinates

Triple integrals are sometimes expressed in terms of cylindrical coordinates.

One such case occurs if a function of r, θ and z is continuous throughout a region

$$R = \{(r,\theta,z) : a \le r \le b,\ c \le \theta \le d,\ k \le z \le \ell\}.$$

If a point in the region R is represented by (r_i, θ_i, z_i), then the triple integral of the function over the region R is

$$\iiint_R f(r,\theta,z)dv = \lim_{\|P\|\to 0} \sum_{i=1}^{n} f(r_i,\theta_i,z_i,)\,\Delta v_i,$$

where $\|P\|$ represents the length of the longest diagonal in the region.

$$\iiint_R f(r,\theta,z)dv = \int_k^{\ell}\int_c^{d}\int_a^{b} f(r,\theta,z)r\,dr\,d\theta\,dz$$

By using inner partitions, triple integrals may be defined over complicated regions.

Thus,

$$\iiint_R f(r,\theta,z)dv = \int_{\alpha}^{\beta}\int_{p(\theta)}^{q(\theta)}\int_{k_1(r,\theta)}^{k_2(r,\theta)} f(r,\theta,z)r\,dz\,dr\,d\theta$$

Problem Solving Examples:

Find the volume bounded by the cylinder $z = \dfrac{4}{(y^2 + 1)}$ and the planes $y = x$, $y = 3$, $x = 0$, and $z = 0$.

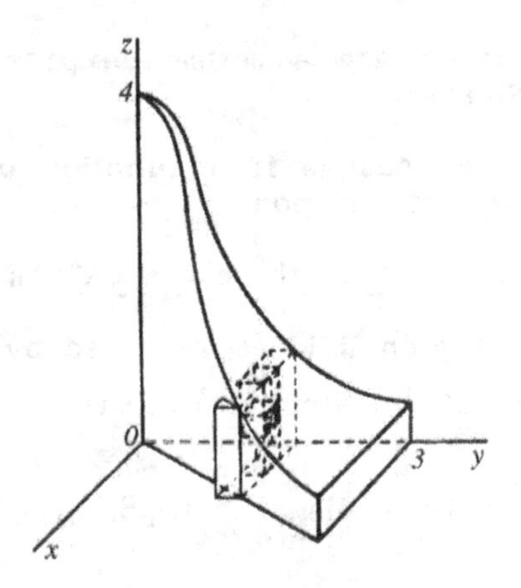

The figure indicates that the volume may be found by integrating first with respect to z between 0 and $\dfrac{4}{(y^2 + 1)}$, then with respect to x between 0 and y, and finally with respect to y, between 0 and 3. Thus we obtain:

$$V = \int_0^3 \int_0^y \int_0^{\frac{4}{(y^2+1)}} dz\, dx\, dy$$

$$= \int_0^3 \int_0^y \frac{4}{y^2 + 1}\, dx\, dy$$

$$= \int_0^3 \frac{4}{y^2 + 1}\, dy = 2\left(\ln (y^2 + 1) \right|_0^3$$

$$= 2 (\ln 10 - \ln 1) = 2 \ln 10.$$

16.6.2 Spherical Coordinates

Triple integrals may also be expressed using spherical coordinates. If a function f of ρ, ϕ and θ is continuous throughout a region

$$R = \{(\rho, \phi, \theta) = a \leq \rho \leq b,\ c \leq \theta \leq d,\ k \leq \phi \leq \ell\},$$

then

$$\iiint_R f(\rho, \phi, \theta)\,dv = \int_c^d \int_k^\ell \int_a^b f(\rho, \phi, \theta)\rho^2 \sin\phi \, d\rho \, d\phi \, d\theta .$$

The letter ρ does not denote density in the equation of spherical coordinates.

Problem Solving Examples:

Find the volume cut from the cone: $x^2 + y^2 - z^2 = 0$, by the sphere: $x^2 + y^2 + (z-2)^2 = 4$

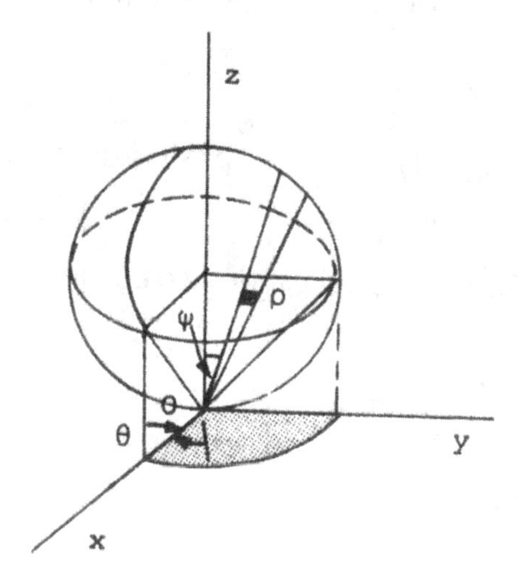

The calculation can be straightforward if we reformulate the equations of the cone and the sphere in the spherical coordinate system, based on the transformation

$$x = \rho \sin \phi \cos \theta$$

$$y = \rho \sin \phi \sin \theta$$

$$z = \rho \cos \phi$$

The equation of the sphere is:

$$\rho = 4 \cos \phi,$$

and that of the cone is:

$$\phi = \frac{1}{4} \pi .$$

Due to the symmetry, we find the volume in the first octant, and the required volume (over the xy-plane) is four times this. Hence

$$V = 4 \int_0^{\frac{\pi}{2}} \int_0^{\frac{\pi}{4}} \int_0^{4 \cos \phi} \rho^2 \sin \phi \, d\rho \, d\phi \, d\theta$$

$$= \frac{256}{3} \int_0^{\frac{\pi}{2}} \int_0^{\frac{\pi}{4}} \cos^3 \phi \sin \phi \, d\phi \, d\theta$$

$$= \frac{16}{3} \int_0^{\frac{\pi}{2}} d\theta$$

$$= \frac{8\pi}{3} .$$

16.7 Surface Area A

Let f represent a function which is defined as $f(x,y) \geq 0$ throughout a region R in the xy plane and which has continuous first derivatives on R.

T is defined as the part of the graph of f for which a projection on the xy-plane is R.

The area of T is defined by the formula

$$A = \iint_R \sqrt{(f_x(x,y))^2 + (f_y(x,y))^2 + 1}\; dA.$$

This formula is also applicable when $f(x,y) \leq 0$ on the region R.

Problem Solving Examples:

Find the area of the surface generated by revolving the lemniscate: $r^2 = a^2 \cos 2\theta$, about the line: $\theta = \frac{\pi}{2}$.

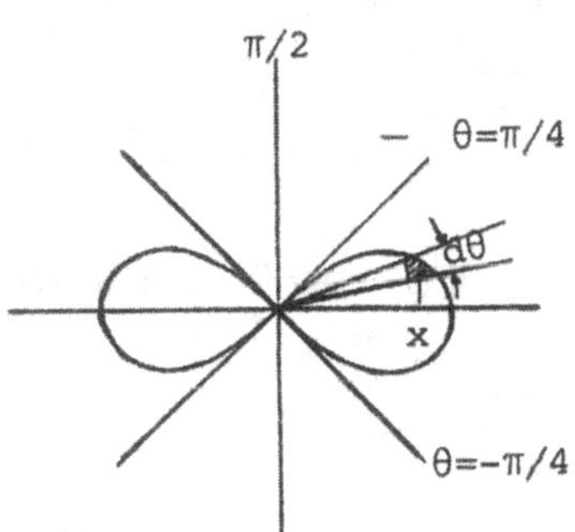

As shown in the figure, the area generated by the shaded region, when revolved about the $\frac{\pi}{2\text{-axis}}$, is:

$2\pi x ds$,

where x is the radius of generation, and ds is the length of the arc subtended by $d\theta$. Since, at any point P on the edge of the lemniscate,

$$x = r\cos\theta,\ rd\theta = ds,$$

and due to symmetry,

$$A = 4\pi \int_0^{\frac{\pi}{4}} r\cos\theta\,(rd\theta)$$

$$= 4\pi a^2 \int_0^{\frac{\pi}{4}} \cos\theta\cos 2\theta\, d\theta = 4\pi a^2 \int_0^{\frac{\pi}{4}} \cos\theta\,(1-2\sin^2\theta)\, d$$

$$= 4\pi a^2 \left[\sin\theta - \frac{2\sin^3\theta}{3}\right]_0^{\frac{\pi}{4}}$$

$$= 4\pi a^2 \left(\frac{1}{\sqrt{2}} - \frac{2}{3}\frac{1}{2\sqrt{2}}\right)$$

$$= 4\pi a^2 \frac{2}{3}\left(\frac{1}{\sqrt{2}}\right) = 4\frac{\pi a^2\sqrt{2}}{3}$$

Note, that multiplying both sides of the equation by r^2, we obtain:

$$r^4 = r^2a^2\cos 2\theta = a^2(r^2\cos^2\theta - r^2\sin^2\theta) = a^2\, x^2 - y^2\ ,$$

On the other hand,

$$r^2 = x^2 + y^2,$$

Thus,

$$(x^2 + y^2)^2 = a^2\,(x^2 - y^2).$$

Therefore, insertion of $x = r\cos\theta$ does not affect the calculation.

Find the area of the surface formed by revolving about the x-axis the parabola: $y^2 = 2x - 1$, from $y = 0$ to $y = 1$.

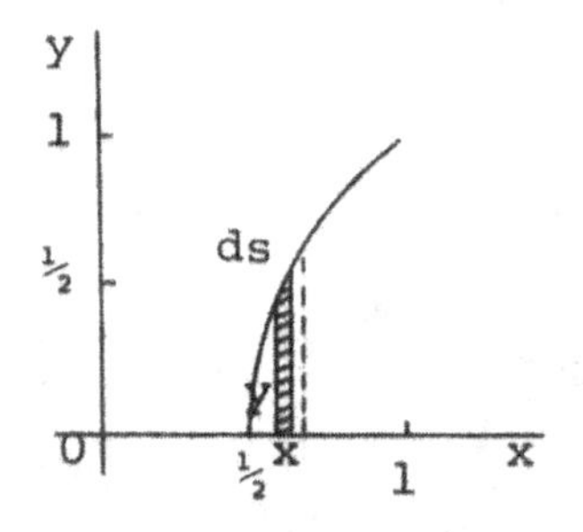

We consider the shaded strip as shown in the figure. Its length is y units and width Δx units. If this strip is rotated about the x-axis, it sweeps a volume that looks like a disk with radius y units, and thickness of Δx units. The circumference $2\pi r = 2\pi y$. Hence its surface area is

$$\Delta S = 2\pi y \, \Delta s.$$

Noting that the parabola is made up of such disks for

$$\frac{1}{2} \le x \le 1,$$

the required surface area is the sum of all surface areas of all the disks which compose the entire given solid. Mathematically, if S is the surface area required, then

$$S \cong \sum_{i=1}^{n} 2\pi y_i \, \Delta s_i,$$

or

$$S = \lim_{n \to \infty} \sum_{i=1}^{n} 2\pi y_i \, \Delta s_i \, , \; \frac{1}{2} \le x_i \le 1$$

Since $2y\,dy = 2\,dx$, and $\frac{dx}{dy} = y$, the differential of arc length is given by:

$$(ds)^2 = (dx)^2 + (dy)^2$$

$$= \left(1+\left(\frac{dx}{dy}\right)^2\right)(dy)^2.$$

$$ds = \sqrt{1+\left(\frac{dx}{dy}\right)^2} \cdot dy.$$

Furthermore, $y \geqq 0$, so that $|y| = y$, and hence:

$$S = 2\pi \int_0^1 y\sqrt{1+y^2}\,dy$$

$$= \left(\frac{2\pi}{3}(1+y^2)^{\frac{3}{2}}\right)_0^1$$

$$= \frac{2\pi}{3}(2\sqrt{2}-1).$$

16.8 Improper Integrals

$$\int_a^{\infty} f(x)dx = \lim_{t\to\infty} \int_a^t f(x)dx,$$

if the limit exists

$$\int_{-\infty}^{a} f(x)dx = \lim_{t\to -\infty} \int_t^a f(x)dx,$$

if the limit exists.

The above two expressions are improper integrals since one of their limits is not a real number.

Improper integrals either converge or diverge

They converge when the limit on the right-hand side of the equation exists as t goes to infinity.

They diverge when the limit on the right-hand side of the equation does not exist as t goes to infinity.

$$\int_{-\infty}^{+\infty} f(x)dx = \int_{-\infty}^{a} f(x)dx + \int_{a}^{\infty} f(x)dx$$

The above integral converges if and only if both integrals on the right-hand side of the equation are convergent.

Another form of an improper integral is one in which the integrand does not exist for some value on the closed interval of integration.

Example: If f is continuous on the half-open interval [a,b) and becomes infinite or undefined at b, then, by definition,

$$\int_{a}^{b} f(x)dx = \lim_{t \to b^-} \int_{a}^{t} f(x)dx,$$

provided the limit exists.

Likewise, if f is continuous on the half-open interval (a,b] and becomes infinite or undefined at a, then by definition,

$$\int_{a}^{b} f(x)dx = \lim_{t \to a^+} \int_{t}^{b} f(x)dx,$$

provided the limit exists.

As previously stated, the above integral converges if the limit on the right-hand side of the expression exists, and diverges otherwise.

Problem Solving Examples:

Find the area under the curve: $y = 3x^{-2}$, from $x = 1$ to $x = \infty$.

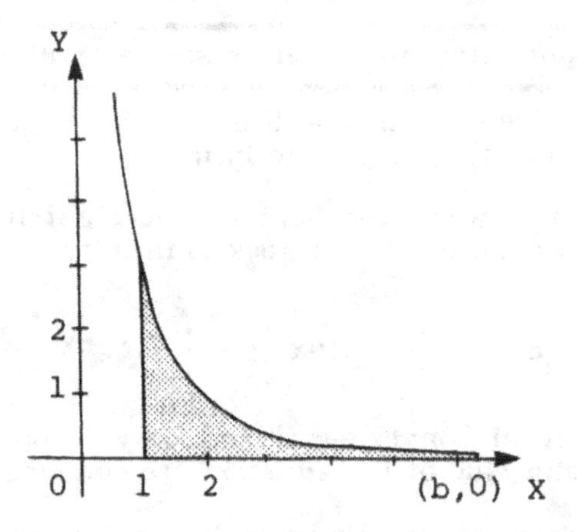

A figure is very important in all such problems as it shows the places where the area is not bounded. The curve: $y = 3x^{-2}$, has $y = 0$ as a horizontal asymptote. It is then necessary to find the value of the shaded area in the diagram and evaluate the limit of that area as $b \to \infty$, <u>if possible</u>. The area is evaluated as the integral of the upper function, $3x^{-2}$, minus the lower function, $y = 0$ (the x-axis). The limits are given as $x = 1$ and $x = \infty$. We can evaluate such an integral (an integral involving infinity as one of the limits), by the method of improper integrals. Thus,

$$A = \int_1^{\infty} (3x^{-2} - 0)\,dx = \int_1^{\infty} 3x^{-2}\,dx = \lim_{b \to \infty} \int_1^{b} 3x^{-2}\,dx$$

$$= 3 \lim_{b \to \infty} \left(-\frac{1}{x} \right]_1^b \Bigg)$$

$$= 3 \lim_{b \to \infty} \left(-\frac{1}{b} + 1 \right) = 3,$$

since $\frac{1}{b} \to 0$ as $b \to \infty$. Hence this area exists and equals 3 square units.

Find the area under the curve: $y = x^{-\frac{1}{2}}$ from x = 0 to x = 2.

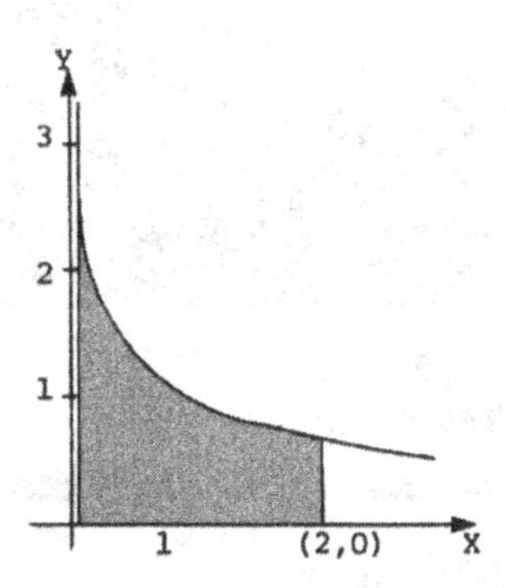

The curve: $y = x^{-\frac{1}{2}}$ has x = 0 as a vertical asymptote. It is then necessary to find the value of the shaded area in the diagram and evaluate the limit of that area as $h \to 0$, if possible. The area is

evaluated as the integral of the upper function which is $y = x^{-\frac{1}{2}}$, minus the lower function, y = 0 (the x-axis). The limits are given as x = 0 and x = 2.

We evaluate such an integral (one involving zero as one of the limits) by the method of improper integrals. Thus,

$$A = \int_0^2 \left(x^{-\frac{1}{2}} - (0) \right) dx$$

$$= \int_0^2 x^{-\frac{1}{2}} \, dx = \lim_{h \to 0} \int_h^2 x^{-\frac{1}{2}} \, dx = \lim_{h \to 0} \left(2x^{\frac{1}{2}} \Big|_h^2 \right)$$

$$= \lim_{h \to 0} \left(2\sqrt{2} - 2\sqrt{h} \right) = 2\sqrt{2} \text{ sq. units.}$$

Without analysis one might think that the area under such a curve is always infinite since its boundary recedes to infinity along the y-axis. But for this curve such is not true, and it is therefore advisable to evaluate the integral in each case before a conclusion is reached.

CHAPTER 17

Vector Fields

17.1 Vector Fields

Definition:

All the vectors in a region such that a unique vector having initial point P is associated with each point P in the region is called a vector field.

Examples of vector fields

1) Electric fields
2) Magnetic fields

A vector field that is the gradient of a scalar function is called a conservative vector field.

In the rectangular coordinate system the vector associated with the point (x,y,z) may be denoted by F(x,y,z), where

$$F(x,y,z) = M(x,y,z)i + N(x,y,z)j + P(x,y,z)k.$$

Every equation of the type given above determines a vector field. The function F(x,y,z) is called a vector function.

Problem Solving Examples:

Compute the integral of the vector field $\vec{F}(x,y,z) = (y,-x,z^2)$ over the paraboloid $z = x^2 + y^2$ with $0 \leq z \leq 1$.

A Given that X(t,u) parametrizes a surface, and letting $\vec{n}$ represent the outward normal unit vector to the surface, the integral of a vector field over a surface is defined by

$$\int\int_S \vec{F} \cdot \vec{n}\, dA = \int\int_R \vec{F} \cdot \vec{n} \left\| \frac{\partial X}{\partial t} \cdot \frac{\partial X}{\partial u} \right\| dt\, du$$

where $\vec{F}$ is a vector field in some open set in R^3 . Furthermore, using the fact that $\frac{\vec{N}}{\|\vec{N}\|} = \vec{n}$, where $\vec{N}$ is any normal in the same direction as $\vec{n}$(unit normal) and $\|\vec{N}\|$ is its norm,

$$\vec{n} \left\| \frac{\partial X}{\partial t} \cdot \frac{\partial X}{\partial u} \right\| = \pm \frac{\partial X}{\partial u} \cdot \frac{\partial X}{\partial u} .$$

Thus

$$\int\int_S \vec{F} \cdot \vec{n}\, dA = \int\int_R \vec{F}(X(t,u)) \cdot \left[\pm \left(\frac{\partial X}{\partial t} \cdot \frac{\partial X}{\partial u} \right) \right] dt\, du. \qquad (1)$$

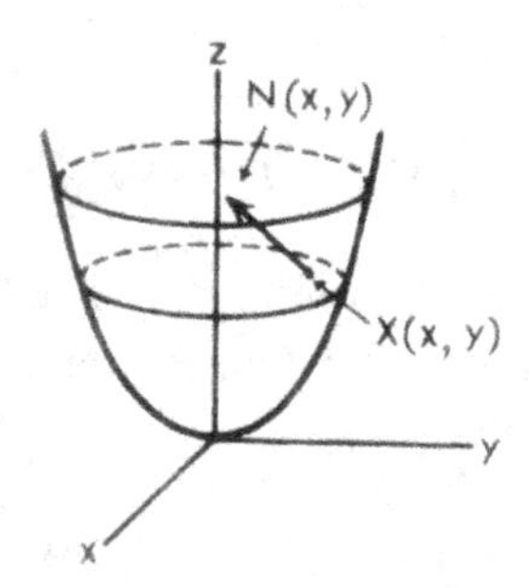

In this problem the surface $z = x^2 + y^2$ with $0 \le z \le 1$ is given. Parametrize the surface by $X(x,y) = (x,y,x^2 + y^2)$. Computing

$$\vec{N}(x,y) = \left(\frac{\partial x}{\partial x}, \frac{\partial y}{\partial x}, \frac{\partial z}{\partial x} \right) \cdot \left(\frac{\partial x}{\partial y}, \frac{\partial y}{\partial y}, \frac{\partial z}{\partial y} \right)$$

we find $\vec{N}(x,y) = (1,0,2x) \cdot (0,1,2y)$

$$= \begin{vmatrix} \hat{i} & \hat{j} & \hat{k} \\ 1 & 0 & 2x \\ 0 & 1 & 2y \end{vmatrix} = -2x\hat{i} - 2y\hat{j} + \hat{k}$$

$$= (-2x, -2y, 1)$$

Note that $\vec{N}(x,y)$ points inward and hence the outward normal direction is that of $-\vec{N}$. Next compute

$$\vec{F}(X(x,y)) \cdot \vec{N}(x,y) = (y, -x, z^2) \cdot (-2x, -2y, 1)$$

$$= -2xy + 2xy + z^2 = z^2$$

$$= (x^2 + y^2)^2 .$$

Substituting this value into (1) along the negative sign due to the inward orientation

$$\iint_S \vec{F} \cdot \vec{n}\, dA = -\iint_R (x^2+y^2)^2\, dx\, dy$$

where R is the disc $x^2 + y^2 \leq 1$ in the xy-plane. To evaluate this integral change to polar coordinates, letting $x = r \cos \theta$, $y = r \sin \theta$ and $dx\, dy = r\, dr\, d\theta$ to obtain

$$-\int_0^{2\pi} \int_0^1 (r^2 \cos^2\theta + r^2 \sin^2\theta)^2\, r\, dr\, d\theta$$

$$= -2\pi \int_0^1 r^4 r\, dr = -2\pi \int_0^1 r^5\, dr$$

$$= -2\pi \frac{r^6}{6} \Big|_0^1 = -\frac{2\pi}{6} = -\frac{\pi}{3} .$$

Determine whether the following vector fields have potential functions and if so find them:

a) $F(x,y) = (3xy, \sin xy^3)$

b) $F(x,y) = (2xy, x^2 + 3y^2)$

a) We use the theorem: Given that f,g are differentiable functions having continuous partial derivatives on an open set V in R^2 if $\frac{\partial f}{\partial y} \neq \frac{\partial g}{\partial x}$ then the vector field, $F(x,y) = (f(x,y), g(x,y))$, does not have a potential function.

Since $F(x,y) = (3xy, \sin xy^3)$, we have $\frac{\partial f}{\partial y} = 3x$ and $\frac{\partial g}{\partial x} = y^3 \cdot \cos xy^3$. Hence, since $\frac{\partial f}{\partial y} \neq \frac{\partial g}{\partial x}$ the vector field does not have a potential function.

b) We use the theorem: Given that f,g are differentiable functions having continuous first partial derivatives at all points of some open rectangle, if $\frac{\partial f}{\partial y} = \frac{\partial f}{\partial x}$ then the vector field, $F(x,y) = (f(x,y), g(x,y))$, has a potential function.

We are given $F(x,y) = (2xy, x^2 + 3y^2)$. Let

$$f(x,y) = 2xy \text{ and } g(x,y) = x^2 + 3y^2$$

Then $\frac{\partial f}{\partial y} = 2x$; $\frac{\partial g}{\partial x} = 2x$ and so, $\frac{\partial f}{\partial x} = \frac{\partial g}{\partial y}$. Hence, a potential function does exist. We want to find $\varphi(x,y)$ such that

$$\frac{\partial \varphi}{\partial x} = 2xy \text{ and } \frac{\partial \varphi}{\partial y} = x^2 + 3y^2. \tag{1}$$

Solving the first equation with respect to x we have $\partial\varphi = 2xy\ \partial x$ and

$$\varphi(x,y) = \int 2xy\, dx + h(y) = x^2y + h(y) \tag{2}$$

where h(y) is some function of y. Now taking the partial derivative of φ(x,y) with respect to y, we have

$$\frac{\partial \varphi}{\partial y} = x^2 + \frac{dh}{dy} \qquad (3)$$

and upon equating (3) with (1) we have

$$\frac{dh}{dy} = 3y^2$$

which gives $h = y^3 + C$, upon integrating with respect to g. Then (2) gives $\varphi(x,y) = x^2y + y^3 + C$. This is a potential function for F(x,y).

17.2 Line Integrals

Consider a plane curve given by the parametric equations

$$x = h(t) \text{ and } y = g(t) \quad (a \leq t \leq b),$$

where g and h are smooth on the interval [a,b].

Let it be given that the norm $||\Delta||$ of the subdivision of c is, by definition, the largest of the Δs_i, where Δs_i denotes the length of the subarc $\overline{P_{i-1}P_i}$.

Then the line integral of function f(x,y) along c from A →B is given by

$$\int_C f(x,y)ds = \lim_{||\Delta|| \to 0} \sum_{i=1} f(u_i v_i)\Delta s_i.$$

If f is continuous on the interval, then the above limit exists and we can rewrite the above as:

$$\int_C f(x,y)ds = \int_{a_1}^{b_2} f(h(t),g(t)) \sqrt{[h'(t)]^2+[g'(t)]^2}\, dt.$$

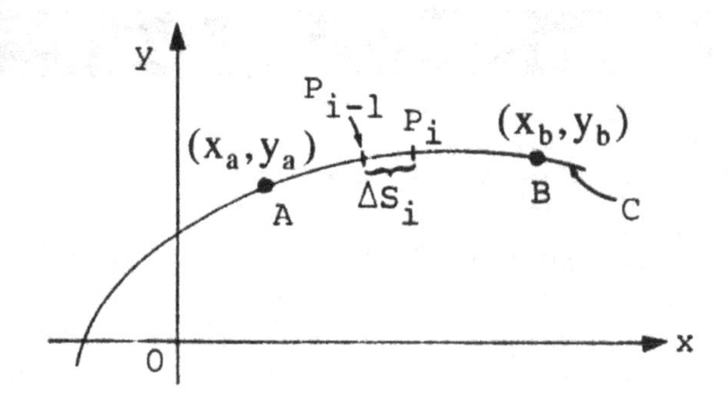

17.3 Green's Theorem

A smooth curve is one such that if represented parametrically by $x = g(t)$ and $y = h(t)$, g' and h' must be continuous on the closed interval $[a,b]$ and should not be zero simultaneously except possibly at a or b.

A smooth closed curve is one such that

$$A = (g(a),h(a)) = (g(b),h(b)) = B.$$

A simple curve is one such that for all numbers t_1 and t_2,

$$t_1 \neq t_2 \text{ and } (g(t_1),h(t_1)) \neq (g(t_2),h(t_2)).$$

GREEN'S THEOREM

Consider c, a piecewise smooth simple closed curve and R, the region consisting of c and its interior. If M and N are continuous function of x and y with continuous first partial derivatives in R, then

$$\int_c Mdx + Ndy = \iint_R \left(\frac{\partial N}{\partial x} - \frac{\partial M}{\partial y} \right) dA.$$

Problem Solving Examples:

Let C be the ellipse $x^2 + 4y^2 = 4$. Compute

$$\oint_C (2x - y)dx + (x + 3y)dy \text{ by Green's Theorem}$$

A Green's Theorem states: Let p,q, be functions on a region R, which is the interior of a closed path C (parametrized counterclockwise). Then

$$\int_C pdx + qdy = \iint_R \left(\frac{\partial q}{\partial x} - \frac{\partial p}{\partial y}\right) dxdy.$$

In the given problem,

$$\oint_C (2x - y)dx + (x + 3y)dy \qquad (1)$$

where C is the ellipse $x^2 + 4y^2 = 4$, is the integral. To use Green's theorem let $p = (2x - y)$ and $q = (x + 3y)$. Then $\frac{\partial p}{\partial y} = -1$ and $\frac{\partial q}{\partial x} = 1$ so that (1) equals

$$\iint_R (1 + 1)dxdy = 2\iint_R dxdy = 2 \cdot (\text{area of the ellipse}). \qquad (2)$$

Hence rewriting the ellipse as $\frac{x^2}{4} + y^2 = 1$, and using the formula

Area $= \pi ab$ where $\frac{x^2}{a^2} + \frac{y^2}{b^2} = 1$, the area of the given ellipse is 2π.

Thus, (2) becomes $2 \cdot 2\pi = 4\pi$. Thus the value of the line integral is 4π.

Use Green's Theorem to find:

a) $\int_C y^2dx - xdy$ clockwise around the triangle whose vertices are at (0,0), (0,1), (1,0).

b) The integral of the vector field $\vec{F}(x,y) = (y+3x, 2y-x)$ counterclockwise around the ellipse $4x^2 + y^2 = 4$.

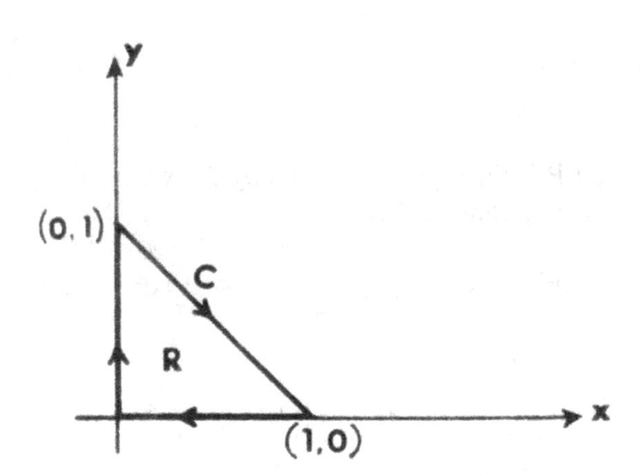

Green's Theorem states:

$$\oint_C pdx + qdy = \iint_R \left(\frac{\partial q}{\partial x} - \frac{\partial p}{\partial y}\right) dxdy \qquad (1)$$

where C is the boundary of the region R.

a) From the figure the region lies to the right of the path. Therefore use C^- instead of C so that the region lies to the left. Then use the fact that

$$\int_C \vec{F} = \int_{C^-} \vec{F}. \text{ Hence}$$

$$\int_C y^2dx - xdy = -\int_{C^-} y^2dx - xdy. \qquad (2)$$

Now apply Green's Theorem to the right hand side of equation (2).

Let $p = y^2$ and $q = -x$ so that $\frac{\partial q}{\partial x} = -1$ and $\frac{\partial p}{\partial y} = 2y$ to obtain

$$\int_C y^2dx - xdy = -\iint_R -(1 - 2y)dxdy.$$

From figure 1 R is the region bounded by the x-axis, the y-axis, and the line $x + y = 1$. Thus (2) becomes

$$\int_C y^2dx - xdy = -\int_0^1\int_0^{1-y} -(1 + 2y)\,dxdy$$

$$= \int_0^1 \left(x + 2yx \Big|_0^{1-y}\right)dy = \int_0^1 [1 - y + 2y(1 - y)]dy$$

$$= \int_0^1 (1 + y - 2y^2)dy = y + \frac{y^2}{2} - \frac{2y^3}{3}\Big|_0^1 = 1 + \frac{1}{2} - \frac{2}{3}$$

$$= \frac{5}{6}.$$

b) Since $\vec{F}(x,y)$ is being integrated counterclockwise over the ellipse $4x^2 + y^2 = 4$, we have the region R to the left of C with C being a closed curve. Given $\vec{F}(x,y) = (y+3x, 2y-x)$; if we let $p = y+3x$ and $q = 2y-x$ then Green's Theorem may be applied. First calculate:

$\frac{\partial p}{\partial y} = 1$and $\frac{\partial q}{\partial x} = -1$, then substitute into (1) to obtain

$$\int_C (y + 3)dx + (2y - x)dy \iint_R (-1 - 1)\,dxdy \qquad (3)$$

$$= -2 \iint_R dxdy = -2 \text{ (area of the ellipse).}$$

If the ellipse is rewritten in the form $\frac{x^2}{a^2} + \frac{y^2}{b^2} = 1$ we could use the equation for area of the ellipse, $A = \pi ab$. Since the ellipse is given as $4x^2 + y^2 = 4$ rewrite it as $\frac{x^2}{1^2} + \frac{y^2}{2^2} = 1$ to obtain $A = \pi\,(1)\,(2) = 2\pi$. Thus

$$\int_{\text{ellipse}} \vec{F} = -2 \cdot 2\pi = -4\pi.$$

17.4 Divergence and Curl

In rectangular coordinates

$$\nabla = i\frac{\partial}{\partial x} + j\frac{\partial}{\partial y} + k\frac{\partial}{\partial z}$$

$$\text{grad } f = \nabla f = \frac{\partial f}{\partial x}i + \frac{\partial f}{\partial y}j + \frac{\partial f}{\partial z}k$$

The curl of F, denoted by curl F or $\nabla \times F$, is defined by

$$\text{curl } F = \nabla \times F = \begin{vmatrix} i & j & k \\ \frac{\partial}{\partial x} & \frac{\partial}{\partial y} & \frac{\partial}{\partial z} \\ M & N & P \end{vmatrix}$$

where $F(x,y,z) = M(x,y,z)i + N(x,y,z)j + P(x,y,z)k$.

The divergence of F, denoted by div F or $\nabla \cdot F$, is defined by

$$\text{div } F = \nabla \cdot F = \frac{\partial M}{\partial x} + \frac{\partial N}{\partial y} + \frac{\partial P}{\partial z}$$

The ratio of the closed-surface integral of a vector field taken about a small (vanishing) closed surface to the small volume enclosed by the closed surface is known as the divergence of the vector field.

The ratio of the closed-line integral of a field taken about a small (vanishing) closed path to the small area enclosed expressed as a vector is known as the curl of the field.

Problem Solving Examples:

Let $\vec{F} = (F_1, F_2, F_3)$ be a vector field that satisfies the following conditions

$$\frac{\partial F_2}{\partial z} = \frac{\partial F_3}{\partial y}\ ;\ \frac{\partial F_3}{\partial x} = \frac{\partial F_1}{\partial z}\ ;\ \frac{\partial F_2}{\partial x} = \frac{\partial F_1}{\partial y}\ ,$$

on a region bounded by a curve c. (See the figure below.) Prove, using Stokes's Theorem that $\oint_C \vec{F} \cdot d\vec{s} = 0$.

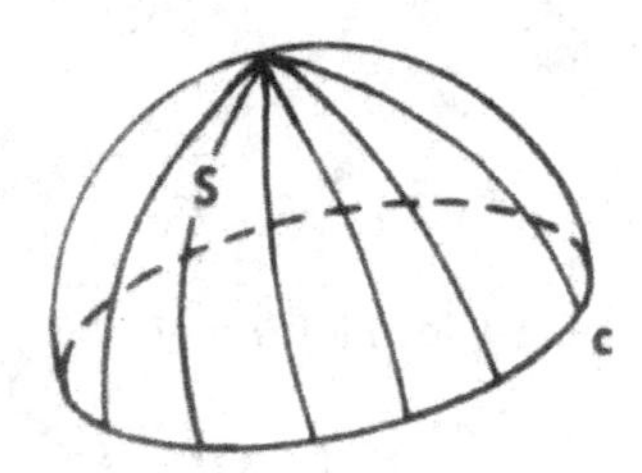

By Stokes's Theorem $\oint_C \vec{F} \cdot d\vec{s} = \int_S \Delta x \vec{F} \cdot \vec{n}\, da$. Let us compute $\Delta x \vec{F}$ (the curl of F).

$$= \begin{vmatrix} \hat{i} & \hat{j} & \hat{k} \\ \frac{\partial}{\partial x} & \frac{\partial}{\partial y} & \frac{\partial}{\partial z} \\ F_1 & F_2 & F_3 \end{vmatrix}$$

$$= \left(\frac{\partial F_3}{\partial y} - \frac{\partial F_2}{\partial z}, \frac{\partial F_1}{\partial z} - \frac{\partial F_3}{\partial x}, \frac{\partial F_2}{\partial x} - \frac{\partial F_1}{\partial y} \right).$$

This is the zero vector since each component is zero by our as–sumptions. Therefore $\oint_C \vec{F} \cdot d\vec{s} \int_S \Delta x \vec{F} \cdot \vec{n}\, dA = \int_S 0\, dA = 0.$

Q Verify Stokes's Theorem for $\vec{F}(x,y,z) = (3y, -xy, yz^2)$ where S is the surface. $2z = x^2 + y^2$ bounded by $z = 2$ and C is its boundary.

Stokes's Theorem states:

$$\oint_C \vec{F} \cdot d\vec{r} = \iint_S (\text{curl } \vec{F}) \cdot \vec{n}\, dA.$$

To verify Stokes's Theorem first compute

$$\oint_C \vec{F} \cdot d\vec{r} \tag{1}$$

Since $d\vec{r} = (dx, dy, dz)$ and $\vec{F}(x,y,z) = (3y, -xz, yz^2)$, $\vec{F} \cdot d\vec{r} = 3y\, dx - xz\, dy + yz\, dz$. Hence (1) becomes

$$\oint_C 3y\, dx - xz dy + yz^2\, dz\ . \tag{2}$$

The boundary C of S is $x^2 + y^2 = 4$, $z = 2$. Let $x = 2\cos\theta$; $y = 2\sin\theta$; $z = 2$. Substituting into (2)

$$\oint_{2\pi}^{0} [3(2\sin\theta)(-2\sin\theta) - (2\cos\theta)(2)(2\cos\theta) + (2\sin\theta)\ (2)\, 0]\, d\theta$$

(Integral goes from 2π to 0 since the boundary curve is oriented so that the surface lies to the left of the curve. See figure).

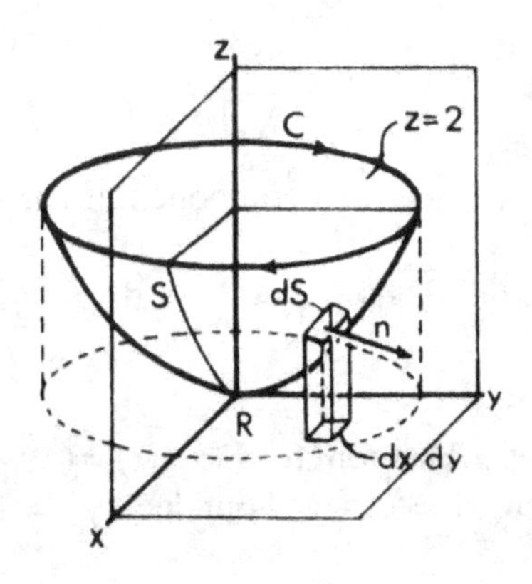

$$= \int_{2\pi}^{0} (-12 \sin^2\theta - 8 \cos^2\theta)\, d\theta$$

$$= \int_{0}^{2\pi} 12 \sin^2\theta\, d\theta + \int_{0}^{2\pi} 8 \cos^2\theta\, d\theta \qquad (3)$$

The first integral can be evaluated by using the equality

$\sin^2\theta = \dfrac{1 - \cos 2\theta}{2}$. Therefore

$$12\int_{0}^{2\pi} \sin^2\theta\, d\theta = 12\int_{0}^{2\pi} \frac{1}{2}\, d\theta - \int_{0}^{2\pi} \cos 2\theta\, d\theta = 12\pi + 0 = 12\pi .$$

To evaluate the second integral use $\cos^2\theta = \dfrac{1 + \cos 2\theta}{2}$;

$$8\int_{0}^{2\pi} \cos^2\theta\, d\theta = 8\int_{0}^{2\pi} \frac{d\theta}{2} + \int_{0}^{2\pi} \cos 2\theta\, d\theta = 8\pi + 0 = 8\pi.$$

Hence (3) equals 20π. So $\oint_C \vec{F} \cdot d\vec{r} = 20\pi$ (4)

To verify Stokes's Theorem we now need to show

$$\iint_S (\text{curl } \vec{F} \cdot \vec{n})\, dA = 20\pi .$$

Using the definition

$$\nabla x\vec{F} = \begin{vmatrix} \hat{i} & \hat{j} & \hat{k} \\ \frac{\partial}{\partial x} & \frac{\partial}{\partial y} & \frac{\partial}{\partial z} \\ f_1 & f_2 & f_3 \end{vmatrix}$$

where $\nabla x\vec{F}$ represents the curl $\vec{F}$ and $\vec{F} = (f_1,f_2,f_3)$,

$$\nabla xF = \begin{vmatrix} \hat{i} & \hat{j} & \hat{k} \\ \frac{\partial}{\partial x} & \frac{\partial}{\partial y} & \frac{\partial}{\partial z} \\ 3y & -xz & yz^2 \end{vmatrix}$$

$$= \left(\frac{\partial(yz^2)}{\partial y} - \frac{\partial(-xz)}{\partial z}\right)\hat{i} - \left(\frac{\partial(yz^2)}{\partial x} - \frac{\partial(3y)}{\partial z}\right)\hat{j} + \left(\frac{\partial(-xz)}{\partial x} - \frac{\partial(3y)}{\partial y}\right)\hat{k}$$

$$= (z^2 + x)\hat{i} - (z + 3)\hat{k} = (z^2 + x, 0, -z-3) .$$

Since S is the paraboloid $2z = x^{2+}y^2$ bounded by $z = 2$ this is a surface in the form $g(x,y,z) = z - f(x,y) = 0$. The unit normal can be expressed as

$$\frac{\nabla(x^2 + y^2 - z)}{||\nabla(x^2 + y^2 - 2z)||}$$

where ∇ is the gradient. Therefore

$$\vec{n} = \frac{(2x,2y,-2)}{\sqrt{4x^2 + 4y^2 + 4}} = \frac{(x, y, -1)}{\sqrt{x^2 + y^2 + 1}} .$$

But since $\iint_S (\text{curl } \vec{F}) \cdot \vec{n}\, dA = \iint_R (\text{curl } \vec{F}) \cdot \vec{N}$ and $\vec{n} = \frac{\vec{N}}{||\vec{N}||}$

$$\iint_S (\text{curl } \vec{F}) \cdot \vec{n}\, dA = \iint_R (z^2 + x, 0, -z-3) \cdot (x, y, -1)\, dx\, dy \qquad (5)$$

$$= \iint_R (xz^2 + x^2 + z + 3)\, dx\, dy,$$

which, when changed to polar coordinates, becomes

$$\int_0^{2\pi}\int_0^2 \left((r\cos\theta)\frac{r^4}{4} + r^2\cos^2\theta + \frac{r^2}{2} + 3\right) r\, dr\, d\theta$$

$$= \int_0^{2\pi}\int_0^2 \frac{r^6}{4}\cos\theta\, dr\, d\theta + \int_0^{2\pi}\int_0^2 r^3\cos^2\theta\, dr\, d\theta + \int_0^{2\pi}\int_0^2 \left(\frac{r^3}{2} + 3r\, dr\right) d\theta.$$

The first integral equals 0 since $\int_0^{2\pi} \cos\theta\, d\theta = 0$. The second integral can be computed by the equality

$$\cos^2\theta = \frac{1 + \cos 2\theta}{2} \text{ so that } \int_0^{2\pi}\int_0^2 r^3\cos^2\theta\, dr\, d\theta$$

$$= 4\int_0^{2\pi} \left(\frac{1}{2} + \frac{\cos 2\theta}{2}\right) d\theta = 4\pi$$

The last integral $\int_0^{2\pi} \left(\frac{r^3}{2} + 3r\right) dr\, d\theta$

$$= 2\pi\int_0^2 \left(\frac{r^3}{2} + 3r\right) dr = \left[\frac{16}{8} + \frac{12}{2}\right] 2\pi = 16\pi.$$

So (5) equals $4\pi + 16\pi = 20\pi$ which checks with (4) and verifies Stokes's Theorem.

Quiz: Multiple Integration and Vector Fields

1. The volume, V, of the region in space bounded above by the surface $x^2 + y^2 + z^2 = 4$ and below by $z = -\sqrt{x^2 + y^2}$ is represented by a triple integral in spherical coordinates as

$$\iiint_V \rho^2 \sin\phi d\rho\, d\phi d\theta$$

find the upper limit of integration for ϕ.

(A) π

(B) $\frac{3\pi}{4}$

(C) $\frac{\pi}{2}$

(D) $\frac{\pi}{4}$

(E) $\frac{\pi}{6}$

2. At the point (2, –1, 2) on the surface $z = xy^2$, find a direction vector for the greatest rate of decrease of z.

(A) $\hat{i} - 2\hat{j}$

(B) $\hat{i} - 4\hat{j}$

(C) $\frac{(\hat{i} - 4\hat{j})}{\sqrt{17}}$

(D) $-\hat{i} - 4\hat{j}$

(E) $\hat{i} - \hat{j}$

3. Find the curl of $\bar{u} = xyz\,\vec{i} + xy^2\vec{j} + yz\,\vec{k}$

(A) $xz\,\vec{i} + (x - yz)\,\vec{j} + 2y\,\vec{k}$

(B) $(x - z)\,\vec{i} - yz\,\vec{j} + xyz\,\vec{k}$

(C) $z\,\vec{i} + xy\,\vec{j} + (y^2 - xz)\,\vec{k}$

(D) $xy\,\vec{i} - (z - y)\,\vec{j} + (xy - yz)\,\vec{k}$

(E) $(xy - yz)\,\vec{i} - yz\,\vec{j} + x\,\vec{k}$

4. Find Green's function for $y'' + 5y' + 6y = \sin x$

(A) $2e^{2(t-x)} + 3e^{3(t-x)}$

(B) $e^{2(t-x)} - e^{3(t-x)}$

(C) $e^{(t+x)} - e^{(t-x)}$

(D) $2e^{(t-x)} - 3e^{(t-x)}$

(E) $e^{3(x-t)} - e^{2(x-t)}$

5. The moment with respect to the yz plane of the volume in the first octant bounded by the paraboloid $z = x^2 + y^2$ and the plane $z = 4$ is written as an integral in cylindrical coordinates as $\iiint_V F(z, r, \theta)\, dz\, dr\, d\theta$. Find $F(z, r, \theta)$.

(A) $r \cos \theta$

(B) $r \sin \theta$

(C) $rz \cos \theta$

(D) $r^2 \sin \theta$

(E) $r^2 \cos \theta$

6. Given $x^2z - 2yz^2 + xy = 0$, find $\frac{\partial x}{\partial z}$ at $(1, 1, 1)$.

(A) 0

(B) $\frac{4}{3}$

(C) -1

(D) 1

(E) None of these

7. The partial derivative $\frac{\partial}{\partial y}\left[\int_0^1 e^{y \sin x}\, dx\right]$ is equal to

(A) $\int_0^1 e^{y \sin x}\, dx$

(B) $\int_0^1 \cos x\, e^{y \sin x}\, dx$

(C) $\int_0^1 \sin x\, e^{y \sin x}\, dx$

(D) $\int_0^1 e^{y \sin x} (\sin y)\, dy$

(E) $\int_0^1 y \sin x\, e^{y \sin x}\, dx$

8. If $f(x, y) = \frac{2}{x^2} + 3xy$ for $x \neq 0$, and the gradient of f at (r, s) has length r, then which of the following equations is satisfied by r and s?

(A) $16 + 24\,r^3s + 9r^6s^2 + 8\,r^8 = 0$

(B) $24 + 9\,r^3s + r^6s^2 + 8\,r^8 = 0$

(C) $16 - r^3s + 9r^6s^2 + 8\,r^8 = 0$

(D) $16 - 24\,r^3s + 9r^6s^2 + r^8 = 0$

(E) $16 - 24\,r^3s + 9r^6s^2 + 8\,r^8 = 0$

9. The iterated integral $\int_0^1 \int_{\frac{y}{2}}^1 e^{x^2}\,dx\,dy$ can be expressed as

(A) $\int_0^1 \int_0^{2x} e^{x^2}\,dx\,dy$

(B) $\int_{\frac{y}{2}}^1 \int_0^1 e^{x^2}\,dx\,dy$

(C) $\int_0^1 \int_0^{2y} e^{x^2}\,dx\,dy$

(D) $\int_0^1 \int_0^{2y} e^{x^2}\,dx\,dy$

(E) $\int_0^1 \int_y^1 e^{x^2}\,dx\,dy + \int_0^1 \int_0^y e^{x^2}\,dx\,dy$

10. Evaluate $\int_{-\infty} e^x\,dx$

(A) Diverges

(B) 0

(C) e

(D) $e - 1$

(E) 1

ANSWER KEY

1. (B)
2. (D)
3. (C)
4. (B)
5. (E)
6. (D)
7. (C)
8. (E)
9. (A)
10. (E)

CHAPTER 18

Infinite Series

18.1 Indeterminate Forms

An indeterminate form is a ratio $\frac{f(x)}{g(x)}$ in which $f(c) = g(c) = 0$ or $f(c) = g(c) = \infty$ for some c. Even though $\frac{f(c)}{g(c)}$ is meaningless, $\lim_{x \to c} \frac{f(x)}{g(x)}$ may exist in such cases.

The indeterminate form $\left(\frac{0}{0}\right)$ and $\left(\frac{\infty}{\infty}\right)$.

Problem Solving Examples:

The function $f(x) = \frac{2x^2 - x - 3}{x + 1}$ is defined for all values of x except $x = -1$, since at $x = -1$ both numerator and denominator vanish. Does

$$\lim_{x \to -1} f(x)$$

exist, and, if so, what is its value?

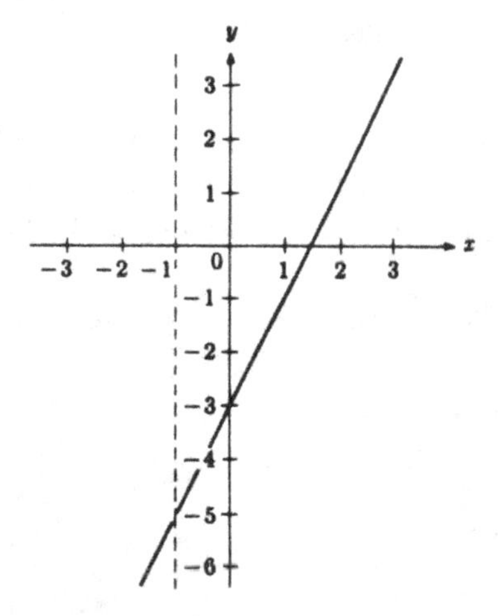

A As a visual aid, let us sketch f(x). This appears to be a straight line with a "hole" at the point (–1, –5). From the diagram we see that

$$\lim_{x \to -1} f(x) = -5.$$

However, a more systematic method of obtaining limits, without relying on pictorial representation and intuition is desirable. By means of factoring, we can write f(x) in the form

$$f(x) = \frac{(2x - 3)(x + 1)}{x + 1}.$$

Now if $x \neq -1$, we are allowed to divide both the numerator and denominator by $(x + 1)$. Then

$$f(x) = 2x - 3, \text{ if } x \neq -1.$$

This function tends to –5 as x tends to –1, since simple substitution now works. Therefore,

$$\lim_{x \to -1} f(x) = -5.$$

Note that we never substituted the value x = –1 in the original expression.

In the above solution, it was possible to factor the numerator and then proceed with direct substitution to obtain a definite answer. However, factoring is not always possible, and when the function

f(x) takes the form $\frac{0}{0}$, as in this case, the preferable approach is to

apply L'Hôpital's rule as follows:

$$\lim_{x \to -1} \frac{2x^2\ x - 3}{x + 1} = \lim_{x \to -1} \frac{4x - 1}{1} = -5.$$

Evaluate $\lim_{x \to 2} \frac{(2x^2 - 4x)}{x - 2}$.

The function takes the form $\frac{0}{0}$, and therefore we can apply L'Hôpital's rule to obtain:

$$\lim_{x \to 2} \frac{4x - 4}{1} = 4$$

We can also solve the problem in a different way by noting that the numerator can be factored.

$$\lim_{x \to 2} \frac{2x^2 - 4x}{x - 2} = \lim_{x \to 2} \frac{2x(x - 2)}{x - 2}$$

$$= \lim_{x \to 2} 2x$$

$$= 4.$$

18.1.1 The Mean Value Theorem

Suppose that f and g are differentiable functions on the interval [a,b] and that they are also continuous. If g' is never 0 in (a,b), then there is a number c in (a,b) such that

$$\frac{f'(c)}{g'(c)} = \frac{f(b)-f(a)}{g(b)-g(a)}$$

Problem Solving Examples:

If $f(x) = 3x^2 - x + 1$, find the point x_o at which $f'(x)$ assumes its mean value in the interval [2,4].

Recall the mean value theorem. Given a function f(x) which is continuous in [a,b] and differentiable in (a,b), there exists a point x_o where $a < x_o < b$ such that:

$$\frac{f(b) - f(a)}{b - a} = f'(x_o),$$

where x_o is the mean point in the interval.

In our problem, $3x^2 - x + 1$ is continuous, and the derivative exists in the interval (2,4). We have:

$$\frac{f(4) - f(2)}{4 - 2} = \frac{[3(4)^2 - 4 + 1] - [3(2)^2 - 2 + 1]}{4 - 2}$$

$$= f'(x_o),$$

or

$$\frac{45 - 11}{2} = 17 = f'(x_o) = 6x_o - 1.$$

$$6x_o = 18$$

$$x_o = 3.$$

x_o is the point where $f'(x)$ asumes its mean value.

18.1.2 L'Hôpital's Rule $\left(\frac{0}{0}\right)$

Suppose that

$$\lim_{x \to c} f(x) = 0, \lim_{x \to c} g(x) = 0, \lim_{x \to c} \frac{f'(x)}{g'(x)} = \ell,$$

and that the Mean Value Theorem holds in some interval about c. Then,

$$\lim_{x \to c} \frac{f(x)}{g(x)} = \lim_{x \to c} \frac{f'(x)}{g'(x)} = \ell$$

Problem Solving Examples:

Evaluate $\lim\limits_{x \to a} \dfrac{a - x}{\ln \frac{x}{a}}$.

The function takes the form $\frac{0}{0}$, and therefore it is possible to apply L'Hôpital's rule. Now

$$\lim_{x \to a} \frac{a - x}{\ln \frac{x}{a}} = \lim_{x \to a} \frac{a - x}{\ln x - \ln a}.$$

Differentiating numerator and denominator gives:

$$\lim_{x \to a} \frac{-1}{\frac{1}{x}} = -a.$$

18.1.3 L'Hôpital's Rule $\left(\frac{\infty}{\infty}\right)$

Suppose that

$$\lim_{x \to c} f(x) = \infty, \quad \lim_{x \to c} g(x) = \infty, \quad \text{and} \lim_{x \to c} \frac{f'(x)}{g'(x)} = \ell$$

Then,

$$\lim_{x \to c} \frac{f(x)}{g(x)} = \lim_{x \to c} \frac{f'(x)}{g'(x)} = \ell$$

L'Hôpital's rule is used for one - sided limits as well as for ordinary limits. You may have to apply L'Hôpital's rule more than once.

Problem Solving Examples:

Evaluate $\lim\limits_{x\to\infty} \dfrac{3x^3 - x + 7}{x^3 + 4x^2 + x - 3}$.

The function takes the form $\frac{\infty}{\infty}$, and therefore L'Hôpital's rule may be applied. However, the rule would have to be applied several times to arrive at the final answer.

The shorter method is to note that $x \to \infty$, and divide through by the highest power of x.

$$\lim_{x\to\infty} \frac{3x^3 - x + 7}{x^3 + 4x^2 + x - 3} = \lim_{x\to\infty} \frac{3 - \frac{1}{x^2} + \frac{7}{x^3}}{1 + \frac{4}{x} + \frac{1}{x^2} - \frac{3}{x^3}}$$

$$= \frac{3}{1}$$

$$= 3.$$

18.2 Infinite Sequence

Definition:

An infinite sequence is a function whose domain is the set of positive integers.

Definition:

A sequence $\{a_n\}$ has the limit ℓ (denoted by $\lim\limits_{n\to\infty} a_n = \ell$) if for every $c > 0$, there exists a positive number N such that if $n > N$, then $|a_n - \ell| < c$.

If the limit $(\lim_{n\to\infty} a_n)$ does not exist, then the sequence $\{a_n\}$ has no limit.

The statement $\lim_{n\to\infty} a_n = \infty$ means that for every positive real number s, there exists a number N such that if $n > N$, then $a_n > s$.

Theorem:

1) $\lim_{n\to\infty} r^n = 0$, if $|r| < 1$.

2) $\lim_{n\to\infty} |r^n| = \infty$, if $|r| > 1$.

Problem Solving Examples:

Prove that the following given sequences $\{s_n\}$ and $\{t_n\}$ converge to s and t respectively:

a) $\lim_{n\to\infty} (s_n + t_n) = s + t$;

b) $\lim_{n\to\infty} cs_n = cs$, for constant c ;

c) $\lim_{n\to\infty} (c + s_n) = c + s$, for constant c ;

d) $\lim_{n\to\infty} s_n t_n = st$;

e) $\lim_{n\to\infty} \frac{1}{s_n} = \frac{1}{s}$, provided

$s_n \neq 0$ $(n = 1,2, \dots)$, $s \neq 0$.

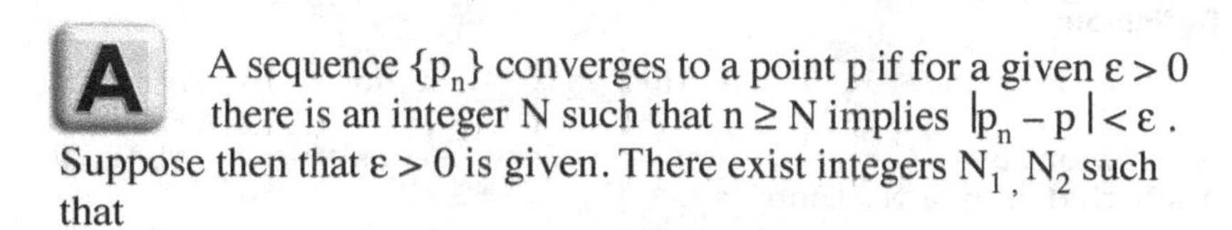

A sequence $\{p_n\}$ converges to a point p if for a given $\varepsilon > 0$ there is an integer N such that $n \geq N$ implies $|p_n - p| < \varepsilon$. Suppose then that $\varepsilon > 0$ is given. There exist integers N_1, N_2 such that

$$|s_n - s| < \varepsilon \text{ for } n \geq N_1$$

and

$$|t_n - t| < \varepsilon \text{ for } n \geq N_2$$

a) Let $N = \max(N_1, N_2)$, then $n \geq N$ implies

$$|(s_n + t_n) - (s + t)| \leq |s_n - s| + |t_n - t| < 2\varepsilon.$$

This proves (a).

b)

$$|(cs_n) - cs| = |c|\,|s_n - s| < |c|\,\varepsilon \text{ for } n \geq N_1 .$$

Hence (b) follows.

c) Similarly (c) follows from

$$|(c + s_n) - (c + s)| = |s_n - s| < \varepsilon \text{ for } n \geq N_1 .$$

d) Since

$$s_n t_n - st = (s_n - s)(t_n - t) + s(t_n - t) + t(s_n - s) ,$$

if

$$N = \max(N_1 1, N_2) ,$$

then

$n \geq N$ implies

$$|s_n t_n - st| \leq |s_n - s|\,|t_n - t| + |s|\,|t_n - t| + |t|\,|s_n - s|$$

$$< \varepsilon^2 + |s|\varepsilon + |t|\varepsilon = \varepsilon(\varepsilon + |s| + |t|)$$

and so (d) follows.

e) For $n \geq N_1$, since

$$|s| - |s_n| \leq |s_n - s| < \varepsilon$$

implies

$$|s_n| > |s| - \varepsilon,$$

$$\left| \frac{1}{s_n} - \frac{1}{s} \right| = \left| \frac{s_n - s}{s_n s} \right| < \frac{\varepsilon}{|s|(|s| - \varepsilon)} .$$

Therefore, (e) holds.

Note that the final result is not bounded by ε in particular cases, but rather by something which is also arbitrarily small for arbitrarily small $\varepsilon > 0$.

18.3 Convergent and Divergent Series

Definition:

An infinite series $\sum_{k=0}^{\infty} a_k$, sometimes written $a_0 + a_1 + a_2 \ldots$, is the sequence $\{s_0, s_1, s_2, s_3, \ldots\}$ of partial sums.

An infinite series converges if $\lim_{n \to \infty} s_n = s$ for some real number s. The series diverges if the sequence of the partial sums diverges (the limit does not exist).

Geometric series have the form $a + ar + ar^2 + \ldots + ar^{n-1}$, where a and r are real numbers and $a \neq 0$.

Theorem:

The geometric series

$$a + ar + ar^2 + \ldots + ar^{n-1},$$

with $a \neq 0$:

1) converges and has the sum $\frac{a}{1-r}$ if $|r| < 1$.

2) diverges if $|r| \geq 1$.

If an infinite series Σa_n is convergent, then $\lim_{n\to\infty} a_n = 0$. The infinite series Σa_n is divergent if $\lim_{n\to\infty} a_n \neq 0$. If $\lim_{n\to\infty} a_n = 0$, this does not mean that the series is convergent.

Theorem:

For every $c > 0$, if there exists an integer N such that $|s_k - s_\ell| < c$ whenever $k, \ell > N$, then the infinite series Σa_n is convergent.

Theorem:

If Σa_n and Σb_n are infinite series such that $a_i = b_i$ for all $i > k$, where k is a positive integer, then both series converge or both series diverge.

Theorem:

If Σa_n and Σb_n are convergent series with the sums A and B, respectively, then:

1) $\Sigma(a_n + b_n)$ converges and has the sum A + B.

2) If c is a real number, Σca_n converges and has the sum cA.

3) $\Sigma(a_n - b_n)$ converges and has the sum A - B.

Problem Solving Examples:

Show that the series $\sum_{k=1}^{\infty} \frac{1}{k^p}$ is convergent for p> 1 and is divergent for $p \leq 1$.

A To determine if the series is convergent or divergent the following theorem (called the integral test) is used: Let f(x) be a function which is positive, continuous and non-increasing as x increases for all values of $x \geq N$, where N is some fixed positive integer. Let the terms of an infinite series be given by $u_n = f(n)$ when $n \geq N$. If

$$\lim_{R\to\infty} \int_N^R f(x)\,dx < \infty \quad (= \infty)$$

then

$$\sum_{n=N}^{\infty} u_n < \infty \ (= \infty)$$

The proof of this theorem starts with the given relationship:

$$f(k+1) \leq f(x) \leq f(k),\ k \geq N,\ k < x \leq k+1$$

(since f(x) is nonincreasing).

Integrating f(x) from k to k + 1 preserves the inequalities:

$$f(k+1) \leq \int_k^{k+1} f(x)\,dx \leq f(k)$$

where

$$k = N, N+1, \ldots, n.$$

Adding these inequalities gives the result

$$\sum_{k=N+1}^{n+1} f(k) \leq \int_N^{n+1} f(x)\,dx \leq \sum_{k=N}^{n} f(k)\,. \tag{1}$$

If the integral

$$\int_N^{\infty} f(x)\,dx$$

is convergent, then from (1)

$$\sum_{k=N+1}^{\infty} f(k) \leq \int_N^{\infty} f(x)\,dx\,.$$

This shows that the partial sums of the series

$$\sum_{n=N+1}^{\infty} u_n$$

are bounded, and hence that this series is convergent (this from a theorem which states: If $u_n \geq 0$ for every n, then the series

$$\sum_{n=0}^{\infty} u_n$$

is convergent if and only if the sequence $\{S_n\}$ of partial sums is bounded.) The series

$$\sum_{n=1}^{\infty} u_n$$

is then convergent also. For the second case, suppose the integral

$$\int_N^{\infty} f(x)\,dx$$

is divergent. Since $f(x) > 0$ this can happen only if

$$\int_N^{n+1} f(x)\,dx \to +\infty$$

as $n \to \infty$.

It then follows, from the second inequality in (1) that $f(N) + \ldots + f(n) \to +\infty$, and hence that the series is divergent. Thus, the integral test is proven.

For the series

$$\sum_{k=1}^{\infty} \frac{1}{k^p}$$

consider the integral

$$\int_1^b \frac{dx}{x^p} = \left.\frac{x^{-p+1}}{-p+1}\right|_1^b = \frac{1}{1-p}\left(\frac{1}{b^{p-1}} - 1\right) \quad (\text{for } p \neq 1).$$

Then, for $p > 1$

$$\lim_{b\to\infty} \int_1^b \frac{dx}{x^p} = \lim_{b\to\infty} \frac{1}{1-p}\left(\frac{1}{b^{p-1}} - 1\right) = \frac{1}{p-1}.$$

Therefore, the integral converges and by the theorem, so does the series.

For $p < 1$,

$$\lim_{b\to\infty} \int_1^b \frac{dx}{x^p} = \lim_{b\to\infty} \frac{1}{1-p}\left(\frac{1}{b^{p-1}} - 1\right)$$

is unbounded. Therefore, the integral diverges and so does the series.

For $p = 1$,

$$\lim_{b\to\infty} \int_1^b \frac{dx}{x^p} = \lim_{b\to\infty} \int_1^b \frac{dx}{x} = \lim_{b\to\infty} \log b.$$

This grows without bounds. Therefore the integral and thus the series diverges. This shows that the series, called the p-series, converges for $p > 1$ and diverges for $p \leq 1$. note: If $p = 1$, the series is called the harmonic series.

Test the following series for convergence:

a)

$$\sum_{n=2}^{\infty} \frac{1}{n(\log n)^2}$$

b)

$$\sum_{n=2}^{\infty} \frac{1}{n(\log n)}$$

c)

$$\sum_{n=4}^{\infty} \frac{1}{n(\log n)\,[\log(\log n)]^2}$$

a) To test this series for convergence, apply the integral test. Since the series is

$$\sum_{n=2}^{\infty} \frac{1}{n(\log n)^2},$$

set up the integral

$$\int_2^b \frac{dx}{x(\log x)^2}.$$

Then, if the limit of the integral as $b \to \infty$ exists, it follows that the series converges. If the limit of the integral as $b \to \infty$, goes to infinity, then the series diverges. Therefore, to solve

$$\int_2^b \frac{dx}{x(\log x)^2}$$

let $u = \log x$, $du = \frac{dx}{x}$. This yields

$$\int_{\log 2}^{\log b} \frac{du}{u^2} = -u^{-1}\Big|_{\log 2}^{\log b} = \frac{1}{\log 2} - \frac{1}{\log b}$$

Hence

$$\lim_{b\to\infty} \int_2^b \frac{dx}{x(\log x)} = \lim_{b\to\infty}\left(\frac{1}{\log 2} - \frac{1}{\log b}\right) = \frac{1}{\log 2}$$

and thus the series converges.

b) To test the series

$$\sum_{n=2}^{\infty} \frac{1}{n(\log n)},$$

again use the integral test. Therefore set up the integral

$$\int_2^b \frac{dx}{x(\log x)} dx .$$

To solve this integral let $u = \log x$, $du = \frac{dx}{x}$. This yields

$$\int_{\log 2}^{\log b} \frac{du}{u} = \log u\Big|_{\log 2}^{\log b} = \log(\log b) - \log(\log 2) .$$

Hence

$$\lim_{b\to\infty} \int_2^b \frac{dx}{x(\log x)} = \lim_{b\to\infty} (\log(\log b) - \log(\log 2))$$

which is unbounded and thus the series diverges.

c) To test the series,

$$\sum_{n=4}^{\infty} \frac{1}{n \log n\, [\log(\log n)]^2}$$

also use the integral test. Therefore to solve

$$\int_4^b \frac{dx\,dx}{x \log x\,[\log(\log x)]^2} \quad \text{let } u = \log(\log x),\ du = \frac{dx}{x \log x}.$$

Since $\int \frac{du}{u^2} = -u^{-1}$, this yields

$$-\frac{1}{\log(\log x)}\Bigg|_4^b = -\frac{1}{\log(\log b)} + \frac{1}{\log(\log 4)}.$$

Hence

$$\lim_{b\to\infty} \int_4^b \frac{dx}{x \log x\,[\log(\log x)]^2}$$

$$= \lim_{b\to\infty}\left(\frac{-1}{\log(\log b)} + \frac{1}{\log(\log 4)}\right)$$

$$= \frac{1}{\log(\log 4)},$$

and the series converges.

18.4 Positive Term Series

A positive term series is a series for which every term is positive.

Theorem:

If Σa_n is a positive term series and if there exists a number m such that $s_n < m$ for every n, then the series converges and has a sum $s \leq m$. If no such m exists, the series diverges.

18.4.1 The Integral Test

If a function f is positive, continuous and decreasing on the interval $[1,\infty)$, then the infinite series:

1) converges if $\int_1^\infty f(x)dx$ converges.

2) diverges if $\int_{1}^{\infty} f(x)dx$ diverges.

Problem Solving Examples:

For $b_n = \frac{1}{\{(n-1)\log(n-1)\}}$ prove that $\frac{b_{n+1}}{b_n}$ equals

$1 - \frac{1}{n} - (n \log n)^{-1} + 0\,((n^2 \log n)^{-1})$, and then deduce Gauss's test:

If a series of positive terms a_n has

$$\frac{a_{n+1}}{a_n} = 1 - n^{-1} + 0\,(n^{-q})$$

for some $q > 1$, then Σa_n is divergent.

Let

$$\sum_{n=1}^{\infty} a_n$$

denote a series of positive terms. It is well known that if

$$\frac{a_{n+1}}{a_n} < k < 1$$

for all large n or if

$$a_n^{\frac{1}{n}} < k < 1$$

for all large n, then the series converges. On the other hand if

$$\frac{a_{n+1}}{a_n} > k > 1$$

or if $\sqrt[n]{a_n} > k > 1$

for all large n, then the series diverges. However, these two tests fail if a_{n+1} and a_n are of the same order

i.e., if
$$\frac{a_{n+1}}{a_n} \to 1 \text{ as } n \to \infty .$$

In this case we must try to estimate the difference

$$\frac{a_{n+1}}{a_n - 1}$$

and see what the next term looks like if we wish to make some statements regarding convergence on the divergence of the series. The problem at hand asks us to carry out this procedure.

$$\frac{b_{n-1}}{b_n} = \frac{(n-1)\log(n-1)}{n \log n}$$

so

$$\frac{b_{n+1}}{b_n} - 1 = \frac{\{(n-1)\log(n-1) - n\log n\}}{n \log n}$$

$$= \frac{1}{n \log n}\{n \log(n-1) - n \log n - \log(n-1)\}$$

$$= \frac{1}{n \log n}\{n(\log(n-1) - \log n) - \log(n-1)\}$$

$$= \frac{1}{n \log n}\{n \log(1-\tfrac{1}{n}) - \log(n(1-\tfrac{1}{n}))\}$$

$$= \frac{1}{n \log n}\{n \log(1-\tfrac{1}{n}) - \log n - \log(1-\tfrac{1}{n})\}$$

$$= -\frac{1}{n} + \frac{1}{n \log n}\{n \log(1-\tfrac{1}{n}) - \log 1 - \tfrac{1}{n})\} \quad (1)$$

Recall now the power series expansion for log (1 – x),

$$\log(1-x) = -x - \frac{x^2}{2} - \frac{x^3}{3} - \dots \qquad |x| < 1.$$

If $x = \frac{1}{n}$, then for large n, x is very small and

$$\log\left(1-\frac{1}{n}\right) \approx -\frac{1}{n}.$$

so

$$n \log\left(1-\frac{1}{n}\right) \approx -1. \text{ This implies}$$

i.e.,

$$\frac{b_{n+1}}{b_n} - 1 \approx -\frac{1}{n} - \frac{1}{n \log n} + \frac{1}{n^2 \log n},$$

$$\frac{b_{n+1}}{b_n} \approx 1 - \frac{1}{n} - \frac{1}{n \log n} + \frac{1}{n^2 \log n}.$$

This approximate identity will now be made precise as follows.

Consider

$$\log(1+x) = \int_0^x \frac{d}{dy} \log(1+y)\, dy \qquad |x| < 1.$$

Integrate by parts to get

$$\log(1+x) = x - \int_0^x y(1+y)^{-1}\, dy \tag{2}$$

The second term on the right side of (2) is less than or equal to

$$\int_0^{|x|} |y|\, |1+y|^{-1}\, dy$$

let $x = -\frac{1}{n}$ and let $n \geq 2$ so that if

$$|y| \leq |x| \leq \frac{1}{n}$$

$$1 + y \geq 1 - |y| \geq \frac{1}{2} \quad \text{and}$$

$$\left|\int_0^x y(1+y)^{-1}\, dy\right| \leq |x| \int_0^{|x|} \frac{1}{2}\, dy = \frac{|x|^2}{2} = \frac{1}{2n^2}$$

Thus

$$\log\left(1-\frac{1}{n}\right) = -\frac{1}{n} + R_n$$

where

$$|R_n| \le \frac{1}{2n^2},$$

so

$$n\log\left(1-\frac{1}{n}\right) = -1 + R'_n$$

where

$$|R'_n| = n|R_n| \le \frac{1}{2n}.$$

Substituting these into (1) gives

$$\frac{b_{n+1}}{b_n} = 1 - \frac{1}{n} + \frac{1}{n\log n}\left\{-1 + R'_n + \frac{1}{n} - R_n\right\}$$

$$= 1 - \frac{1}{n} - \frac{1}{n\log n} + \frac{1}{n^2\log n} + \frac{(R'_n - R_n)}{n\log n}$$

$$= 1 - \frac{1}{n} - \frac{1}{n\log n} + Q_n$$

where

$$Q_n = \frac{1}{n^2\log n} + \frac{(R'_n - R_n)}{n\log n}$$

and

$$(n^2\log n)\ |Q_n| \le 1 + n(|R'n| + |Rn|)$$

$$\le 1 + \frac{1}{2} + \frac{1}{2n}$$

$$\le 2.$$

This implies that $Q_n = 0((n^2 \log n)^{-1})$
and one can write

$$\frac{b_{n+1}}{b_n} = 1 - \frac{1}{n} - \frac{1}{n \log n} + 0((n^2 \log n)^{-1}).$$

This proves the first part of the problem.

In order to prove Gauss's test write

$$b_{n+1} = \beta_n b_n$$

$$a_{n+1} = \alpha_n a_n$$

where

$$\alpha_n = 1 - \frac{1}{n} + \frac{M_n}{n^q} \quad |M_n| \leq C \text{ for all n}$$

$$\beta_n = 1 - \frac{1}{n} - \frac{1}{n \log n} + \frac{k_n}{n^2 \log n} \quad |kn| \leq C \text{ for all n.}$$

Then

$$\alpha_n - \beta_n = \frac{1}{n \log n} + \frac{M_n}{n^q} - \frac{k_n}{n^2 \log n}$$

or

$$\alpha_n - \beta_n = \frac{1}{(n \log n)}\left[1 + \frac{M_n}{n^{q-1}} \log n - \frac{k_n}{n}\right]$$

$$\geq \frac{1}{(n \log n)}\left[1 - \frac{C}{n^{q-1}} \log n - \frac{C}{n}\right]$$

since

$$q > 1 \quad \frac{\log n}{n^{q-1}} \to 0 \quad \text{as } n \to \infty \quad \text{by}$$

L'Hopital's rule. Hence there exists an integer N_o so that for every $n \geq N_o$,

$$1 - \frac{C}{n^{q-1}} \log n - \frac{C}{n} > 0.$$

Then

$$\alpha_n > \beta_n \text{ for every } n \geq N_o.$$

Now

$$a_{N_o+k+1} = \alpha_{N_o+k}\,\alpha_{No+k-1} \ldots \alpha_{N_o}\, a_{N_o}$$

$$\geq \beta_{N_o+k}\,\beta_{N_o+k-1} \ldots \beta_{N_o} \left(\frac{a_{N_o}}{b_{N_o}}\right) b_{N_o}$$

$$= \left(\frac{a_{N_o}}{b_{N_o}}\right) b_{N_o+k+1} \quad \text{for all } k \geq 0.$$

Therefore

$$\sum_{k=0}^{\infty} a_{N_o+k+1} \geq \left(\frac{a_{N_o}}{b_{N_o}}\right) \sum_{k=0}^{\infty} b_{N_o+k+1} = \infty$$

since

$$\Sigma \frac{1}{n \log n}$$

diverges by the integral test.

18.4.2 The P-Series

The p-series expressed as $\sum_{n=1}^{\infty} \frac{1}{n^p}$, converges if $p > 1$ and diverges if $p \leq 1$.

Problem Solving Examples:

Establish the convergence or divergence of the series:

$$\frac{1}{1+\sqrt{1}}+\frac{1}{1+\sqrt{2}}+\frac{1}{1+\sqrt{3}}+\frac{1}{1+\sqrt{4}}+\cdots.$$

To establish the convergence or divergence of the given series we first determine the nth term of the series. By studying the law of formation of the terms of the series we find the

n[th] term to be $\frac{1}{1+\sqrt{n}}$. To determine whether this series is convergent or divergent we use the comparison test. We choose $\frac{1}{n}$, which is a known divergent series since it is a p-series, $\frac{1}{n^p}$, with $p = 1$. If we can show $\frac{1}{1+\sqrt{n}} > \frac{1}{n}$, then $\frac{1}{1+\sqrt{n}}$ is divergent. But we can see this is true, since $1 + \sqrt{n} < n$ for $n > 1$. Therefore the given series is divergent.

18.4.3 Comparison Test

Let Σa_n and Σb_n represent positive term series.

1) If $a_n \le b_n$ and the series Σb_n converges, then Σa_n converges.
2) If $a_n \ge b_n$ and the series Σb_n diverges, then Σa_n also diverges.

Problem Solving Examples:

Test for convergence:

a) $\int_1^\infty \frac{x^2 dx}{2x^4-x+1}$ b) $\int_1^\infty \frac{x\,dx}{3x^4+6x^2+1}$.

To test this integral for convergence, use the following test, known as the Quotient test: Suppose $\int_a^\infty f(x)\,dx$ and $\int_b^\infty g(x)\,dx$ are integrals of the first kind with nonnegative integrands:

i) If the limit $\lim\limits_{x\to\infty} \dfrac{f(x)}{g(x)} = L$ exists (finite) and is not zero, then either both integrals are convergent or both are divergent.

ii) If $L = 0$ in (i) and $\int_b^\infty g(x)\,dx$ converges, then $\int_a^\infty f(x)\,dx$ converges.

iii) If $L = \infty$ in (i) and $\int_b^\infty g(x)\,dx$ diverges, then $\int_a^\infty f(x)\,dx$ diverges.

To apply the test, first observe that for large values of x the integrand is approximately $\dfrac{1}{2x^2}$, since $2x^4 - x + 1$ will be approximately equal to $2x^4$. Specifically, taking

$$f(x) = \frac{x^2}{2x^4 - x + 1}, \quad g(x) = \frac{1}{x^2}$$

we find $\lim\limits_{x\to\infty} \dfrac{f(x)}{g(x)} = \lim\limits_{x\to\infty} \dfrac{x^4}{2x^4 - x + 1} = \dfrac{1}{2}$.

Now the integral $\int_1^\infty \frac{1}{x^2}\,dx$ is convergent since $\int_1^\infty \frac{1}{x^p}\,dx$ is convergent for $p > 1$. Therefore, since both integrals are either convergent or divergent by (i), and because it has been shown that $\int_1^\infty \frac{1}{x^2}\,dx$ is convergent, then both are convergent. Thus, $\int_1^\infty \frac{x^2\,dx}{2x^4 - x + 1}$ is convergent.

b) $\int_1^\infty \frac{x}{3x^4 - 6x^2 + 1}\,dx$. This integral will be tested for convergence by two methods.

Method 1: The Comparison Test.

For large x, the integrand is approximately equal to $\frac{x}{3x^4} = \frac{1}{3x^3}$.

Since $\frac{x}{3x^4 - 6x^2 + 1} \le \frac{1}{3x^3}$ and $\frac{1}{3}\int_1^\infty \frac{1}{x^3}\,dx$ converges (because $\int_1^\infty \frac{1}{x^p}\,dx$ converges for p>1), by the comparison test $\int_1^\infty \frac{x}{3x^4 - 6x^2 + 1}\,dx$ converges.

Method 2: The Quotient Test.

Let $f(x) = \frac{x}{3x^4 - 6x^2 + 1}$, $g(x) = \frac{1}{x^3}$. Then $\lim_{x\to\infty} \frac{f(x)}{g(x)}$

$= \lim_{x\to\infty} \frac{x^4}{3x^4 - 6x^2 + 1} = \frac{1}{3}$. Now, the integral $\int_1^\infty \frac{1}{x^3}\,dx$ is convergent (p integral with p = 3). Therefore, since $\lim_{x\to\infty} \frac{f(x)}{g(x)} = \frac{1}{3}$ and $\int_1^\infty g(x)\,dx$ converges, $\int_1^\infty f(x)\,dx = \int_1^\infty \frac{x}{3x^4 - 6x^2 + 1}\,dx$ converges by the quotient test.

18.5 Alternating Series: Absolute and Conditional Convergence

An alternating series is an infinite series in which successive terms have opposite signs. An alternating series is usually expressed as

$$a_1 - a_2 + a_3 - a_4 + \ldots + (-1)^{n-1} a_n + \ldots,$$

or

$$-a_1 + a_2 - a_3 + a_4 - \ldots + (-1)^n a_n + \ldots,$$

where each $a_i > 0$.

18.5.1 Alternating Series Test

Let $\{a_k\}$ represent a decreasing series of positive terms. If $a_k > 0$, then $\sum_{k=0}^{\infty}$ and $\lim a_k = 0$ as $(-1)^k a_k$ then $\sum_{k=0}^{\infty} (-1)^k a_k$ converges.

Theorem:

If the condition $a_k > a_{k+1} > 0$ for every positive integer k holds for the alternating series $\sum (-1)^{k-1} a_k$ and $\lim_{k \to \infty} a_k = 0$, then the error in approximating the sum S of the series by the kth partial sum, S_k, is numerically less then a_{k+1}.

Problem Solving Examples:

Test the alternating series:

$$\frac{1+\sqrt{2}}{2} - \frac{1+\sqrt{3}}{4} + \frac{1+\sqrt{4}}{6} - \frac{1+\sqrt{5}}{8} + \cdots$$

for convergence.

A An alternating series is convergent if (a) the terms, after a certain n^{th} term, decrease numerically, i.e., $u_{n+1} < u_n$, and (b) the general term approaches 0 as n becomes infinite. Therefore, we determine the nth term of the given alternating series. By discovering the law of formation, we find that the general term is

$\pm\frac{1=\sqrt{n+1}}{2n}$. Therefore, the preceding term is $\pm\frac{1+\sqrt{n}}{2(n-1)}$. To satisfy condition (a) stated above, we must show that:

$$\frac{1+\sqrt{n+1}}{2n} < \frac{1+\sqrt{n}}{2(n-1)}.$$

Obtaining a common denominator for both these terms,

$$\frac{1}{2}\cdot\frac{(1+\sqrt{n+1})(n-1)}{n(n-1)} < \frac{(1+\sqrt{n})(n)}{2(n-1)}\cdot\frac{1}{2}.$$

Since the denominators are the same, to prove condition (a) we must show,

$$(1+\sqrt{n+1})(n-1) < (1+\sqrt{n})(n),$$

which is obvious, since subtracting 1 from n has a greater effect than adding 1 to $\sqrt{n}$. Since $u_{n+1} < u_n$, we have the first condition for convergence.

Now we must show that

$$\lim_{n\to\infty} \frac{1+\sqrt{n}}{2n-2} = 0.$$

We find that $\lim_{n\to\infty} \frac{(1+\sqrt{n})}{2n-2} = \frac{\infty}{\infty}$, which is an indeterminate form.

We therefore apply L'Hôpital's Rule obtaining:

$$\lim_{n\to\infty} \frac{\frac{1}{2}n^{-\frac{1}{2}}}{2} = \lim_{n\to\infty} \frac{1}{4\sqrt{n}} = 0 .$$

Since both conditions hold, the given alternating series is convergent.

18.5.2 Absolute Convergence

The series Σa_k is absolutely convergent if the series $\Sigma |a_k| = |a_1| + |a_2| + \ldots + |a_n|$, obtained by taking the absolute value of each term, is convergent.

Problem Solving Examples:

Evaluate: $\int_0^1 e^{-x^2}dx$.

A We use the theorem: If the series f(x) converges for $|x| < r$, the integral of f(x) may be found by integrating the series term by term, and the integral series converges for $|x| < r$. We first attempt to obtain the region of convergence of the given integrand.

The function $f(x) = e^{-x^2}$ can be expanded using Maclaurin's Formula:

$$f(x) = \sum_{n=0}^{\infty} \frac{f^{(n)}(0)}{n!} x_n ,$$

where $f^{(n)}(0)$ is the nth derivative of the function evaluated at $x = 0$. However, to minimize the effort, since e^u is well-known, (for example, from the Euler Formula) we have:

$$e^u = 1 + u + \frac{u^2}{2!} + \ldots + \frac{u^{n-1}}{(n-1)!} + \ldots ,$$

substituting $-x^2$ for u,

$$e^{-x^2} = 1 - x^2 + \frac{x^4}{4!} - \frac{x^6}{3!} + \ldots + \frac{(-1)^{n-1}x^{2(n-1)}}{(n-1)!} + \ldots .$$

This kind of series is called an alternating series, since the real constant terms are alternately positive and negative. This is a power series, the convergence of which can be determined, using one of the following interrelated methods:

i) if $\lim_{n\to\infty} \frac{a_{n-1}}{a_n} = r$, ($a_n$ is the coefficient of x to the nth degree), then the series is absolutely convergent for $|x| < r$ and divergent for $|x| > r$;

ii) Cauchy's Ratio Test, states that where the ratio:

$$\frac{u_{n+1}}{u_n}$$ (u_n is the nth sequence of the function).

exists as n becomes infinite, and is less than unity, the given series converges absolutely. If this limit does not exist, or if it is greater than unity, the series diverges. The behavior of the series cannot be determined for x = r, in the first case, and the limit equals unity in the second one. The series is thus absolutely convergent for all values of x, $|x| < \infty$. Integration, under the given limit yields

$$\int_0^1 e^{-x^2}\, dx = \int_0^1 \left\{1 - x^2 + \frac{x^4}{4!} - \frac{x^6}{6!} + \ldots\right\} dx$$

$$\simeq 0.747.$$

Note that the integrated terms are also absolutely convergent for all x.

18.5.3 Conditional Convergence

The series Σa_k is conditionally convergent if $\Sigma |a_k|$ diverges while Σa_k converges.

Problem Solving Examples:

Determine if the following integrals converge absolutely:

a) $\int_0^{\infty} \frac{\cos x}{\sqrt{1+x^3}}\, dx$ b) $\int_0^{\infty} \sin x^2\, dx$.

a) To test this integral for absolute convergence, apply the following theorem:

If f(x) is continuous for $a \le x < \infty$ and if $\lim_{x\to\infty} x^p f(x) = A$ for $p > 1$, then the integral $\int_a^\infty |f(x)|\,dx$ converges, from which it follows that $\int_a^\infty f(x)dx$ is absolutely convergent.

To apply the theorem to the given problem take p such that $1 < p < \frac{3}{2}$. Therefore, taking $p = \frac{5}{4}$ yields

$$\lim_{x\to\infty} x^p f(x) = \lim_{x\to\infty} x^{\frac{5}{4}} \frac{\cos x}{\sqrt{1+x^3}}$$

$$= \lim_{x\to\infty} \frac{x^{\frac{5}{4}} \cos x}{x^{\frac{3}{2}} (1+x^{-3})^{\frac{1}{2}}} = \lim_{x\to\infty} \frac{\cos x}{x^{\frac{1}{4}} (1+x^{-3})^{\frac{1}{2}}} = 0 .$$

Hence since $p = \frac{5}{4} > 1$ and $A = 0$, by the theorem, $\int_0^\infty \left| \frac{\cos x}{\sqrt{1+x^3}} \right| dx$ converges. Thus, it follows that $\int_0^\infty \frac{\cos x}{\sqrt{1+x^3}} dx$ converges absolutely.

b) $\int_0^\infty \sin x^2\,dx$. This integral has previously been shown to be convergent. To test it for absolute convergence, again as before, begin by setting $x = \sqrt{t}$, so that $dx = \frac{1}{2\sqrt{t}} dt$. Then

$$\int_0^\infty |\sin x^2|\,dx = \frac{1}{2}\int_0^\infty \frac{|\sin t|}{\sqrt{t}} dt, \tag{1}$$

which is an improper integral of the first kind.

To proceed, apply the following theorem: If g(x) is continuous and is a nonincreasing function, and if $\lim_{x\to\infty} g(x) = 0$, then the integral

$\int_a^\infty g(x)\,|\sin x|\,dx$ converges if $\int_a^\infty g(x)dx$ converges. Similarly, $\int_a^\infty g(x)\,|\sin x|\,dx$ diverges if $\int_a^\infty g(x)dx$ diverges. To apply the theorem, let $a = 1$, $g(t) = t^{-1/2}$. Then $g(t)$ is a nonincreasing function and

$$\lim_{t\to\infty} g(t) = \lim_{t\to\infty} \frac{1}{t^{\frac{1}{2}}} = 0.$$

Furthermore

$$\int_a^\infty \frac{1}{t^{\frac{1}{2}}}\,dt = \lim_{b\to\infty} \int_1^b t^{-1/2}\,dt = \lim_{b\to\infty} 2b^{1/2} - 2 = \infty.$$

Hence,

$$\int_1^\infty \frac{|\sin t|}{\sqrt{t}}\,dt$$

diverges by the theorem, since $\int_a^\infty g(x)dx$ diverges. Thus by (1), $\int_a^\infty |\sin x^2|\,dx$ is divergent and because the given integral was proven to be convergent, this shows it to be conditionally convergent.

18.5.4 Ratio Test

Suppose that in the series Σa_k, every $a_k \neq 0$, and

$$\lim_{k\to\infty} \left|\frac{a_{k+1}}{a_k}\right| = \rho \quad \text{or} \quad \lim_{k\to\infty} \left|\frac{a_{k+1}}{a_k}\right| = +\infty .$$

Then:

1) If $\rho < 1$, the series Σa_k converges absolutely.

2) If $\rho > 1$, or if $\lim_{k\to\infty} \frac{|a_{k+1}|}{|a_k|} = +\infty$, the series diverges.

3) If $\rho = 1$, the test gives no information.

Problem Solving Examples:

Test the series:

$$1+\frac{2!}{2^2}+\frac{3!}{3^3}+\frac{4!}{4^4}+\ldots$$

by means of the ratio test. If this test fails, use another test.

To make use of the ratio test, we find the nth term of the given series, and the $(n+1)^{th}$ term. If we let the first term, $1 = u_1$, then $\frac{2!}{2^2} = u_2$, $\frac{3!}{3^3} = u_3$, etc., up to $u_n + u_{n+1}$. We examine the terms of the series to find the law of formation, from which we conclude:

$$u_n = \frac{n!}{n^n} \text{ and, } u_{n+1} = \frac{(n+1)!}{(n+1)^{n+1}} .$$

Forming the ratio $\frac{u_n+1}{u_n}$ we obtain:

$$\frac{(n+1)!}{(n+1)^{n+1}} \times \frac{n^n}{n!}$$

$$\frac{(n+1)(n!)}{(n+1)^n(n+1)} \times \frac{n^n}{n!} = \frac{n^n}{(n+1)^n} .$$

Now, we find $\lim_{n\to\infty} \left|\frac{n^n}{(n+1)n}\right|$. This can be rewritten as:

$$\lim_{n\to\infty} \frac{n^n}{\left[n\left[1+\frac{1}{n}\right]\right]^n} = \lim_{n\to\infty} \frac{n^n}{n^n \bullet \left[1+\frac{1}{n}\right]^n} = \lim_{n\to\infty} \frac{1}{\left[1+\frac{1}{n}\right]^n}$$

We now use the definition:

$$e = \lim_{x\to 0} (1+x)^{\frac{1}{x}}.$$

If we let $x = \frac{1}{n}$ in this definition, we have:

$$\lim_{\frac{1}{n}\to 0}\left[1+\frac{1}{n}\right]^{\frac{1}{\frac{1}{n}}} = \lim_{n\to\infty}\left[1+\frac{1}{n}\right]^n,$$ which is what we have previously.

Therefore, $\lim_{n\to\infty}\dfrac{1}{\left(1+\frac{1}{n}\right)^n}=\dfrac{1}{e}$. Since $e \approx 2.7, \dfrac{1}{e}\approx\dfrac{1}{2.7}$ which is less than 1.

Hence, by the ratio test, the given series is convergent.

18.5.5 Root Test

Let Σa_k represent an infinite series.

1) If $\lim_{k\to\infty}\sqrt[k]{|a_k|} = \ell < 1$, the series is absolutely convergent.

2) If $\lim_{k\to\infty}\sqrt[k]{|a_k|} = \ell > 1$ or $\lim_{k\to\infty}\sqrt[k]{|a_k|} = \infty$, the series is divergent.

3) If $\lim_{k\to\infty}\sqrt[k]{|a_k|} = 1$, the series may be absolutely convergent, conditionally convergent, or divergent.

Problem Solving Examples:

Determine if the series

$$\frac{1}{2}+\frac{1}{3}+\frac{1}{2^2}+\frac{1}{3^2}+\frac{1}{2^3}+\frac{1}{3^3}+\ldots$$

is convergent or divergent by applying the ratio and the root tests.

The series can be rewritten as the sum of the sequence of numbers given by

$$a_n = \begin{cases} \dfrac{1}{2^{(n+1)/2}} & \text{if n is odd } (> 0) \\ \dfrac{1}{3^{n/2}} & \text{if n is even } (n > 0) \end{cases}$$

Now the ratio test states: if $a_k >$ and $\lim\limits_{k\to\infty} \dfrac{a_k+1}{a_k} = \ell < 1$,

then $\sum\limits_{k=1}^{\infty} a_k$

converges. Similarly, if

$\lim\limits_{k\to\infty} \dfrac{a_k+1}{a_k} = \ell \;\; (1 < \ell \;\; \le \infty)$ then

$\sum\limits_{k=1}^{\infty} \;\; a_k$ diverges.

If $\ell = 1$, the test fails. Therefore applying this test gives:

If a_n is odd,

$$\lim_{n\to\infty} \frac{a_n+1}{a_n}$$

$$= \lim_{n\to\infty} \frac{\dfrac{1}{3^{n/2}}}{\dfrac{1}{2^{(n+1)/2}}} = \lim_{n\to\infty} \frac{2^{(n+1)/2}}{3^{n/2}} = \lim_{n\to\infty} \left[\frac{2}{3}\right]^{n/2} 2^{1/2} = 0.$$

If a_n is even,

$$\lim_{n\to\infty} \frac{a_n+1}{a_n} = \lim_{n\to\infty} \frac{\dfrac{1}{2^{(n+1)/2}}}{\dfrac{1}{3^{n/2}}} = \lim_{n\to\infty} \left[\frac{3}{2}\right]^{n/2} 2^{1/2}$$

and no limits exists.

Hence, the ratio test gives two different values, one < 1 and the other > 1, therefore the test fails to determine if the series is convergent. Thus, another test, known as the root test is now applied. This test states: let

$$\sum_{k=1}^{\infty} a_k$$

be a series of nonnegative terms, and let

$$\lim_{n\to\infty} (\sqrt[n]{a_n}) = S, \text{ where } 0 \le S \le \infty. \text{ If:}$$

1) $0 \le S < 1$, the series converges

2) $1 < S \le \infty$, the series diverges

3) $S = 1$, the series may converge or diverge.

Applying this test yields, if a_n is odd

$$\lim_{n\to\infty} \sqrt[n]{a_n} = \lim_{n\to\infty} \sqrt[n]{\frac{1}{2^{(n+1)/2}}} = \lim_{n\to\infty} \sqrt[n]{\frac{1}{2^{n/2}}}\ \sqrt[n]{\frac{1}{2^{1/2}}}$$

$$= \lim_{n\to\infty} \frac{1}{\sqrt{2}}\frac{1}{2^{1/2n}} = \frac{1}{\sqrt{2}} < 1.$$

If a_n is even

$$\lim_{n\to\infty} \sqrt[n]{a_n} = \lim_{n\to\infty} \sqrt{\frac{1}{3^{n/2}}} = \frac{1}{\sqrt{3}} < 1.$$

Thus, since for both cases, the

$$\lim_{n\to\infty} \sqrt[n]{a_n} < 1,$$

the series converges.

18.6 Power Series

A power series is a series of the form

$$c_0 + c_1(x-a)+c_2(x-a)^2 +\ldots+ c_n(x-a)^n$$

in which a and c_i, i = 1,2,3, etc., are constants.

The notations

$$\sum_{n=0}^{\infty} c_n(x-a)^n \text{ and } \sum_{n=0}^{\infty} c_n x^n$$

are used to describe power series.

A power series $\Sigma c_n x^n$ is said to converge:

1) at x_1 if and only if $\Sigma c_n x^n$ converges.

2) on the set S if and only if $\Sigma c_n x^n$ converges for each $x \in S$.

If $\Sigma c_n x^n$ converges at $x_1 \neq 0$, then it converges absolutely whenever $|x| < |x_1|$. If $\Sigma c_n x^n$ diverges at x_1, then it diverges for $|x| > |x_1|$.

Problem Solving Examples:

Find a power series in x for:

a) tan x

b) $\dfrac{\sin x}{\sin 2x}$ $(x \neq 0)$

To find the power series of the given functions, we need the following theorem:

Given the two power series

$$\sum_{n=0}^{\infty} a_n x^n = a_0 + a_1 x + a_2 x^2 + \dots + a_n x^n + \dots$$

and

$$\sum_{n=0}^{\infty} b_n x^n = b_0 + b_1 x + b_2 x^2 + \dots + b_n x^n + \dots ,$$

where $b_0 \neq 0$, and where both of the series are convergent in some interval $|x| < R$, let f be a function defined by

$$f(x) = \frac{a_0 + a_1 x + a_2 x^2 + \dots + a_n x^n + \dots}{b_0 + b_1 x + b_2 x^2 + \dots + b_n x^n + \dots} .$$

Then for sufficiently small values of x the function f can be represented by the power series

$$f(x) = c_0 + c_1 x + c_2 x^2 + \dots + c_n x^n + \dots ,$$

where the coefficients $c_0, c_1, c_2, \dots , c_n , \dots$
are found by long division or equivalently by solving the following relations successively for each c_i ($i = 0$ to ∞):

$$b_0 c_0 = a_0$$

$$b_0 c_1 + b_1 c_0 = a_1$$

$$\vdots$$

$$b_0 c_n + b_1 c_{n-1} + \dots + b_n c_0 = a_n$$

$$\vdots$$

a) To find the power series expansion of tan x we need the Taylor's series for sin *x* and cos *x*. That is

$$\sin x = x - \frac{x^3}{3!} + \frac{x^5}{5!} - \dots$$

and

$$\cos x = 1 - \frac{x^2}{2!} + \frac{x^4}{4!} - \dots$$

Then,

$$\tan x = \frac{\sin x}{\cos x} = \frac{x - \frac{x^3}{3!} + \frac{x^5}{5!} - \dots}{1 - \frac{x^2}{2!} + \frac{x^4}{4!} - \dots} \tag{1}$$

Therefore, by the theorem we can find the power series expansion of tan x by dividing the numerator by the denominator on the right side of (1). Hence, using long division we have

$$\begin{array}{r|l}
 & x + \frac{1}{3}x^3 + \frac{2}{15}x^5 + \dots \\
\hline
1 - \frac{1}{2}x^2 + \frac{1}{24}x^4 - \dots & x - \frac{1}{6}x^3 + \frac{1}{120}x^5 - \dots \\
 & x - \frac{1}{2}x^3 + \frac{1}{24}x^5 - \dots \\
\cline{2-2}
 & \quad \frac{1}{3}x^3 - \frac{1}{30}x^5 + \dots \\
 & \quad \frac{1}{3}x^3 - \frac{1}{6}x^5 + \dots \\
\cline{2-2}
 & \qquad \frac{2}{15}x^5 - \dots \\
 & \qquad \frac{2}{15}x^5 - \dots \\
\cline{2-2}
\end{array}$$

Thus

$$\tan x = x + \frac{1}{3}x^3 = \frac{2}{15}x^5 + \dots$$

b) Since

$$\sin x = x - \frac{x^3}{3!} + \frac{x^5}{5!} - \dots$$

we have

$$\sin(2x) = 2x - \frac{(2x)^3}{3!} + \frac{(2x)^5}{5!} - \ldots,$$

so that

$$\frac{\sin x}{\sin 2x} = \frac{x - \frac{x^3}{3!} + \frac{x^5}{5!} - \ldots}{2x - \frac{(2x)^3}{3!} + \frac{(2x)^5}{5!} - \ldots}. \tag{2}$$

Now multiplying the numerator and denominator on the right side of (2) by 1/x yields

$$\frac{\sin x}{\sin 2x} = \frac{1 - \frac{x^2}{6} + \frac{x^4}{120} - \ldots}{2 - \frac{4}{3}x^2 + \frac{4}{15}x^4 - \ldots}.$$

Now by long division

$$\begin{array}{r|l}
 & \frac{1}{2} + \frac{1}{4}x^2 + \frac{5}{48}x^4 + \ldots \\
\hline
2 - \frac{4}{3}x^2 + \frac{4}{15}x^4 & 1 - \frac{1}{6}x^2 + \frac{1}{120}x^4 - \ldots \\
 & \underline{\frac{1}{2} - \frac{4}{6}x^2 + \frac{4}{30}x^4 - \ldots} \\
 & \frac{1}{2}x^2 - \frac{15}{120}x^4 + \ldots \\
 & \underline{\frac{1}{2}x^2 - \frac{1}{3}x^4 + \ldots} \\
 & \frac{25}{120}x^4 - \ldots \\
 & \underline{\frac{25}{120}x^4 - \ldots}
\end{array}$$

Thus, for $x \neq 0$,

$$\frac{\sin x}{\sin(2x)} = \frac{1}{2} + \frac{1}{4}x^2 + \frac{5}{48}x^4 + \dots .$$

18.6.1 Calculus of Power Series

If the series $\Sigma c_n x^n$ converges on the interval $(-a,a)$, then

$$\Sigma \frac{d}{dx}(c_n x^n) = \Sigma n c_n x^{n-1}$$

also converges on $(-a,a)$.

Problem Solving Examples:

a) Find an expansion in powers of x of the function

$$f(x) = \int_0^1 \frac{1 - e^{-tx}}{t} dt .$$

b) Use the result from part (a) to find $f\left(\frac{1}{2}\right)$ approximately.

a) Using the fact that for all values x the series representation for e^x is

$$e^x = \sum_{n=0}^{\infty} \frac{x^n}{n!} = 1 + x\frac{x^2}{2!} + \dots + \frac{x^n}{n!} + \dots$$

we have

$$e^{-tx} = 1 - tx + \frac{t^2x^2}{2!} - \frac{t^3x^3}{3!} + \dots .$$

Hence,

$$1 - e^{-tx} = tx - \frac{t^2x^2}{2!} + \frac{t^3x^3}{3!} - \dots$$

so that

$$\frac{1-e^{-tx}}{t} = x - \frac{tx^2}{2!} + \frac{t^2x^3}{3!} - \ldots + (-1)^{n-1}\frac{t^{n-1}x^n}{n!} + \ldots .$$

Now this series representation is valid for all values of x and t. In addition, the radius of convergence of the power series in t is $R = \infty$. This is because,

$$\lim_{n\to\infty} \frac{a_{n+1}}{a_n} = \lim_{n\to\infty} |tx| \frac{1}{n+1} = 0$$

so that $R = \infty$.

Therefore, we can integrate the series term by term (this by the theorem in the previous problem) to obtain,

$$f(x) = xt - \frac{t^2x^2}{2 \cdot 2!} + \frac{t^3x^3}{3 \cdot 3!} - \ldots + (-1)^{n-1}\frac{t^n x^n}{n \cdot n!} + \ldots \Big|_0^1$$

or

$$f(x) = x - \frac{x^2}{2 \cdot 2!} + \frac{x^3}{3 \cdot 3!} - \ldots + (-1)^{n-1}\frac{x^n}{n \cdot n!} + \ldots . \tag{1}$$

b) From (1) we have,

$$f\left(\frac{1}{2}\right) = \int_0^1 \frac{1-e^{\frac{-t}{2}}}{t}\,dt = \frac{1}{2} - \frac{1}{2 \cdot 2!}\left(\frac{1}{2}\right)^2 + \frac{1}{3 \cdot 3!}\left(\frac{1}{2}\right)^3 - \ldots$$

which approximately equals 1.13.

18.6.2 The Differentiation of Power Series

If

$$f(x) = \sum_{n=0}^{\infty} c_n x^n \quad \text{for all x in } (-a,a),$$

then f is differentiable on $(-a,a)$, and

$$f'(x) = \sum_{n=1}^{\infty} n c_n x^{n-1} \quad \text{for all x in } (-a,a).$$

A power series defines an infinite differentiable function in the interior of its interval of convergence.

The derivatives of this function may be obtained by differentiating term by term.

Problem Solving Examples:

a) Using power series, show that

$$\frac{d(\sin x)}{dx} = \cos x \text{ and } \frac{d(\cos x)}{dx} = -\sin x.$$

b) Then show that sin a cos b + cos a sin b = sin (a+b) and

$$\cos a \cos b - \sin a \sin b = \cos(a+b).$$

The power series expansion for sin x and cos x are for all values x

$$\sin x = x - \frac{x^3}{3!} + \frac{x^5}{5!} - \frac{x^7}{7!} + \ldots \quad (1)$$

$$\cos x = 1 - \frac{x^2}{2!} + \frac{x^4}{4!} - \frac{x^6}{6!} + \ldots . \quad (2)$$

Since, by theorem, a power series can be differentiated term-by-term in any interval lying entirely within its radius of convergence, we have (1) and (2), for all values x

$$\frac{d(\sin x)}{dx} = 1 - \frac{3x^2}{3!} + \frac{5x^4}{5!} - \frac{7x^6}{7!} + \ldots$$

$$= 1 - \frac{x^2}{2!} + \frac{x^4}{4!} - \frac{x^6}{6!} + \ldots = \cos x.$$

and
$$\frac{d(\cos x)}{dx} = \frac{-2x}{2!} + \frac{4x^3}{4!} - \frac{6x^5}{5!} + \ldots$$

$$= -x + \frac{x^3}{3!} - \frac{x^5}{5!} + \ldots = -\sin x$$

b) Using the result from part (a) we have

$$\{\sin x \cos(h-x) + \cos x \sin(h-x)\}'$$
$$= \cos x \cos(h-x) + \sin x \sin(h-x)$$
$$- \sin x \sin(h-x) - \cos x \cos(h-x)$$
$$= 0$$

Thus sin x cos(h–x) + cos x sin(h–x) is constant and this constant equals the value sin h (this was found by letting x = 0). Hence

$$\sin x \cos(h-x) + \cos x \sin(h-x) = \sin h.$$

Now replacing x by a and h by a+b yields

$$\sin a \cos b + \cos a \sin b = \sin(a+b)$$

From which differentiation with respect to a yields

$$\cos a \cos b - \sin a \sin b = \cos(a+b).$$

Note that from this result we obtain

$$\cos^2 x - \sin^2 x = \cos 2x \ .$$

18.7 Taylor Series

$$f(x) = \sum_{n=0}^{\infty} \frac{f^{(n)}(a)}{n!}(x-a)^n$$

This formula represents the Taylor series for f about the point a or the expansion of f into a power series about a.

For the special case where a = 0, the Taylor series is represented by the formula

$$f(x) = \sum_{n=0}^{\infty} \frac{f^{(n)}(0)}{n!} x^n$$

This is called the Maclaurin series.

Problem Solving Examples:

Give a Taylor expansion of $f(x,y) = e^x \cos y$ on some compact convex domain E containing (0,0).

Let $f(x,y) = \phi(x)\,\psi(y)$ where $\phi(x) = e^x$ and $\psi(y) = \cos y$. Also let $E = E_x \times E_y$ where E_x and E_y are compact convex subsets of R, i.e., closed and bound intervals. Note that

$$\sup_{x \in E_x} \left\| \frac{d^k\phi(x)}{dx^k} \right\| = e^{\max E_x}$$

and

$$\sup_{y \in E_y} \left\| \frac{d^k\psi(y)}{dy^k} \right\| \leq 1.$$

Hence $\phi(x)$ and $\psi(y)$ are real analytic on E.

Furthermore, note that

$$e^x = \sum_{i=0}^{\infty} \frac{x^i}{i!} \frac{d^i e^x}{dx^i}\bigg|_{x=0} = \sum_{i=0}^{\infty} \frac{x^i}{i!} \qquad (1)$$

and

$$\cos y = \sum_{i=0}^{\infty} \frac{x^i}{i!} \frac{d^i \cos y}{dy^i}\bigg|_{y=0} = \sum_{i=0}^{\infty} (-1)^i \frac{y^{2i}}{(21)!} . \qquad (2)$$

Since $f = \phi\psi$ and ϕ and ψ are real analytic on E, f is real analytic on E:

$$\sup_{x,y\in E} \left\| \frac{\partial^k f(x,y)}{\partial x^i \partial y^{k-1}} \right\| \leq e^{\max E_x} .$$

hence, from (1) and (2)

$$e^x \cos y = \left(\sum_{i=0}^{\infty} \frac{x^i}{i!} \right) \left(\sum_{j=0}^{\infty} (-1)^j \frac{y^{2j}}{(2j)!} \right)$$

$$= \sum_{k=0}^{\infty} \left(\sum_{\substack{i+j=k \\ i,j\geq 0}} (-1)^j \frac{x^i y^{2j}}{i!(2j)!} \right) . \qquad (2)$$

The first three terms corresponding to k=0, k=1, and k=2 give the approximation:

$$e^x \cos y \approx 1 + x - \frac{1}{2} y^2 + \frac{1}{2} x^2 - \frac{1}{2} xy^2 + \frac{1}{24} y^4$$

which is known to be accurate near (0,0).

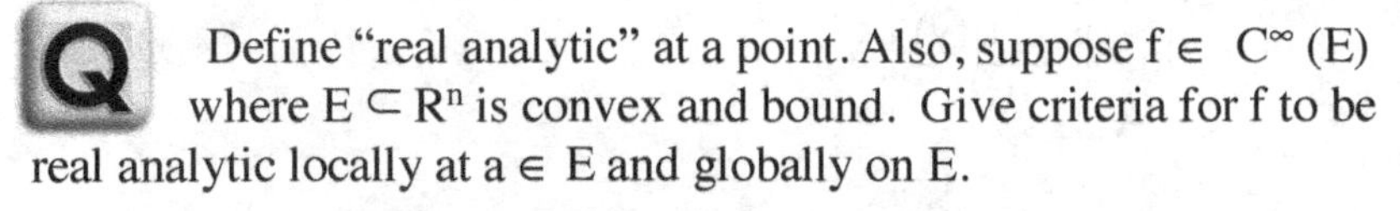

Q Define "real analytic" at a point. Also, suppose $f \in C^\infty(E)$ where $E \subset R^n$ is convex and bound. Give criteria for f to be real analytic locally at $a \in E$ and globally on E.

A A function $f \in C^\infty(E)$ (that is, all partial derivatives exist and are continuous on E) is real analytic at $a \in E$ if the Taylor series about a converges in some neighborhood of a. Furthermore, if the Taylor series about every $a \in E$ converges on E, then f is real analytic on E. The Taylor series expansion of f about $a \in E$ is

$$f(x) = f(a) + \sum_{i=1} \frac{\partial t}{\partial x_i}(a)(x_i - a_i) + \ldots +$$

$$\frac{1}{k!} \sum_{\substack{i_j=1 \\ 1 \le j \le k}}^{n} \frac{\partial^k f}{\partial x_{i_1} \ldots \partial x_{i_k}}(a)(x_{i_1} - a_{i_1}) \ldots (x_{i_k} - a_{i_k}) + \ldots$$

$$= f(a) + J_f(a)(x-a) + \frac{1}{2} < H_f(a)(x-a), x-a > + \ldots \tag{1}$$

Where $J_f(a)$ and $H_f(a)$ are the Jacobian and Hessian matrices, respectively, and $<, >$ denotes the Euclidean inner product. To show that (1) converges note that, by Taylor's Theorem,

$$f(x) = f(a) + J_f(a)(x-a) + \frac{1}{2} H_f(a)(x-a), x-a > + \ldots + R_k(x,a)$$

where

$$R_k(x,a) = \frac{1}{k!} \sum_{\substack{i_j=1 \\ 1 \le j \le k}}^{n} \frac{\partial^k f}{\partial x_{i} \ldots \partial x_{i_k}}(\bar{x})(x_{i_1} - a_{i_1}) \ldots (x_{i_k} - a_{i_k}), \tag{2}$$

for some $\bar{x}$ such that $0 < ||\bar{x} - a|| < ||x\text{-}a||$, and that (1) converges if and only if $R_k(x,a) \to 0$ as $k \to \infty$. Suppose that

$$\sup_{x \in E} \frac{\partial^k f(x)}{\partial x_{i_1} \ldots \partial x_{i_k}} \le c\,\alpha^k k! \tag{3}$$

for non-negative real constants c and α. Then, from (2) and (3),

$$\left| R_k(x,a) \right| \le \frac{1}{k!} \left| \sum_{\substack{i_j=1 \\ 1 \le j \le k}}^{n} c\,\alpha^k k! \, ||x-a||^k \right.$$

$$= \frac{c\,\alpha^k\,k!}{k!}\,\|x-a\|^k\,n^k$$

$$= c\,(n\,\alpha\,\|x-a\|)^k$$

Therefore, when $\|x-a\| < \frac{1}{n\alpha}$, $R_k\,(x-a) \to 0$ as $k \to \infty$.

Hence, (3) is a sufficient criterion for f to be real analytic locally at any $a \in E$. Now suppose that

$$\sup_{x \in E} \left| \frac{\partial^k f(x)}{\partial x_{i_1} \ldots \partial x_{i_k}} \right| \leq c\,\alpha^k \tag{4}$$

for non-negative real constants c and α. Then (2) and (4) give

$$|R_k\,(x,a)| \leq \frac{1}{k!} \sum_{\substack{i_j=1 \\ 1 \leq j \leq k}}^{n} c\,\alpha^k\,\|x-a\|^k$$

$$= \frac{c\,\alpha^k\,\|x-a\|^k\,n^k}{k!}$$

$$= c\,\frac{(n\,\alpha\,\|x-a\|)^k}{k!}\,. \tag{5}$$

Since the power series whose k^{th} term is the right hand side of (5) converges by the ratio test, $R_k\,(x-a) \to 0$ as $k \to \infty$. Hence, (4) is a sufficient criterion for f to be real analytic globally on E.

18.7.1 Validity of Taylor's Expansion and Computations with Series

The Taylor series for e^x about $x = a$ converges to e^x for any a and x. This is exemplified by the formula

$$e^x = 1 + (x-a) + \frac{(x-a)^2}{2!} + \ldots + \frac{(x-a)^n}{n!} + \ldots$$

This formula is applicable for all values of a and x.

The following functions are obtained from the Maclaurin series:

$$\sin x = \sum_{n=0}^{\infty} \frac{(-1)^n x^{2n+1}}{(2n+1)!},$$

$$\cos x = \sum_{n=0}^{\infty} \frac{(-1)^n x^{2n}}{(2n)!}$$

Problem Solving Examples:

Find the Maclaurin series and the interval of convergence for the function $f(x) = \cos x$.

To find the Maclaurin series for the given function, we determine $f(0), f'(x), f'(0), f''(x), f''(0)$, etc. We find:

$f(x) = \cos x$	$f(0) = 1$
$f'(x) = -\sin x$	$f'(0) = 0$
$f''(x) = -\cos x$	$f''(0) = -1$
$f'''(x) = \sin x$	$f'''(0) = 0$
$f^4(x) = \cos x$	$f^4(0) = 1$
$f^5(x) = -\sin x$	$f^5(0) = 0$
$f^6(x) = -\cos x$	$f^6(0) = -1.$

We develop the series as follows:

$$f(x) = f(0) + f'(0)x + \frac{f''(0)}{2!}x^2 + \frac{f'''(0)}{3!}x^3 + \frac{f^4(0)}{4!}x^4$$

$$+ \frac{f^5(0)}{5!}x^5 + \frac{f^6(0)}{6!}x^6 + \ldots$$

By substitution:

$$\cos x = 1 + 0 - \frac{x^2}{2!} + 0 + \frac{x^4}{4!} + 0 - \frac{x^6}{6!} + \ldots$$

$$= 1 \frac{x^2}{2!} + \frac{x^4}{4!} - \frac{x^6}{6!} + \ldots .$$

We examine the terms of this series to determine the law of formation. We find the nth term of the series to be $\frac{x^{2n-2}}{(2n-2)!}$. Then the (n+1)th term is $\frac{x^{2n}}{(2n)!}$. Therefore, the Maclaurin series is:

$$\cos x = 1 - \frac{x^2}{2!} + \frac{x^4}{4!} - \frac{x^6}{6!} + \ldots \pm \frac{x^{2n-2}}{(2n-2)!} \pm \frac{x^{2n}}{(2n)!} \ldots .$$

To find the interval of convergence we use the ratio test. We set up the ratio $\frac{u_{n+1}}{u_n}$, obtaining:

$$\frac{x^{2n}}{(2n)!} \times \frac{(2n-2)!}{x^{2n-2}} = \frac{x^{2n}}{(2n)(2n-2)!} \times \frac{(2n-2)!}{x^{2n-2}} = \frac{x^2}{2n}.$$

Now, we find $\lim_{n \to \infty} \left| \frac{x^2}{2n} \right| = |0| = 0$. By the ratio test we know that if $\lim_{n \to \infty} \left| \frac{u_{n+1}}{u_n} \right| < 1$, the series converges. Since 0 is always less than 1, the series converges for all values of x.

18.7.2 Binomial Theorem

For each real number m, we have

$$(1+x)^m = 1 + \sum_{n=1}^{\infty} \frac{m(m-1)(m-2)\ldots(m-n+1)}{n!} x^n \text{ for } |x| < 1.$$

Problem Solving Examples:

Show that:

a) $$\lim_{n\to\infty} \frac{n^4 + n^3 - 1}{(n^2+2)(n^2 - n - 1)} = 1$$

b) $$\lim_{n\to\infty} \left(1 + \frac{C}{n^2}\right)^n = 1,$$

C a constant.

By definition,

$$\lim_{n\to\infty} p_n = p$$

if for all $\varepsilon > 0$, there exists N such that $n \geq N$ implies that

$$|p_n - p| < \varepsilon.$$

a)

$$\lim_{n\to\infty} \frac{n^4 + n^3 - 1}{(n^2+2)(n^2 - n - 1)}$$

$$= \lim_{n\to\infty} \frac{n^4 (1 + 1/n - 1/n^4)}{n^2(1 + 2/n^2)\, n^2 (1 - 1/n - 1/n^2)}$$

$$= \lim_{n\to\infty} \frac{(1 + 1/n - 1/n^4)}{(1 + 2/n^2)(1 - 1/n - 1/n^2)}$$

$$= \frac{1+0-0}{(1+0)(1-0-0)} = 1$$

since

$$\lim_{n \to \infty} \frac{1}{n^p} = 0 \text{ for } p > 0.$$

b) We want to show that

$$\lim_{n \to \infty} \left(1 + \frac{C}{n^2}\right)^n = 1,$$

However, from the definition of a convergent sequence, this is equivalent to showing that

$$\left| \left(1 + \frac{C}{n^2}\right)^n - 1 \right| < \varepsilon$$

where $\varepsilon > 0$, and $n \geq N$.

To do this we make use of the binomial theorem,

$$(a+b)^m = \sum_{r=0}^{m} \binom{m}{r} a^{m-r} b^r ,$$

where

$$\binom{m}{r} = \frac{m!}{r!(m-r)!} .$$

Hence,

$$\left| \left(1 + \frac{C}{n^2}\right)^n - 1 \right| = \left| \sum_{r=0}^{n} \binom{n}{r} \left(\frac{C}{n^2}\right)^r - 1 \right|$$

$$= \left| \sum_{r=1}^{n} \binom{n}{r} \left(\frac{C}{n^2}\right)^r \right|$$

$$\leq \sum_{r=1}^{n} \frac{n^r}{r!} \left(\frac{|C|}{n^2}\right)^r \text{ since } \binom{n}{r} \leq \frac{n^r}{r!} .$$

$$\leq \sum_{r=1}^{n} \left(\frac{|C|}{n}\right)^r \text{ since } \frac{n^r}{r!n^r} < 1.$$

But

$$\sum_{r=1}^{n} \left(\frac{|C|}{n}\right)^r = \frac{1 - \left(\frac{|C|}{n}\right)^{n+1}}{1 - \frac{|C|}{n}} - 1 = \frac{\frac{|C|}{n} - \left(\frac{|C|}{n}\right)^{n+1}}{1 - \frac{|C|}{n}}$$

$$= \frac{\frac{|C|}{n}\left(1 - \left(\frac{|C|}{n}\right)^n\right)}{1 - \frac{|C|}{n}}$$

$$< \frac{\frac{|C|}{n}}{1 - \frac{|C|}{n}}$$

$$\left(\text{if } n > |C| \text{ since } 1 - \left(\frac{|C|}{n}\right)^n < 1\right).$$

Therefore,

$$\left|\left(1 + \frac{C}{n^2}\right)^n - 1\right| < \frac{\frac{|C|}{n}}{1 - \frac{|C|}{n}} = \frac{|C|}{n - |C|}$$

which $\to 0$ as $n \to \infty$.

Thus,

$$\lim_{n \to \infty} \left(1 + \frac{C}{n^2}\right)^n = 1.$$

Quiz: Infinite series

1. Which of the following series are convergent?

I. $1 - \frac{1}{2^2} + \frac{1}{3^2} - \ldots + \frac{1}{n^2} + \ldots$

II. $1 + \frac{1}{\sqrt{2}} + \frac{1}{\sqrt{3}} + \ldots + \frac{1}{\sqrt{n}} + \ldots$

III. $\frac{1}{5} + \frac{1}{6} + \frac{1}{7} + \ldots + \frac{1}{n+4} + \ldots$

(A) I only

(B) II only

(C) I, II, and III

(D) II and III only

(E) none

2. The coefficient of x^3 in the Taylor series for $f(x) = \ln x$ about $x = 1$ is:

(A) $\frac{1}{6}$

(B) $\frac{2}{3}$

(C) $\frac{1}{2}$

(D) $\frac{1}{3}$

(E) $\frac{1}{4}$

3. For $-1 < x \leq 1$, if $f(x) = \sum_{m=1}^{\infty} \frac{(-1)^{m+1} x^{3m-2}}{3m-2}$, then $f'(x) =$

(A) $\sum_{m=1}^{\infty} (-1)^{m+1} x^{3m}$

(B) $\sum_{m=1}^{\infty} (-1)^{m+1} x^{3m-3}$

(C) $\sum_{m=1}^{\infty} (-1)^{3m} x^{3m}$

(D) $\sum_{m=1}^{\infty} (-1)^{m} x^{3m-3}$

(E) $\sum_{m=1}^{\infty} (-1)^{m} x^{3m}$

4. What are the values of x for which the series $x - \frac{x^2}{2^2} + \frac{x^3}{3^2} - \frac{x^4}{4^2} + \ldots$ converges?

(A) All values of x

(B) $0 \leq x \leq 2$

(C) $0 < x < 2$

(D) $0 < x < 1$

(E) $-1 \leq x \leq 1$

5. Given that $\cos x = \sum_{n=0}^{\infty} \frac{(-1)^n x^{2n}}{(2n)!}$ what is $\cos (3x^2)$?

(A) $\sum_{n=0}^{\infty} \frac{x^{4n}}{(2n)!}$

(B) $\sum_{n=0}^{\infty} \frac{(-9x)^{2n}}{(2n)!}$

(C) $\sum_{n=0}^{\infty} \frac{(-1)^n x^{2n}}{(2n)!}$

(D) $\sum_{n=0}^{\infty} \frac{-9x^{4n}}{(2n)!}$

(E) $\sum_{n=0}^{\infty} \frac{(-9)^n x^{4n}}{(2n)!}$

6. If $f(x) = \sum_{n=0}^{\infty} a_n (x-1)^n$ for some constants a_n, which of the following could be the interval of convergence for f?

(A) $[0, 1]$

(B) $\left[\frac{1}{2}, 2\right]$

(C) $[0, 2]$

(D) $\left(0, \frac{3}{2}\right)$

(E) $[0, 4]$

7. If $e^x = \sum_{n=0}^{\infty} \frac{x^n}{n!}$, then $\sum_{n=1}^{\infty} \frac{(-1)^n x^{2n}}{n!}$ must be

(A) e^{-x^2}

(B) $\cos x$

(C) $1 - e^{-x^2}$

(D) $e^{-x^2} - 1$

(E) none of the above

8. For $-1 < x < 1$, if $f(x) = \sum_{n=1}^{\infty} \frac{(-1)^{n+1} x^{5n-3}}{5n-3}$, $f'(x) =$

(A) $\sum_{n=1}^{\infty} (-1)^{5n} x^{5n}$

(B) $\sum_{n=1}^{\infty} (-1)^n x^{5n}$

(C) $\sum_{n=1}^{\infty} (-1)^{n+1}x^{5n}$

(D) $\sum_{n=1}^{\infty} (-1)^{n+1}x^{5n-4}$

(E) $\sum_{n=1}^{\infty} (-1)^{n}x^{5n-4}$

9. If $f(x) = \sum_{n=0}^{\infty} \frac{(-1)^n x^{2n+1}}{(2n+1)!}$, then $f'(x) =$

(A) $\sum_{n=0}^{\infty} \frac{(-1)^n x^{2n+1}}{(2n-1)!}$

(B) $\sum_{n=0}^{\infty} \frac{(-1)^n x^{2n}}{(2n)!}$

(C) $\sum_{n=0}^{\infty} (-1)^n x^{2n}$

(D) $\sum_{n=0}^{\infty} \frac{(-1)^n x^{2n}}{(2n-1)!}$

(E) $\sum_{n=0}^{\infty} (-1)^n x^{2n-1}$

10. Which of the following series converge?

I. $\frac{1}{5} + \frac{1}{10} + \frac{1}{15} + \frac{1}{20} + \ldots + \frac{1}{5n} \ldots$

II. $\frac{1}{5} + \frac{1}{25} + \frac{1}{125} + \frac{1}{625} + \ldots + \frac{1}{n^5} \ldots$

III. $\frac{1}{5} - \frac{1}{6} + \frac{1}{10} - \frac{1}{11} + \frac{1}{15} - \frac{1}{16} \ldots$

$+ \frac{1}{5n} - \frac{1}{1+5n} + \ldots$

(A) I only

(B) II only

(C) III only

(D) I and II only

(E) II and III only

ANSWER KEY

1. (A)
2. (D)
3. (B)
4. (E)
5. (E)
6. (C)
7. (D)
8. (D)
9. (B)
10. (E)

NOTES

NOTES

NOTES

NOTES

NOTES

NOTES